Automation
in Biotechnology

Automation in Biotechnology

**A Collection of Contributions Based on Lectures
Presented at the Fourth Toyota Conference, Aichi, Japan,
21-24 October 1990**

Edited by

ISAO KARUBE

*Research Center for Advanced Science & Technology,
Tokyo University, 4-6-1 Komaba, Meguro-ku,
Tokyo 153, Japan*

ELSEVIER

Amsterdam – London – New York – Tokyo   1991

ELSEVIER SCIENCE PUBLISHERS B.V.
Molenwerf 1
P.O. Box 211, 1000 AE Amsterdam, Netherlands

Distributors for the United States and Canada:

ELSEVIER SCIENCE PUBLISHING COMPANY INC.
655, Avenue of the Americas
New York, NY 10010, U.S.A.

ISBN 0-444-88767-9

Printed in The Netherlands

TOYOTA CONFERENCES

The First Toyota Conference
Molecular Conformation and Dynamics of Macromolecules in Condensed Systems, 28 September – 1 October 1987 (Aichi, Japan), edited by
M. Nagasawa

The Second Toyota Conference
Organization of Engineering Knowledge for Product Modelling in Computer Integrated Manufacturing, 2–5 October 1988 (Aichi, Japan), edited by
T. Sata

The Third Toyota Conference
Integrated Micro-Motion Systems – Micromachining, Control and Applications, 22–25 October 1989 (Aichi, Japan), edited by F. Harashima

The Fourth Toyota Conference
Automation in Biotechnology, 21–24 October 1990 (Aichi, Japan), edited by
I. Karube

The Fifth Toyota Conference
Nonlinear Optical Materials, 6–9 October 1991 (Aichi, Japan), edited by
S. Miyata

The Sixth Toyota Conference
Turbulence and Molecular Processes in Combustion, 11–14 October 1992 (Japan)
edited by T. Takeno

CONTENTS

Preface   ix

Organizing Committee   xi

List of Contributors   xiii

Automation in Biotechnology
    I. Karube   1

Automated DNA Sequence Analysis and the Human Genome
    W.R. McCombie, A.R. Kerlavage, A. Martin-Gallardo, J. Gocayne, M. FitzGerald,   19
    S. Trapp, T. Onai, S. Arakawa and M. Dubnick

Automated DNA Sequencing
    D.A. Amorese   29

An Automated System for Sequencing of the Human Genome
    E. Soeda, Y. Murakami, K. Nishi and I. Endo   57

Automated Laboratory Instrumentation for the Preparation and Analysis of Biotechnology
Samples
    J.N. Little   69

Possible Contribution of STM to Biotechnology
    M. Ono   81

STM Investigation of Macromolecules
    B. Michel   91

Handling of Biological Molecules and Membranes in Microfabricated Structures
    M. Washizu   113

Recognition and Counting of Cells by Image Processing
    T. Fukuda, S. Shiotani, F. Arai, H. Asama, T. Nagamune and I. Endo   127

Innovative Actuators for Micromanipulation and Microinjection in Biotechnology
    T. Higuchi   145

Application of A.C. Electrical Fields to the Manipulation and Characterisation of Cells
    R. Pethig   159

Micromechanical Silicon Devices Applied for Biological Cell Fusion Operation
    K. Sato, Y. Kawamura and S. Tanaka   187

VIII

Transgenesis of Animals
    H. Kondoh, K. Agata and K. Ozato   203

Automation of Plant Tissue Culture Process
    Y. Miwa   217

Trends in Automation for Clonal Propagation by Tissue Culture
    J. Aitken-Christie   235

Robotic Workcell for Flexibly Automated Handling of Young Transplants
    K.C. Ting   261

Environmental Control and Automation in Micropropagation
    T. Kozai   279

The Ruthner Container System
    E. Ruthner   305

Requirements and Technologies for Automated Plant Growth Systems on Space Bases
    T.W. Tibbitts, R.J. Bula, R.C. Morrow, R.B. Corey and D.J. Barta   325

Automation in Space Life Sciences
    M. Yamashita   337

Automated Measurements from Laboratory to Industrial Scale. A Tool for Better
Understanding of Fermentation Processes
    J. Villadsen   349

Trend of Automation in Bio-Industry
    M. Hachiya   367

PREFACE

Biotechnology, which is hardly fifteen years old, has emerged as a new breed of high technology. It has, to a large extent, been promoted by a number of breakthroughs in basic techniques. However, more sophisticated procedures and techniques are required for further improvement of basic research and practical development. Recently, intense research efforts have been expended to automate experimental procedures such as measurement, analysis and manipulation, and to expand the application of robotic technology to research and development in bio-engineering fields.

In response to rapid expansion of the life sciences it has become desirable to automate the principal techniques. Work has focussed on areas such as genetic engineering, bioreactor control systems, cell technology and laboratory instrumentation. Compared with automation in existing industries, these are primarily future technology and further improvement is required to reach the level of practical use.

With this situation in mind, the Fourth Toyota Conference was organized under the title of "Automation in Biotechnology". The meeting was held at Nisshin, Aichi, Japan on October 21-24, 1990. Twenty-two scientists and engineers from all over the world were invited to present their work. Sixty researchers joined the conference to discuss the state of the art and to exchange their thoughts and views on the future of this emerging field. The titles of the regular sessions were:

1. Laboratory Automation
2. Automatic Instrumentation
3. Bioindustry Instrumentation

This book includes all the papers, most of which have been revised with reference to the discussions at the conference. We believe this conference and the proceedings will represent a milestone in this key technology.

The Fourth Conference was financially supported by Toyota Motor Corporation, as were the previous ones. We would like to express our sincere gratitude to all of the committee members, speakers and those people who took part in discussions at the conference. We are also thankful to Toyota Motor Corporation for its financial support and kind offering of the company premises for the meeting. Finally, we would like to express our hearty appreciation to the members of Toyota Central Research & Development Laboratories, Inc. for their helpful assistance in the preparation and execution of the conference.

Organizing Committee of
the 4th Toyota Conference

Isao Karube (Chairman)
Takeyoshi Dohi
Masahiko Hachiya
Yoshiyuki Miwa
Hideaki Matsuoka

ORGANIZING COMMITTEE
of the FOURTH TOYOTA CONFERENCE

ISAO KARUBE (Chairman)
Research Center for Advanced Science and Technology
University of Tokyo
4-6-1 Komaba, Meguro-ku, Tokyo 153, Japan

TAKEYOSHI DOHI
Department of Precision Machinery Engineering, Faculty of Engineering
University of Tokyo
7-3-1 Hongo, Bunkyo-ku, Tokyo 113, Japan

MASAHIKO HACHIYA
Systems Engineering Division
Hitachi, Ltd.
4-6 Kanda-Surugadai, Chiyoda-ku, Tokyo 101, Japan
(present address, below)
Research Institute for Innovative Technology for the Earth
No.7 Toyo Kaiji Bld. 8F
11-8-2 Nishishinbashi, Minato-ku, Tokyo 105, Japan

YOSHIYUKI MIWA
Department of Mechanical Engineering
School of Science and Engineering
Waseda University
3-4-1 Okubo, Shinjuku-ku, Tokyo 169, Japan

HIDEAKI MATSUOKA
Department of Biotechnology, Faculty of Technology
Tokyo University of Agriculture and Technology
2-24-16 Naka-cho, Koganei-city, Tokyo 184, Japan

Working Committee

EIICHI TAMIYA
Research Center for Advanced Science and Technology
University of Tokyo
4-6-1 Komaba, Meguro-ku, Tokyo 153, Japan

IZUMI KUBO
Institute of Life Science
Soka University
1-236 Funaki-cho, Hachioji-city, Tokyo 192, Japan

TAKASHI HORIUCHI
Department of Precision Machinery Engineering, Faculty of Engineering
University of Tokyo
7-3-1 Hongo, Bunkyo-ku, Tokyo 113, Japan

HIROSHI SONODA
Systems Development Department, Systems Engineering Division
Hitachi, Ltd.
4-6 Kanda-Surugadai, Chiyoda-ku, Tokyo 101, Japan

ATSUO TAKANISHI
Department of Mechanical Engineering
Waseda University
3-4-1 Okubo, Shinjuku-ku, Tokyo 169, Japan

Auditor of Organizing Committee

AKIRA HOSONO
Toyota Central Research and Development Laboratories, Inc.
41-1 Yokomichi, Nagakute, Nagakute-cho, Aichi-gun,
Aichi-ken 480-11, Japan

Secretary General

HIKOMARO SANO
Information and Patent Division
Toyota Central Research and Development Laboratories, Inc.
41-1 Yokomichi, Nagakute, Nagakute-cho, Aichi-gun,
Aichi-ken 480-11, Japan

List of Contributors

Agata, Kiyokazu
National Institute for Basic Biology, Okazaki, Aichi-ken 444, Japan

Aitken-Christie, Jenny
Forest Health and Improvement Division, Forest Research Institute, Private Bag, Rotorua, New Zealand

Amorese, Douglas A.
Medical Products Department, E. I. du Pont de Nemours and Company (Inc.), Glasgow Site, Wilmington, DE 19898, U.S.A.

Arai, Fumihito
Department of Mechanical Engineering, Nagoya University, Furo-cho, Chikusa-ku, Nagoya, Aichi-ken 464-01, Japan

Arakawa, Shoji
Section of Receptor Biochemistry and Molecular Biology, Laboratory of Cellular and Molecular Neurobiology, National Institute of Neurological Disorders and Stroke, National Institutes of Health, Bethesda, MD 20892, U.S.A.
(present address: Department of Pediatrics, Kyoto Prefectural University of Medicine, Kyoto, Japan)

Asama, Hajime
Riken, 2-1 Hirosawa, Wako, Saitama-ken 351-01, Japan

Barta, D. J.
Wisconsin Center for Space Automation and Robotics, University of Wisconsin-Madison, Madison, WI 53715, U.S.A.

Bula, R. J.
Wisconsin Center for Space Automation and Robotics, University of Wisconsin-Madison, Madison, WI 53715, U.S.A.

Corey, R. B.
Department of Soil Science, University of Wisconsin-Madison, Madison, WI 53706, U.S.A.

Dubnick, Mark
Section of Receptor Biochemistry and Molecular Biology, Laboratory of Cellular and Molecular Neurobiology, National Institute of Neurological Disorders and Stroke, National Institutes of Health, Bethesda, MD 20892, U.S.A.

Endo, Isao
Laboratory of Chemical Engineering, The Institute of Physical and Chemical Research (Riken), 2-1 Hirosawa, Wako, Saitama-ken 351-01, Japan

FitzGerald, Michael
Section of Receptor Biochemistry and Molecular Biology, Laboratory of Cellular and Molecular Neurobiology, National Institute of Neurological Disorders and Stroke, National Institutes of Health, Bethesda, MD 20892, U.S.A.

Fukuda, Toshio
Department of Mechanical Engineering, School of Engineering, Nagoya University, Furo-cho, Chikusa-ku, Nagoya, Aichi-ken 464-01, Japan

Gocayne, Jeannine
Section of Receptor Biochemistry and Molecular Biology, Laboratory of Cellular and Molecular Neurobiology, National Institute of Neurological Disorders and Stroke, National Institutes of Health, Bethesda, MD 20892, U.S.A.

Hachiya, Masahiko
Systems Engineering Division, Hitachi, Ltd., 4-6 Kanda-Surugadai, Chiyoda-ku, Tokyo 101, Japan
(present address: Research Institute of Innovative Technology for the Earth, No.7 Toyo Kaiji Building 8F, 11-8-2 Nishishinbashi, Minato-ku, Tokyo 105, Japan)

Higuchi, Toshiro
Institute of Industrial Science, University of Tokyo, 7-22-1 Roppongi, Minato-ku, Tokyo 106, Japan

Karube, Isao
Research Center for Advanced Science and Technology, University of Tokyo, 4-6-1 Komaba, Meguro-ku, Tokyo 153, Japan

Kawamura, Yoshio
Central Research Laboratory, Hitachi, Ltd., 1-280 Higashi-Koigakubo, Kokubunji, Tokyo 185, Japan

Kerlavage, Anthony R.
Section of Receptor Biochemistry and Molecular Biology, Laboratory of Cellular and Molecular Neurobiology, National Institute of Neurological Disorders and Stroke, National Institutes of Health, Bethesda, MD 20892, U.S.A.

Kondoh, Hisato
Department of Molecular Biology, School of Science, Nagoya University, Furo-cho, Chikusa-ku, Nagoya, Aichi-ken 464-01, Japan

Kozai, Toyoki
Department of Horticulture, Faculty of Horticulture, Chiba University, 648 Matsudo, Chiba 271, Japan

Little, James N.
Zymark Corporation, Zymark Center, Hopkinton, MA 01748, U.S.A.

Martin-Gallardo, Antonia
Section of Receptor Biochemistry and Molecular Biology, Laboratory of Cellular and Molecular Neurobiology, National Institute of Neurological Disorders and Stroke, National Institutes of Health, Bethesda, MD 20892, U.S.A.

McCombie, William Richard
Section of Receptor Biochemistry and Molecular Biology, Laboratory of Cellular and Molecular Neurobiology, National Institute of Neurological Disorders and Stroke, National Institutes of Health, Bethesda, MD 20892, U.S.A.

Michel, Bruno
IBM Research Division, Zurich Research Laboratory, Saumerstrasse 4, 8803 Rüschlikon, Switzerland

Miwa, Yoshiyuki
Department of Mechanical Engineering, School of Science and Engineering, Waseda University, 3-4-1 Okubo, Shinjuku-ku, Tokyo 169, Japan

Morrow, R. C.
Wisconsin Center for Space Automation and Robotics, University of Wisconsin-Madison, Madison, WI 53715, U.S.A.

Murakami, Y.
RIKEN Gene Bank, The Institute of Physical and Chemical Research, 3-1-1 Koyadai, Tsukuba Science City, Ibaraki-ken 305, Japan

Nagamune, Teruyuki
Riken, 2-1 Hirosawa, Wako, Saitama-ken 351-01, Japan

Nishi, K.
Division of Supersensor's Engineering, Instrumentation and Characterization Center, The Institute of Physical and Chemical Research, 2-1 Hirosawa, Wako, Saitama-ken 351-01, Japan

Onai, Takeshi
Section of Receptor Biochemistry and Molecular Biology, Laboratory of
Cellular and Molecular Neurobiology, National Institute of Neurological
Disorders and Stroke, National Institutes of Health, Bethesda, MD 20892,
U.S.A.
(present address: Department of Physiology, 1st Division, Gunma
University, School of Medicine, Maebashi, Japan)

Ono, Masatoshi
Frontier Technology Division, Electrotechnical Laboratory, Agency of
Industrial Science and Technology, 1-1-4 Umezono, Tsukuba-city, Ibaraki-ken
305, Japan

Ozato, Kenjiro
Department of Biology, College of Liberal Arts and Sciences, Kyoto
University, Kyoto 606, Japan

Pethig, Ronald
Institute of Molecular and Biomolecular Electronics, University of Wales,
Dean Street, Bangor, Gwynedd, LL57 1UT, U.K.

Ruthner, Eberhardt
ASG AgroService GmbH, A 1010 Wien, Stubenbastei 12, Austria

Sato, Kazuo
Central Research Laboratory, Hitachi Ltd., 1-280 Higashi-Koigakubo,
Kokubunji, Tokyo 185, Japan

Shiotani, Shigetoshi
Department of Mechanical Engineering, Nagoya University, Furo-cho,
Chikusa-ku, Nagoya, Aichi-ken 464-01, Japan

Soeda, Eiichi
RIKEN Gene Bank, The Institute of Physical and Chemical Research,
3-1-1 Koyadai, Tsukuba Science City, Ibaraki-ken 305, Japan

Tanaka, Shinji
Central Research Laboratory, Hitachi Ltd., 1-280 Higashi-Koigakubo,
Kokubunji, Tokyo 185, Japan

Tibbitts, T. W.
Wisconsin Center for Space Automation and Robotics, University of
Wisconsin-Madison, Madison, WI 53715, U.S.A.
Department of Horticulture, University of Wisconsin-Madison, Madison,
WI 53706, U.S.A

Ting, Kuan-Chong
    Biological and Agricultural Engineering Department, Rutgers University,
    P.O. Box 231, New Brunswick, NJ 08903, U.S.A.

Trapp, Susan
    Section of Receptor Biochemistry and Molecular Biology, Laboratory of
    Cellular and Molecular Neurobiology, National Institute of Neurological
    Disorders and Stroke, National Institutes of Health, Bethesda, MD 20892,
    U.S.A.

Villadsen, John
    Department of Biotechnology and Research Center for Process Biotechnology,
    The Technical University of Denmark, Bygning 223, DK-2800 Lyngby, Denmark

Washizu, Masao
    Department of Electrical Engineering, Faculty of Engineering, Seikei
    University, 3-3-1 Kichijoji-Kitamachi, Musashino-shi, Tokyo 180, Japan

Yamashita, Masamichi
    Space Utilization Research Center, Institute of Space and Astronautical
    Science, 3-1-1 Yoshinodai, Sagamihara, Kanagawa-ken 229, Japan

I. Karube (Ed.) *Automation in Biotechnology*
Proceedings of the 4th Toyota Conference, 21–24 October 1990

1

AUTOMATION IN BIOTECHNOLOGY

Isao Karube
Research Center for Advanced Science and Technology
University of Tokyo

SUMMARY
    Automation in biotechnology have started from the field of bioreactor-
control, to the field of molecular biology, such as genetic- and chromosomal
engineering. In this paper, such novel technniques which are being
developed   are described.

INTRODUCTION
    Biotechnology has emerged as a new breed of high technology, and is
hardly fifteen years old.   The biotechnology has, to a large extent, been
promoted by a number of breakthroughs of basic techniques.   However, the
complicated procedures and techniques are still required for basic research
and   development.
    Recently, there are intense research efforts to automatize experimental
procedures, to measure and to expand the applications of robotic technology
into biotechnological research and development.
    Automation in biotechnology have started from the controlling techniques
in bioreactor system.   Nowadays many bioreactor systems are furnished
with automatic computer-controlling system.   Good sensors are necessary for
effective automatic control.   We have developed various kinds of sensors for
fermentation monitoring.   In the former part of this paper, novel sensors for
the monitoring of viscosity and cell number which we have recently
developed are described.
    Recently, according to the wide-spread of biotechnology and life science,
several approaches have been started to automatize the principal techniques
in molecular biology field, such as genetic engineering, chromosomal
engineering and cellular technology.   Comparing with the automation
technology in bioreactor, they are future technologies and they do not come
up to the level of practical-use.   In the latter part of this paper, these
approaches which are being developed in our group will be described.

MONITORING TECHNIQUES FOR AUTOMATION OF BIOREACTOR

Viscosity Sensor using Piezoelectroric Quartz Crystal

The monitoring of viscosity in a fermentation broth is an important factor, particularly for products with high viscosity produced during a bioprocess. Several methods were reported for bioprocesses, such as using a rotating viscometer or a capillary viscometer. However, these methods require a complicated procedure and were not suitable for *in situ* monitoring. Therefore, the establishment of a simple and rapid *in situ* method has been expected.

On the other hand, the study in piezoelectric quartz crystal has progressed remarkably. A quartz crystal has been used as a mass detector and has been applied to determine gases, ions, and microorganisms. It has also been reported that piezoelectric quartz crystal can oscillate in a liquid.

In our previous study electrical equivalent circuit parameters of quartz crystal in contact with a liquid have been computed by using an impedance analyzer and a microprocesser. The equation for the resonant resistance (R1) was derived as follows:

$$R1 = (2Fs \, \rho\eta)1/2A/_k 2$$

where Fs is the resonant frequency, A is the electrode area, k is an electromechanical coupling factor, and   and   are the density and viscosity, respectively. Resistance R1 and the resonant frequency change of the quartz crystal were found experimentally as good indexes of liquid viscosity. Therefore, it is possible to determine the viscosity by measuring R1 of the quartz crystal. We attempted to apply the principle on-line monitoring of the viscosity of fermentation broth.

We developed a novel viscous sensor utilizing AT-cut quartz crystal for the monitoring of the viscosity of fermentation broth (1). Further, the sensor system was applied to the on-line monitoring of the viscosity of leuconostoc mesenteroides culture which produces dextran during cultivation.

The sensor system was constructed from the piezoelectric quartz crystal fixed to the cell, exposing only one side of the quartz crystal electrode, an oscillating circuit, a peak level meter, and a personal computer (Figure 1). In order to investigate the characteristics of the sensor system, a sensor signal relating to the resonant resistances of the quartz crystal was measured using dextran solutionship was obtained between the sensor signal and the $(\rho\eta)1/2$ of the liquid, where   and   are the density and viscosity, respectively. The sensor signal was dependent not only on the viscosity of the liquid but also

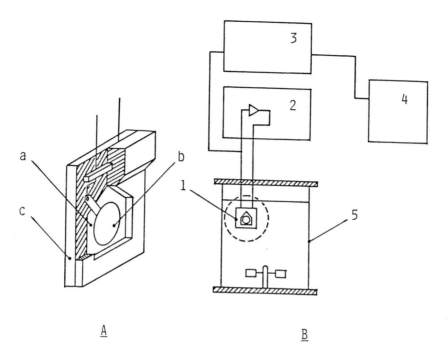

Figure 1. Schematic diagram of the piezoelectric viscosity sensor system. (A) Quartz crystal viscous device: a, quartz plate; b, electrode; c, cell. (B) System for on-line viscosity monitoring of fermentation broth: 1, quartz crystal viscous sensor; 2, oscillating circuit; 3, peak level meter; 4, personal computer; 5, jar fermentor.

on the molecular weight of dextran, because dextran solution shows a non-Newtonian property. The sensor system was applied for the on-line monitoring of the viscosity in dextran fermentation. A good correlation was observed between the sensor signal and the viscosity value measured with a rotational viscometer for the fermentation broth. Little bubbling effect and agitation of the sensor signal were observed, showing that this system can be utilized for viscosity monitoring in a bioprocess.

Piezoelectric gum sensor for the determination of microbial concentration

In a bioprocess, the determination of microbial concentration is very important factor. Turbidimetry and colony counts have been conventionally used for the determination of cell concentration. The turbidimetric method is simple. However, it cannot be applied to colored mixtures and samples sometimes require dilution. Although colony counts are reliable, the method

4

is time-consuming and complicated.

Flow cytometry have currently been applied for process monitoring. However the operations of this device are complicated and technical knowledge is essential.

On the other hand, the sensing devices using ultrasound have been developed in recent years. These devices have been utilized as an ultrasonograph in medical facility, as a fish detector in marine product industry and elsewhere. In the bioindustry, however, very few applications of the ultrasonic techniques have been reported.

Recently, we have developed the piezoelectric gum (NTK Piezo-Rubber) which is a new type of piezoelectric transducer. This transducer possesses a good damping property and can be formed in various shapes.

The piezoelectric gum sensor for the determination of the microbial concentration was developed and the properties of the sensor were investigated (2). Then, the sensor system was applied to the bioprocesses in order to monitor cultivations such as *Saccharomyces cerevisiae* and *Escherichia coli*.

The sensor system was constructed from two piezoelectric gums, a pulse generator and an amplifier (Figure 2). The measurement was based on the attenuation of the sound intensity level in the presence of the

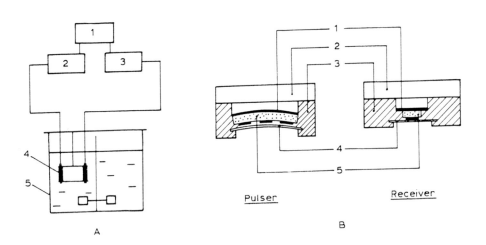

Figure 2. Schematic diagram of the piezoelectric gum sensor system. (A) Monitoring system for the continuous determination of cell concentration: 1, oscilloscope; 2, pulse generator; 3, preamplifier; 4, piezoelectric gum sensor; 5, jar. (B) Piezoelectric gum device: 1, electrode; 2, acrylic acid resin; 3, urethan resin; 4, polyester film; 5, piezoelectric gum.

microorganism. When the sensor system was immersed in the microbial suspension, the output voltage of the sensor decreased with increasing cell concentration. A linear relationship between the output voltage and the cell concentration was obtained. Calibration curves for Escherichia coli and Saccharomyces cerevisiae were linear in the range of 1.0-3.8 g dry cell mass per 1 and 0.5-9.8 g dry cell mass per 1, respectively. The continuous

determination of microbial concentration in a bioprocess was also attempted using this sensor system (Figure 3).

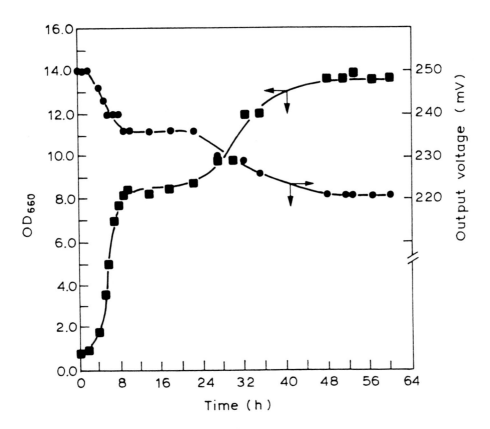

Figure 3. Time course of growth of <u>S. cerevisiae</u>. (●) Output voltage of the sensor; (■) cell concentration obtained by turbidimetric method.

Micro-ethanol sensor

For the on-line monitoring of fermentation processes, the small and selective ethanol sensor is appriciable. Kitagawa et al. developed a micro-alcohol sensor consisting of an immobilized acetic acid bacteria, a gas permeable membrane and an ISFET. This sensor was utilized for the determination of ethanol.

Acetic acid bacteria have been widely used in the industries to make vinegar and ascorbic acid by the oxidative fermentation. In acetic acid fermentation, both alcohol dehydrogenase (ADH) and aldehyde dehydrogenase (ALDH) are involved in oxidization of ethanol to acetic acid via acetaldehyde. When acetic acid bacteria are immobilized on an ISFET, ethanol is oxidized to acetic acid by those bacteria, causing a pH change to occur at the surface of ISFET. Therefore, ethanol concentration can be measured by combining an ISFET and acetic acid bacteria.

Acetobacter aceti IAM 1802 was used for the micro-alcohol sensor. The micro-organisms were immobilized on the gate surface of ISFET, in calcium alginate gel. The ISFET-immobilized micro-organisms and wire-like Ag-AgCl electrode were placed in a small shell

(6x3x22mm) that had a gas permeable menbrane (a porous Teflon membrane, 0.5 µm pore size) fitted to the side. The inside of the shell was filled with inner buffer solution (5 mM tris-HCl buffer containing 0.1 M $CaCl_2$, pH 7.0)

The system consisted of a thermostatic circulating jacketed vessel (3 ml volume), the microbial-FET alcohol sensor, the circuit for measurement of the gate output voltage (Vg), an electrometer and a recorder.

The initial rate change of gate voltage with time, dV/dt, was plotted against the logarithmic value of the ethanol concentration. The minimum detectable response to ethanol was obtained at approximately 0.1 mM. A linear relationship was observed over the range between 3-70 mM.

The response of the micro-alcohol sensor was stable over a wide pH range (pH 2-12), however, at pH's lower than 6, the sensor responded to acetic acid. Volatile organic acids such as acetic acid do not dissociate in lower pH solution, and penetrate through the gas permeable membrane, at pH's higher than 6, on the other hand, the volatile organic acids dissociate to the individual ions, they did not interfere with the measurement of ethanol by sensor. The sensor was stable for 15 h, at 15°C. The sensor showed usefulness for the determination of ethanol, especially in terms of its good selectivity and potential for miniaturization, and it is expected to be integrated into a multi-functional sensor capable of simultaneously determining substrates contained in a solution having a complex composition.

Micro-oxygen electrode and its application for glucose sensor

Several micro-oxygen electrode based on semiconductor technology have been proposed, however, none has yet reached the production line. One reason for this is that they contain a liquid electrolyte solution, making adhension of the gas-permeable membrane to the substrate difficult, even if epoxy resin is used. Therefore, mass production of such a device is impossible. We recently developed a novel disposable oxygen electrode based on conventional semiconductor fabrication technology.   The key improvements were (1) to use a porous material (agarose gel) to support the electrolyte solution and (2) to use a hydrophobic polymer (negative photoresist) as the gas-permeable membrane and directly cast it over the porous material.

The electrode has a U-shaped groove, 300μm deep, and two gold electrodes over the SiO_2 layer that electrically insulates them. Each gold electrode covers about half of the oxygen electrode, that is, the bottom and side walls of the groove and the upper part. Agarose gel containing a 0.1 M potassium chloride aqueous solution was poured into the groove, after which it was covered by the gas-permeable membrane. Several hydrophobic polymers were tested as the gas-permeable membrane, and a negative photoresist was found to be effective. In this oxygen electrode, a negative photoresist was used as the gas-permeable membrane, because the resulting membrane can be selectively formed, its adhesive properties are adequate, it is tolerant of chemical agents, and iti is hard to break. The agarose layer was approximately 300μm thick, and the gas-permeable membrane was 2μm thick when applied at 1500 rpm. Only the pad areas of the two gold electrodes were left exposed, while the other parts were covered with the same hydrophobic polymer used for the gas-permeable membrane. This insulates each electrode when used in an aqueous solution. The cathode and the anode were the same size. Both gold electrodes were 200μm apart in each case.

The response time of the oxygen electrode to an oxygen concentration change from saturation to almost zero was measured by adding sodium sulfite to reduce the dissolved oxygen concentration. The response profiles were similar regardless of the electrode size. It responded as soon as sodium sulfite was added to the buffer solution, a steady-state being achieved 8-10 min later. The 90% response time of the oxygen electrode was approximately 3 min, which is about 3 to 4 times longer than that of conventional oxygen electrodes. This is not considered to be a function of the cathode area but is probably governed mainly by oxygen diffusion through the agarose gel. The distance between the cathode and the gas-permeable membrane can be shortened and will be our next step.

A linear relationship was obtained for an oxygen concentration between about 1 and 7.0 ppm (saturated) when the terminal voltage between the two gold electrodes was 0.8 V, even though it deviated from linearity at lower concentrations. Similar calibration curves were obtained by using other oxygen electrodes. This confirmed that the device can be used as an oxygen sensor.

The stability of the oxygen electrode was tested. When larger oxygen electrodes (4mm wide) were used, their response decreased after they were used successively for a few times in experiments. If they are stored for several hours in a phosphate buffer solution or in distilled water with no voltage applied, however, their sensitivity returns to the initial level. The small oxygen electrode (2mm wide) could be used more than 15 times, although the resoponse decreased slightly as with the larger oxygen electrode. The micro-oxygen sensor is more stable than the conventional oxygen electrode. Poor stability of the larger oxygen electrodes was thought to be due to accumulation of reaction products in the vicinity of each of the two gold electrodes.

We then constructed a micro-glucose sensor utilizing this micro-oxygen electrode. Glucose oxidase was immobilized by cross-linking with bovine serum albumin (BSA) and glutaraldehyde on a sensitive part of the oxygen electrode.

The glucose sensor responsed as soon as the glucose solution was injected into the buffer solution in which the sensitive part of the sensor was dipped, and stabilized 5-10 min after the injection. The time delay of the oxygen electrode seemed to be more important than the thickness of the GOD-immobilized membrane. The response time of the glucose sensor will be shortedned once the oxygen electrode is improved.

The sensor responded almost linearly for glucose concentrations between 0.2 and 2 mM, which is comparable to conventional glucose sensors. The glucose sensor was sensitive to glucose at normal blood glucose concentrations (5 mM), but the sensitivity is easily shifted by adjusting the amount of immobilized enzyme.

The stability of the glucose sensor was evaluated by performing the same experiments. In this experiment, the stability of the 3 mm-wide glucose sensor was evaluated by performing the same experiments. In this experiment, the stability of the 3 mm-wide glicose sensor was evaluated. In subsequent experiments at 30°C, its response gradually decreased, but it returned to the initial level when the sensor was stored with no voltage applied, as can be seen in the case of the oxygen electrodes. The stability seems to be mainly dependent on the stability of the oxygen electrode used as the transducer.

Micro CO_2 sensor and its application

Bacterial CO_2 sensors were fabricated using the oxygen electrode mentioned above and autotrophic bacteria which utilize CO_2 as a C source. The bacteria were immobilized on the gas-permeable membrane of the oxygen electrode or inside of the groove together with the electrolyte. In the former case, another gas-permeable membrane was formed over the immobilized bacteria. In the latter case, a fabrication process that makes immobilization of bacteria on a small area possible was developed. The CO_2 sensors were designed to be used in buffer solutions for dissolved CO_2 determination. If a sample solution containning CO_2 is supplied in the buffer solution, the CO_2 permeates through the gas-permeable membrane and is assimilated by the bacteria. Then, oxygen is consumed following the elevation of the bacterial respiration. By monitoring the oxygen concentration decrease with the oxygen electrode, the CO_2 concentration can be determined.

The response time of the CO_2 sensor was 2 to 3 minutes. A linear relationship for the $NaHCO_3$ concentration was obtained between 0.5 and 3.5 mM at 30°C and pH5.5. The CO_2 sensor can be used up to 10 times.

Furthermore, the CO_2 sensor was applied to make a miniature hybrid L-lysine sensor. L-lysine decarboxylase enzyme was immobilized on the sensitive area of the CO_2 sensor. The sensor works by monitoring CO_2 dissociation by the enzyme with the bacterial CO_2 sensor above. The response time of the L-lysine sensor was 1-3 min. The optimum pH was 6.0 and the optimum temperature was 33°C. The response to L-lysine concentration was linear from 25 to 400μM. Reproducible responses were obtained by adding more than 1μM pyridoxal-5'-phosphate. The sensor had excellent selectivity for L-lysine and a stable response for more than 25 repetitive operations.

AUTOMATIC APPROACHS IN MOLECULAR BIOLOGY

Novel DNA Cleavage Techniques

(A) High Voltage Electric Pulse

Recently we found that DNA cleavage occurred during the application of high electric field pulse to cells. This discovery offers serious new and profound promise in the application to DNA transfomation experiments. Therefore, we investigated DNA cleavage phenomena under pulsed high voltage conditions.

In order to clarify factors which are strongly related to DNA cleavage, we paid special attention to electrochemical effects. According to Palecek's paper on the electrochemistry of DNA, adenine (A) and cytosine (C) are reduced at an electrode. The reduction of A and C occurs due top their protonation, which may take place in the electrode region at neutral pH provided adenosine N-1 or cytosine N-3 is able to accept the proton. Furthermore DNA containing guanine offers an anodic peak in the vicinity of -0.3 V in cyclic voltammetry with mercury electrode. A reduction peak due to adenine appears in a subsequent cycle. An anodic peak due to guanine can be observed on the same voltammogram after subsequent cycles. The guanine residue in the DNA chain represents a chemically reversible redox system because guanine is reduced around -1.8 V to dihydroguanine, which is oxidized back to fuanine around -0.3 V. In this way, DNA can directly react with the electrode. However, no research has been performed on the cleavage of the DNA chain by electrochemical reactions.

On the other hand, reactive oxygen species are known to play an important role in DNA cleavage. For example, hydroxy radical attacks the deoxyribose sugars along the backbone of the DNA molecule and breaks the DNA chain with almost no wequence dependence.

A study of DNA cleavage by pulsed high voltage is presented here to examine the effects of electrical parameters on DNA cleavage activity and to estimate how the active species formed relative to the DNA cleavage reaction (3).

A high voltage electric pulse was applied to phage DNA solution. The DNA cleavage reaction was dependent on the voltage amplitude (Figure 4), pulse number and pulse width. Radical scavengers and ESR data indicated

the possibility that active species such as OH radical were strongly related to DNA cleavage.

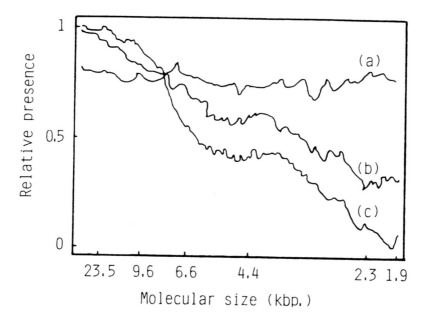

Figure 4. Densitometer scan of electrophoretic patterns with various pulse amplitudes. Forty pulses with various amplitudes were applied to the lamda phage DNA solution. Pulse amplitudes were used as follows: (a) 8 kV/cm; (b) 3 kV/cm; (c) none.

(B) Semiconductor Photocatalysts

Preferential cleavage of DNA with antitumor drugs, such as cisdiamminedichloroplatinum(II), neocartinostatin, metalloporphyrin and bleomycin, has been extensively investigated by a number of laboratories. It is necessary to consider two distinct phases for preferential DNA cleavage: site-specific binding and specific cleavage. Whilst the base specificity of DNA cleavage induced by antitumore drugs is known, the base sequence specificity is, as yet, unknown. Numerous attempts to design a small molecule, for this purpose, have been unsuccessful except for an oligodeoxynucleotide-EDTA Fe(II) probe.

Semiconductor-photocatalysts have many advantages, such as high activity, ease of modification and reusability. We have focussed on these features for the use of semiconductor-photocatalysts as cleaving agents for DNA. We have examined DNA cleavage using semi-conductor particles (4). This is the first attempt on the use of semiconductor-photocatalysts for DNA cleavage. The final aim is to achieve a site-specific cleavage of a double-stranded DNA using photocatalysts.

Semiconductor-photocatalysts were used as double-stranded DNA cleaving agents. In particular, WO_3 had the highest activity in DNA cleavage. DNA cleavage was induced by irradiation for only 30 s using WO_3, and was more inhibited by tryptophan than by histidine. Hydroxy radicals played a more important role than oxygen radicals in the cleavage process, as shown from NBT analysis and the ESR spectrum using DMPO as a radical trap. The inhibiting effect on DNA cleavage increased with increasing recongnition sites of a DNA BINDING PROTEIN.

Cell Fusion using Polymer Membrane

Cell fusion is well understood as a useful technique to develop a new strain of microorganism. The cell fusion was induced by chemical reagents such as polyethyleneglycol (PEG) and polyvinylalcohol, or by viruses such as HVJ and SV5. This reagent, however, was often lethal for cells. Therefore an efficient and useful method was required for the cell fusion of microorganisms.

Here, polymer membranes were used for the cell fusion of different strains of *Saccharomyces cerevisiae* (Table 1)(5).

A porous polymer membrane of nitrocellulose or tetrafluoroethylene (TFE) was employed for fusion of Saccharomyces cerevisiae (AH22 and D13-1A) protoplasts. Protoplasts were adsorbed on the membrane with slight suction. Some part of the protoplasts was trapped in pores of the membrane as observed by electron microscopy. The membrane retaining protoplasts was placed on a selective medium. Several colonies appeared on the medium after 5-7 days incubation at 30 C. The fusion of the two strains was ascertained by DNA content and genetic markers. Fusion frequency was 1.2 x 10-6 in the case of the TFE membrane.

Transformation by High Electric Pulse

DNA-mediated transformation of yeasts provides a new aspec;t of research on eukaryote cells. A key step of the transformation is the injection of DNA fragments through the cell membrane. Since Hinnen et al. found that the use of polyethyleneglycol (PEG) and calcium chloride caused transformation of yeast cells, similar procedures have been performed by many researchers.

Recently, the application of an electric pulse was reported by Neumann et al. , who succeeded in the transformation of mammalian cells. Disadvantages linked to external reagents usch as PEG could be avoided by this method.

Table 1 Fusion frequency by various membranes

Membrane	Physical properties			Fusion frequency (x10-6)
	Pore size (μm)	Thickness (mm)	Porosity ratio (%)	
Tetrafluoro-ethylene	1.5	0.28	80.9	0
	2.6	0.10	80.9	0.5
	3.6	0.11	87.4	1.2
	5.3	0.09	89.3	1.0
	6.7	0.09	89.1	0.3
Nitrocellulose	1.0	0.15	80	0.3
	3.0	0.15	81	0.4
	5.0	0.15	81	0.3
	8.0	0.15	-	0
Polycarbonate	2.0	0.01	<30	0
	3.0	0.01	<30	0
	5.0	0.01	<30	0

Under an electric field above a certain threshold level, a short period field above a certain threshold level, a short period of breakdown occurred in the mammalian cell membrane, and DNA outside the cells flowed in the cells. Incorporation of protein or polystyrene latex particles was also demonstrated under an electric pulse application. These reports show that the cell membrane may be repaired after a puff of electric breakdown.

We applied electric pulse to transformation of yeast cell (6). Yeast cells are smaller than mammalian cells as mentioned above. Therefore, the threshold level is theoretically higher than that for mammalian cells. However, the electric breakdown and its instant repair are also expected for yeast cells if the electric conditions are controlled.

Here, an electric pulse was applied to the spheroplats of yeast cells.

Saccharomyces cerevisiae D13-1A (a his3-532 trpl gal2) was used as a recipient strain for plasmid YRp7. Spheroplasts of S. cerevisiae D13-1A were placed under an electric pulse in the presence of YRp7. Transformants were observed when the pulse height was above 5 kV/cm. The number of transformants increased with increasing pulse height and 945 transformants per g DNA were obtained at 10 kV/cm. Introduction of DNA might be caused by breakdown of the cell membrane under high electric voltage.

DNA Sequence Detection Based on Fluorescence Energy Transfer and Polarization

Recently several studies have described the detection of specific DNA sequences, which have been increasingly more important for diagnosing viral genomes which cause infectious diseases and human gene mutations related to inherited disorders . Most of these methods have relied on immobilization of sample DNA in single stranded form (ss-DNA) to a solid support with DNA restriction fragments often being immobilized directly from gels by the method of Southern. This is followed by binding to a complementary ss-DNA probe, which is labeled using ^{32}P radioactively or non-radioactively with biotin-avidin-enzyme system . Although these techniques are highly effective, they are very time-consuming and expensive processes including electrophoresis, the immobilization of DNA, and so on.

A few simple methods for detection of specific DNA sequences have been reported. Sandwich hybridization and strand displacement technique by branch migration have been developed for diagnostic use. Although these systems have an advantage over conventional method in that it does not require immobilization of sample DNA for each analysis, they still require the lengthy immobilization of probe DNA.

We have reported previously the detection of specific DNA sequence by using a energy transfer detection system. As shown Figure 5, a ss-DNA probe labelled with an acceptor fluorophore formed a complementary duplex to analyte DNA labeled with donor by solution hybridization. In the resultant duplex, energy transfer occured and the subsequent change in fluorescence spectra observed. This technique involved no immobilization stage and produced an exceptionally low background. This detection method using fluorescence energy transfer, however, had several disadvantages including the ionic labeling of both sample and probe DNAs. So, in view of these points, the purpose of the present investigation is to construct a new simple detection system for specific DNA sequences by using a different labeling technique and fluorescence polarization (7).

If a small molecule, whose fluorescent polarization signal is samll, becomes attached to a large one, the polarized anisotropy ratio must increased due to the larger effective volume of complex. In this study, a small probe DNA was first labeled with a fluorophore. Under the appropriate hybridization conditions, the probe DNA was allowed to reassociate with a large analyte DNA to form double strands. This resultant complementary DNA duplex contributed to the slower rotational relaxation

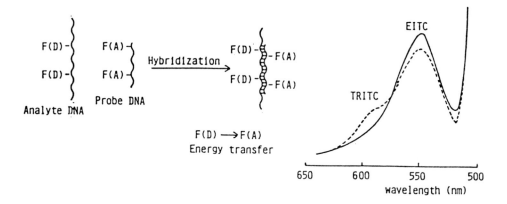

Figure 5. Concept of fluorescence energy transfer and fluorescence spctra after hybridization with probe DNA

[A] Each single-stranded DNA is labelled with a donor fluorophore (F(D)) and an acceptor (F(A)), respectively, and double strands are formed under the hybridization conditions. Because of the existance of both fluorophores in the same vicinity, a resultant energy transfer occures, and consequently the fluorescence spectrum is changed.

[B] The analyte DNA is labelled with eosin isothiocyanate (EITC) as donor, and probe DNA is labelled with tetramethylrhodamine isothiocyanate (TRITC).

(------); complementary, (———); non complementary

for the labeled fluorophore, and an enhancement of fluorescence polarization was observed.   Consequently, the duplex formation between the probe and analyte DNA was confirmed by the increase in the polarization of

fluorophore labeled with the probe DNA.   In contrast, if the analyte DNA contained no nucleotide sequence complementary to the probe DNA, (and thus no probe/analyte DNA duplex formation occured) an unchanged fluorescence polarization was observed.   Thuis specific DNA sequence may be detected by a change in fluorescence polarization without the necessity for sample or probe immobilization.

In order not to sterically hinder the formation of double strands between probe and analyte DNA, a fluorophore was labeled at the 3'-terminus of a oligonucleotide probe DNA by polymerization of amino-deoxynucleotide using teminal deoxynucleotidyl transferase. The small probe DNA was incubated with the analyte DNA and allowed to form double strands with the largen analyte DNA which contained the complementary sequence to the probe DNA. Consequently, the fluorescence polarization singnal was observed to increase due to the slower rotational diffusion of the labeled fluorophore. This method superior because it is more rapid and simpler than energy transfer method using an ionic labeling of both sample and probe DNA, gives satisfactory results for the detection of complementaru sequence in plasmid pBR322.

Direct Monitoring of Gene Expression using Luciferase Gene

The cloning and expression of the special gene in organisms need special techniques and complicated procedures. Gene expression is regulated in a temporal and tissue-specific manner during embryogenesis. It is however not easy to maniqulate tissue specificity of gene expression in transgenic animal because there are a few information about gene expression during embrionic development. We microinjected pRSV DNA containing firefly luciferase gene into the nuclei of MEDAKA (Oryzias latipes) oocytes (8). The luciferese gene is used as a visible indicator for gene expression. LUciferese gene was detected in the microinjected embryos in relation to luciferese activity. The luminescence pattern in transgenic MEDAKA was obtained by photon counting acquisition system. This system is considered to be a potential source in noninvasive and continuous monitoring of gene expression during empryonic development.

REFERENCES

1 H.Endo, K.Sode, I.Karube, and H.Muramatsu, On-line monitoring of the viscosity in dextran fermentation using piezoelectric quartz crystal, Biotechnol. Bioeng., 36 (1990) 636-641.
2 H.Endo, K.Sode, K.Ogura, K.Ohya and I.Karube, Determination of microbial concentration with a piezoelectric gum sensor, J. Biotechnol., 12(1989) 307-316.

3 E. Tamiya, Y.Nakajima, H.Kamioka, M.Suzuki and I.Karube, DNA cleavage based on high voltage electric pulse, FEBS Lett., 234(2)(1988) 357-361.
4 H.Kamioka, M.Suzuki, E.Tamiya and I.Karube, DNA cleavage using semiconductor photocatalysts, J. Molecular Catalysis, 54(1989) 1-8.
5 I.Karube, E.Tamiya and H.Matsuoka, New cell fusion method using polymer membrane, FEBS Lett., 175(1)(1984) 13-15.
6 I.Karube, E.Tamiya and H.Matsuoka, Transformation of Saccharomyces cerevisiae spheroplasts by high electric pulse, FEBS Lett., 182(1)(1985), 90-94.

7 S.Kobayashi, E.Tamiya and I.Karube, Novel DNA sequence detection method based on fluorescent energy transfer and polarization, MRS International Meeting on Advanced Materials, Vol.14 (1989) 95-101.
8 E.Tamiya, T.Sugiyama, K.Masaki, A.Hirose, T.Okoshi and I.Karube, Spatial imaging of luciferase gene expression in transgenic fish, Nucleic Acids Research, 18(4)(1990) 1072.

I. Karube (Ed.) *Automation in Biotechnology*
Proceedings of the 4th Toyota Conference, 21–24 October 1990
© 1991 Elsevier Science Publishers B.V. All rights reserved

AUTOMATED DNA SEQUENCE ANALYSIS AND THE HUMAN GENOME.

W. Richard McCombie, Anthony R. Kerlavage, Antonia Martin-Gallardo, Jeannine Gocayne, Michael FitzGerald, Susan Trapp, Takeshi Onai[#], Shoji Arakawa[*], and Mark Dubnick.

Section of Receptor Biochemistry and Molecular Biology, Laboratory of Cellular and Molecular Neurobiology, National Institute of Neurological Disorders and Stroke, National Institutes of Health, Bethesda, Maryland USA 20892

[#] Present address: Department of Physiology, Ist Division, Gunma University, School of Medicine, Maebashi, Japan.

[*] Present address: Department of Pediatrics, Kyoto Prefectural University of Medicine, Kyoto, Japan

ABSTRACT

We have applied automated DNA sequence analysis to determining the sequence of a number of neurotransmitter receptor cDNA's and genomic clones from human and Drosophila. These include alpha and beta adrenergic, muscarinic cholinergic, octopamine, nicotinic, and GABA/benzodiazepine receptors. These receptors are from two receptor gene super families, the adrenergic/muscarinic/opsin family and the nicotinic/GABA/glycine gene family. The GABA receptors comprise multiple subunits, the genes for which have been localized to human chromosomes 4, 5 and X. The alpha3 subunit gene localized to Xq28 has been isolated as a three cosmid contig from an Xq28 cosmid library. We have also sequenced a three cosmid (58.8kb) contig from chromosome 4p16.3. These lambda and cosmid clones have been sequenced utilizing a variety of strategies including ExoIII deletions, sonication followed by size selection and cloning into M13, double stranded sequencing with forward and reverse primers and by primer walking. Sequencing was performed using fluorescent dye primers and dye terminators on ABI 373A DNA sequencers with Macintosh software. The new software typically allows reads in the range of 450-550 bases per template. Using four ABI 373A sequencers, between 42,000 and 52,000 bases of raw sequence can be produced each day. Automation of the various steps and application of these approaches to human genome sequencing will be discussed.

The standard methods of DNA sequencing are the dideoxy chain termination method developed by Sanger, *et al* (1977) and the chemical degradation method of Maxam and Gilbert (1977). Advances in biological methods to generate single strand templates for chain termination sequencing were made in the early 1980s when the bacteriophage M13 vectors were developed (Messing and Vieira, 1982). The widespread availability of these vectors popularized the chain termination method and by the mid 1980's virtually all DNA sequencing used this general technique.

DNA sequencing using chain termination is carried out in the following manner. A DNA clone containing the sequence of interest, or template, is grown and purified. This DNA is then mixed with a specific single strand DNA primer, usually an M13 "universal" primer. The primer is complimentary to a specific DNA sequence in the template. With the M13 universal primer this sequence is in a position adjacent to the M13 cloning site into which genes or other DNA regions that are to be sequenced are placed. Starting with such a primer template combination, the DNA dependent DNA polymerases, which are enzymes that replicate DNA, can be used to extend the primer and make a new DNA strand that is complimentary to the template by the incorporation of the four deoxynucleotide triphosphates. If however, this *in vitro* DNA replication reaction contains three nucleotide triphosphates and a mix of the fourth nucleotide triphosphate and a dideoxynucleotide analog, the mixture can be used to determine the DNA sequence of the template. This is because in a certain proportion of the primer molecules undergoing extension, a dideoxynucleotide will be incorporated rather than a deoxynucleotide. The dideoxynucleotide, missing a hydroxyl residue, can no longer be used as a substrate and the next base in the chain cannot be added. These basic characteristics can be used to determine the sequence of the template as seen in Figures 1 and 2. An example will illustrate the method. A given primer, template and enzyme are mixed with the three deoxynucleotide triphosphates A, C and G, and a mix of deoxy T and dideoxy T. Using the base pairing rules the strand of DNA beginning with the primer is covalently extended with an A, C or G each time the base T, G or C, respectively, appears in the template. However, if the base A appears in the template, the complimentary deoxy T or dideoxy T is incorporated. The incorporation of deoxy T enables the chain extension to continue with the next base. The incorporation of dideoxy T ends the chain. Thus, a certain number of molecules end at each position where a T should be incorporated. The same reaction is done separately with each of the bases, in turn, being a mix of deoxy and dideoxy nucleotides. This leads to a population of molecules differing in length by one base with the final base in the chain known since it is the particular dideoxy base used in that reaction. These four reaction mixes can then be fractionated in four lanes of a denaturing polyacrylamide gel that is capable of resolving differences in length of one base. The shortest set of molecules (first terminated) has the first base on their end as a dideoxy nucleotide, the next set has the second base and so forth. The sequence is read from the gel as a ladder as shown in Figure 1.

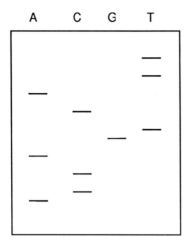

ACCAGTCATT

Fig. 1. A schematic diagram of an autoradigraph of a manual sequencing gel. The results form a ladder from which the sequence can be read. The direction of electroproresis is from top to bottom with the smaller molecules migrating more rapidly through the gel. The letters at the top of the gel represent the dideoxynucleotide present in each reaction. The sequence at the bottom represents the sequence that would be read from the sequencing ladder shown in the figure.

This is made possible by the addition of radioactive bases in the chain to enhance sensitivity. The gel is placed on a piece of filter paper following electrophoresis, dried and subjected to autoradiography. The sequence is then read from the resulting X-ray film by reading the terminated bases from the bottom up (smallest molecules to largest). Automation of this basic biology was the one fundamental technical advance necessary to make large scale sequencing possible. The use of fluorescent primers to initiate the chain termination reactions was described in 1986 (Smith, *et al*) and was the basis of an automated DNA sequencer developed by Applied Biosystems, Inc. (ABI). Rather than using radioisotopes to detect the molecules, this chemistry and instrumentation uses fluorescent primers in the reactions. A schematic of the reactions using such a machine is shown in Figure 2. While somewhat less sensitive than the radioisotope method, this change allows a different fluorochrome to be attached to the primer in each of the four reactions. As a result, all four reactions can be run in a single gel lane. As the sample passes a laser which excites the fluorochrome, each band in the gel is detected and, since a particular fluorescent label corresponds to a particular dideoxynucleotide present in the reaction, the sequence can be determined. The use of one lane per set of reactions greatly minimizes the lane to lane distortion seen on radioisotope sequencing gels. This enabled Applied Biosystems to develop software which automatically determined the sequence from a given template. In addition, the steps of gel removal, drying and autoradiography are eliminated due to the fact that the gel is scanned as it runs and data are analyzed automatically immediately

22

upon completion of the run. We have applied this technology to the scale-up of DNA sequencing

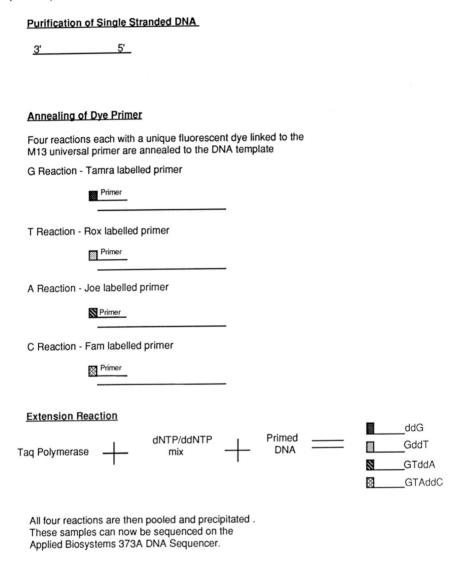

Purification of Single Stranded DNA

3'　　　　　　　5'

Annealing of Dye Primer

Four reactions each with a unique fluorescent dye linked to the
M13 universal primer are annealed to the DNA template

G Reaction - Tamra labelled primer

■ Primer

T Reaction - Rox labelled primer

Primer

A Reaction - Joe labelled primer

Primer

C Reaction - Fam labelled primer

Primer

Extension Reaction

Taq Polymerase ╀ dNTP/ddNTP mix ╀ Primed DNA

■＿＿＿＿＿ ddG
■＿＿＿＿＿ GddT
＿＿＿＿＿ GTddA
＿＿＿＿＿ GTAddC

All four reactions are then pooled and precipitated .
These samples can now be sequenced on the
Applied Biosystems 373A DNA Sequencer.

Fig. 2 Single-stranded sequencing protocol

to the megabase level necessary to carry out analysis of the human genome. The phases of a
large scale, automated sequencing project, as carried out in our laboratory, are shown in Figure
3.

I. **Creation of clones for sequencing**

II. **Sequencing of DNA**

III. **Assembly of sequence data into contigs**

IV. **Verification of final sequence construct**

V. **Computer analysis of sequence data to determine areas of potential interest**

VII. **Biological analysis based upon final sequence and computer findings**

Fig. 3. Phases of a large-scale sequencing project

Automated sequencing began in our laboratory in February of 1987. At that time the sequencing reactions were analyzed with the Applied Biosystems 370A DNA Sequencer using software written for the Hewlett-Packard Vectra computer. The automated sequencer proved itself to be a rapid, highly accurate and efficient device for DNA sequencing (Gocayne, *et al*) Sixteen samples could be run on a sequencing gel yielding about 400 bases of sequence per sample for a total of 6400 raw bases of sequence per day. Using this technology, we sequenced a number of neurotransmitter receptor genes (Fraser, *et al*, 1989; Onai, *et al*, 1989; Arakawa, *et al*, 1990). Current technology has changed the output and ease of use of these machines dramatically. The 373A has been updated with software which runs on the Macintosh and gels can now be run with 24 samples, each yielding 450-500 bases of sequence for a daily output of about 10,800 bases of sequence each day per machine. Through the optimization of procedures which allow us to process samples in parallel, we can now operate four sequencers per day. In order to do this more efficiently, we have been evaluating an engineering prototype of an automated DNA robotic workstation form Applied Biosystems which carries out the DNA sequencing reactions. The robot utilizes a single metal displacement probe which eliminates the need for disposable tips and results in nanoliter accuracy. Our workstation is currently programmed to react 96 templates every 24 hours. When reactions are performed by the workstation, single stranded runs read accurately out to approximately the same number of bases as when the sequencing reactions are done manually. With this automation in place, performing sequencing reactions and running sequencing gels is no longer the bottleneck in large-scale sequencing; rather, the bottleneck has become the processing, assembling and analyzing the data.

During the electrophoresis process, data are collected on Macintosh IIci computers using the ABI Data Collection software. Prior to the run, information about the samples is entered on a Sample Sheet within the program. These fields provide information about the sample or instructions to the software concerning how to analyze the data. A file name is assigned for the reacted template to be run in each of the 24 lanes on the gel. The file names are coded to the templates by a naming scheme which incorporates the project and a unique identifier for each template set in the project. A sample name which relates to the source of the DNA is also given. Comments are added if needed for clarification. The first of the fields which provide instructions to the software indicates the type of primer used in the sequencing reaction. The software uses this information to call the correct files for analysis. The other fields indicate whether the ABI Analysis software should be called automatically, whether a given lane contains sample, and which bases should be called by the Analysis software. The inclusion of this information at the earliest data entry step demonstrates the concept of entering information only once and then passing that information on to all software which requires it. This concept is a driving force in software development for an automated sequencing lab.

Once the Data Collection software is started, the process of collection and analysis is automatic. The length of the run is set in the Data Collection software and when this time has expired, the Analysis software is automatically called and each lane is analyzed in turn. The Analysis software produces a Sample File for each lane, which contains the raw and enhanced chromatogram data, the bases called by the software and annotation including the information entered in the Sample Sheet as well as the date and identification of the sequencing machine. The software will also generate a text file containing the called bases in one of a variety of formats.

Each Sample File generated by the Analysis software is approximately 150 kbytes in size. Thus, the daily output from four sequencers is (24 lanes X 4 sequencers X 150 kb) 14.4 megabytes. We have chosen the Alphatronics Erasable Optical Drive as the storage device for these data. The drive uses two sided disks which can hold approximately 300 megabytes of data per side. Two of these devices are mounted on a Macintosh IIfx which is utilized as a file server (Fig 4). The Sample Files are transferred from each of the 373A Macintosh computers to the optical disks via the network transfer software, TOPS (Sun Microsystems). These data can then be accessed via TOPS from any Macintosh or PC, or via FTP from any Unix device on our network. Routine backups of the working data are made to a second optical disk and data from completed projects are archived on the same media.

The first step in the sequence assembly process is to edit the sequence data using the ABI editing software, SeqEd. This software can be run on any of the Macintosh II family of computers on our network (Fig 4). The software can call the Sample Files on the optical disks over the network. Vector sequence, if present, is identified and removed. Ambiguities at the trailing end of the sequence are likewise removed. The changes are saved to the Sample File which now contains both the edited and unedited versions of the base calls. A text file containing the edited sequence in IntelliGenetics (IG) format is written by the software. Files of this type for each template are transferred to a Sun 4 computer using NCSA Telnet software and the FTP protocol on the Macintosh.

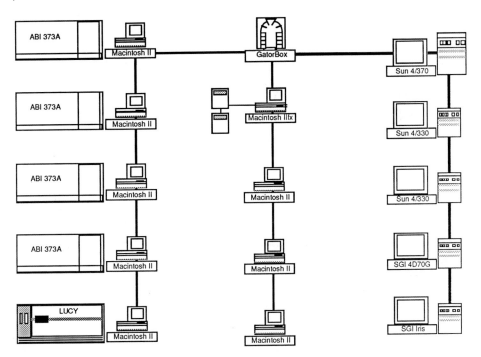

Fig. 4. A schematic diagram of the computer network for DNA sequence data acquisition and analysis. Data from ABI 373A sequencers are collected and analyzed on Macintosh IIci computers. The sample files are transferred over a network to optical disks attached to a Macintosh IIfx file server for working storage and archiving. Several Macintosh IIci computers are used for preliminary editing of the sequences. Edited files are transferred to a Sun4/370 file server where they are assembled into contigs and analyzed using any of five Unix workstations. All computers are on ethernet and file transfer between the Macintosh and Unix computers is handled by a Cayman GatorBox.

The fragments are assembled using the IntelliGenetics GEL software. Once files have been entered into a GEL project, they are searched to assure they do not contain vector sequence. This can be performed automatically in GEL. Each fragment is, in turn, compared with all other fragments in the project and matches are presented to the user, who can either accept or reject the match. If accepted, the two fragments are merged into a contig and this new contig is then searched against the remaining fragments. In this manner, many fragments can be assembled into a relatively small number of contigs. For a typical cosmid size project (40 kilobases), approximately 600 to 700 individual fragments are sequenced and assembled.

Once the sequence of the region has been determined we analyze the sequence to determine potential biological functions. This requires a massive amount of computational power in some instances. Our internal computer network connects with larger, more powerful computers as shown in Figure 5.

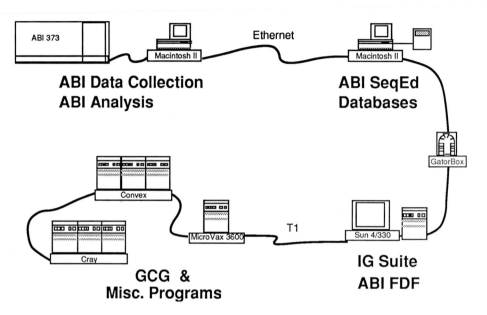

ABI Data Collection
ABI Analysis

ABI SeqEd
Databases

GCG &
Misc. Programs

IG Suite
ABI FDF

Fig. 5. Schematic diagram of DNA data flow. ABI software, including the Data Collection, Analysis, SeqEd, and FDF front end programs, are run on Macintosh IIci computers. The ABI FDF hardware is located in a Sun workstation. The IntelliGenetics Suite is run on Sun workstations for sequence assembly and analysis. Further analysis of finished sequence is carried out using the Genetics Computer Group package on a NIH Convex computer and various other programs are run on a MicroVax or a Cray X-MP computer linked to the laboratory by a T1 connection.

At some point after a sequence is considered "finished", it would be prudent to make the sequence available to the scientific community. Typically, sequences are submitted to GenBank or some other public database (EMBL, DDBJ). The submission of large numbers of sequences or of very large sequences requiring extensive annotation necessitates some automation of the submission process. We are working with Los Alamos National Laboratories to establish their database Schema at our site. The system consists of the GenBank schema in the Sybase relational database format and a tool, the Annotator's Workbench, which simplifies the entry of annotation. Sequences entered via this method can be automatically entered into the public database by conducting an electronic transaction with GenBank.

The current theoretical throughput of our system is roughly 38,400 - 48,000 bases of raw sequence per day. A variety of factors keep us from achieving this rate on a daily basis. While having four sequencers running per day strains the system, using two or three on a daily basis is relatively easy at this time. The output from three sequencers creates a severe data assembly and analysis problem. Current software at all stages is inadequate to handle the amount of data we are generating. The greatest single rate limiting step in the entire process is the assembly of the data into contigs on the computer. For large scale sequencing to advance, new software at all stages of the process must be developed. As part of that process, we are currently evaluating the use of the Fast Data Finder board from ABI which incorporates an array of TRW chips designed for very fast string searching.

Even with these limitations we have generated roughly 2,000,000 bases of raw sequence data since April of this year. The rate at which we can generate both raw and finished sequence data has increased dramatically as our ability to fully utilize existing automated equipment has increased, and our current steady state output is approximately 750,000 raw bases per month. This has brought forth new challenges in the analysis of this data. We are moving to meet this challenge by the development of new strategies for data analysis using the computational powers of supercomputers and the biological power of the polymerase chain reaction.

Acknowledgement

The authors wish to thank Dr. J. Craig Venter, in whose laboratory this work was performed, for his support and encouragement.

References:

Arakawa, S., J. D. Gocayne, W. R. McCombie, D. Urquhart, L. M. Hall, C. M. Fraser and J. C. Venter. 1990. *Drosophila* octopamine receptor: cloning, sequence, chromosomal location and permanent expression in CHO cells. Neuron **2**: 343-354.

Fraser, C. M., S. Arakawa, W. R. McCombie, and J. C. Venter. 1989. Cloning, sequence analysis, and permanent expression of a human alpha$_2$ adrenergic receptor in Chinese hamster ovary cells: evidence for independent pathways of receptor coupling to adenylate cyclase attenuation and activation. J. Biol. Chem. **264**: 11754-11761.

Gocayne J., D.A Robinson, M. G. FitzGerald, F. Z. Chung, A. R. Kerlavage, K. U. Lentes, J. Lai, C. D. Wang, C. M Fraser and J. C. Venter . 1982. Primary structure of rat cardiac beta-adrenergic and muscarinic cholinergic receptors obtained by automated DNA sequence analysis: further evidence for a multigene family. Proc Natl Acad Sci U S A **84**: 8296-300.

Maxam, A.M. and W. Gilbert. 1977. A new method for sequencing DNA. Proc Natl Acad Sci U S A **74**: 560-4

Messing, J. and J. Vieira. 1982. A new pair of M13 vectors for selecting either DNA strand of double-digest restriction fragments. Gene **19**: 269-76.

Onai, T., M. G. FitzGerald, S. Arakawa, J. D. Gocayne, D. A. Urquhart, L. M. Hall, C. M. Fraser, W. R. McCombie and J. C. Venter. 1989. Cloning and sequence analysis and chromosome localization of a Drosophila muscarinic acetylcholine receptor. FEBS Letters **255**: 219-25.

Sanger,F., G. M. Air, B. G. Barrell, N. L Brown, A. R. Coulson, C. A. Fiddes 1977. Nucleotide sequence of bacteriophage phi X174 DNA. Nature **265**: 687-95.

Smith, L.M., J. Z. Sanders, R. J Kaiser, P. Hughes, C. Dodd, C. R. Connell, C. Heiner, S. B. Kent and L. E. Hood. 1986. Fluorescence detection in automated DNA sequence analysis. Nature **321**: 674-9.

I. Karube (Ed.) *Automation in Biotechnology*
Proceedings of the 4th Toyota Conference, 21–24 October 1990
© 1991 Elsevier Science Publishers B.V. All rights reserved

AUTOMATED DNA SEQUENCING

Douglas A. Amorese
E. I. du Pont de Nemours and Company (Inc.)
Medical Products, Glasgow Site
Wilmington, Delaware 19898

ABSTRACT

Traditionally, DNA Sequencing has been a labor-intensive and time-consuming laboratory activity. Each reaction required a precisely controlled, complex reaction mixture and/or reaction condition. The DNA fragments generated in these reactions were typically radiolabeled and detected by autoradiography following separation, based on size, on denaturing polyacrylamide gels. Additionally, the data was manually interpreted and entered into computers for analysis.

Over the last several years, a variety of approaches have been taken to automate various aspects of DNA sequencing. The introduction of fluorescent labels into DNA sequencing fragments has led to the design of systems that control the electrophoresis process and automate data collection and analysis. With the introduction of fluorescent-labeled dideoxynucleotides, the process of generating the sequencing fragments was also simplified. This paper will focus on how fluorescent DNA sequencing systems have simplified various aspects of the sequencing task, from the generation of the sequencing fragments to the entry of the data into computers for analysis.

INTRODUCTION

As our knowledge of the mechanisms of control of cellular processes increases, there is an increased need to understand them at an even greater level. This ultimately requires the determination of the nucleotide sequence of the gene(s) and associated regulatory regions responsible for the processes. This has resulted in an interest in sequencing ever larger amounts of DNA. A consequence of this need is the need to develop efficient and cost-effective methods for DNA sequence determination.

The process of determining the nucleotide sequence can be broken down into three basic tasks; 1) the generation of a set of "DNA sequencing fragments," 2) the fractionation of these fragments based on length, and 3) the detection and ordering of these fragments. DNA sequencing fragments are a nested set of DNA fragments that have one end in common, vary in length, and have a known nucleotide at the opposite end. Typically, multiple reaction sets are performed in which a single type of fragment is generated per set (e.g., all of the fragments of one set end with a guanosine residue, another set, cytosine, etc.). If the last nucleotide on the end of the fragment is known, the sequence of the DNA can be determined by sorting the fragments generated in each reaction by length and determining in which reaction the successively longer fragments were generated.

There are two general methods by which DNA sequencing fragments are typically generated. End-labeled DNAs can be partially degraded through a series of nucleotide-specific chemical modification reactions followed by hydrolysis of the phosphodiester linkage at the position(s) of the modified base (1). The labeled end serves as the "common" end, the partial/random modification of a particular nucleotide species along the length of the DNA strand results in fragments of varying length, and the reaction that the fragment was generated in indicates which base was at the end. Alternatively, DNA sequencing fragments can be generated during the synthesis of DNA from a nucleic acid template by specific elongation terminators (2). Base specific termination is achieved by the enzymatic incorporation of a dideoxynucleotide, a nucleotide analogue that lacks the 3' hydroxyl required for chain elongation. In this case, the primer used to initiate the synthesis serves as the common end, the variability in length is a consequence of the competition between the deoxy and dideoxynucleotide for incorporation by the polymerase at any given point in the template, and again the reaction that the fragment was generated in indicates which base is at the end.

The DNA sequencing fragments generated in these reactions must be fractionated with single-base resolution. That is, it must be possible to distinguish which fragment is longer when comparing fragments that are only one nucleotide different in length. This is achieved by denaturing the fragments generated in each of the reactions from their complementary strands and electrophoresing them in adjacent lanes on a high-resolution urea-containing polyacrylamide gel (DNA sequencing gel).

Sequencing fragments are traditionally labeled with radioisotopes and detected by autoradiography following their electrophoretic separation. The data analysis involves determining the relative lengths of the fragments detected on the autoradiogram and identifying in which lane they are (in which reaction they were generated). The nucleotide sequence is deduced by recording the lane position of each successively longer fragment.

Since the introduction of these DNA sequencing methods, many researchers have sought to improve the quality of data, increase the amount of data obtained in a single experiment (efficiency), and decrease the labor (bench time) associated with this highly repetitive process. As a consequence of this, nearly every facet of DNA sequencing, from the generation of sequencing fragments to their separation and analysis, has seen improvements. New vectors have been introduced (3) that have simplified the task of template preparation. Methods have been developed that make use of alternative types of vectors (4) or even eliminate the need for vectors and cloning in the production of templates. New polymerases have been introduced that have increased the flexibility in the production of DNA sequencing fragments and improved the quality of these fragments. New nucleotide analogues have been developed to reduce secondary structures within fragments. These have resulted in improved resolution and reduced band compressions in the gels, which together, ultimately improve the accuracy of the data. Additionally, new gel systems which permit the determination of a larger number of bases per experiment have also been developed (5,6). Unfortunately, with all of these improvements taking place in the generation of the data, there was little improvement in the tedious task of reading the sequence from the autoradiogram and entering it into a computer. The introduction of digitizing pads simplified the task of recording the data and transferring it to a computer but had little impact on the time or effort required to "read" the data from the films.

To further simplify the task of DNA sequencing, systems were developed that offered "real time detection" of sequencing fragments as they electrophoresed past a detector. These systems eliminated the need for autoradiography - the fixing of the gel, transfer to a solid support, exposure and development of the film, as well as the time required to read the autoradiograph and enter the data into the computer. Within the last 3 to 4 years, a number of these "automated" systems have become commercially available (see reference 7 for a review of instrumentation). This chapter will focus on four fluorescent DNA sequencing systems: the Applied Biosystems, Inc. Model 370A/373A; the

Du Pont Genesis™ 2000 DNA Analysis System; the Hitachi SQ3000 DNA Sequencer; and the Pharmacia Automated Laser Fluorescent (A.L.F.) DNA Sequencer™. In addition to reviewing the basic instruments/detection systems, this paper will discuss how these new sequencing systems have affected the way sequencing fragments are generated and how these systems have led to new approaches to DNA sequence analysis.

SYSTEM OVERVIEW

DNA sequencing technology has undergone significant evolution since it was originally described. The introduction of real time data collection (detection of sequencing fragments as they electrophorese through the gel past a detection zone) was perhaps the next logical step in this progression. While one automated system that does real time data collection with isotopically labeled DNA fragments (the Acugen 402 DNA Sequencer from EG&G) has been commercialized, most of the effort has been directed toward using fluorescent labels. Fluorescent labels provide the high sensitivity required for real time data collection, can be accurately quantitated, are available with a variety of emission wavelengths that can be spectrally distinguished, and eliminate the difficulties associated with using isotopic labels.

Real time data collection differs significantly from the traditional detection system, autoradiography. With autoradiography, the signal is accumulated over a controlled period of time. If the amount of isotope present in the gel is not sufficient to produce a detectable image on the initial film, a new film can be exposed to the gel for a longer time. The background may also increase with longer exposures, so there are practical limits to this detection method. However, this method is quite sensitive, has a broad dynamic range, and is forgiving. In current real time detection systems, the label is detected as it electrophoreses through the gel past a fixed-point detector. Therefore, only a brief period of time is available for capture of the signal. If the signal (actually signal to noise ratio) is too low to be detected or accurately analyzed, the data that is obtained may not be interpretable.

The electrophoretic conditions used to fractionate the sequencing fragments also differ between traditional and real time detection systems. When fragments are going to be detected by autoradiography, they are denatured, applied to a gel, and the gel is typically electrophoresed until a marker dye(s) has migrated a fixed distance. The autoradiograph records the positions of the sequencing fragments at the time the electrophoresis was stopped. The sequence is obtained by determining the relative lengths of the sequencing fragments starting with the smallest fragments and proceeding toward the larger fragments. In this system, the largest fragments (400 to 500nt) may have electrophoresed only a few centimeters. With automated detection systems, fragments are not detected until they have electrophoresed approximately 75% of the length of the gel. At the time when a gel would have finished electrophoresing in a traditional system, only 50 to 70 sets of fragments would have passed by the detection zone of an automated system. This means that the gels used with automated systems must electrophorese for longer times and maintain clean, sharp banding of fragments for greater distances. An additional concern for the automated systems that is of minimal concern in autoradiographic systems is the rate of electrophoresis. With autoradiographic detection, the gels can electrophorese at different rates from day to day because the marker dye(s) is used as the indicator for when to stop electrophoresis. With most of the automated systems, the signal processing is "tuned" to a particular throughput. If the sequencing fragments electrophorese past the detector too rapidly, the data may become difficult to analyze accurately. Thus the gels, in addition to having to perform for longer periods of time, have to perform more consistently in real time detection systems. These distinctions between real time and static detection have provided a challenge to the design of instruments as well as the development of reliable, reproducible protocols.

The commercially available fluorescence-based detection systems are similar in that they all use standard DNA sequencing gels to fractionate the DNA fragments and air-cooled argon lasers to excite the dye molecules. The systems differ in the means used to detect the emitted light and the method used to identify the type of base that has passed through the detection zone. Two general approaches have been taken to determine the base sequence with fluorescent systems. One approach uses four distinguishable fluorescent labels, each corresponding to one of the four sets of reaction products (A, C, G, and T-terminated fragments), that are electrophoresed as a mixture in a single lane (ABI's 370A/373A and Du Pont's Genesis 2000). The alternative method uses a single label in four separate reactions that are then electrophoresed in four separate lanes (Hitachi's SQ3000 and Pharmacia's A.L.F.).

The Applied Biosystems, Inc. (ABI) Models 370A/373A have similar detection systems and for the purposes of this section will be treated as a single system. The system originally described by Smith et al. (8) used four fluorescent dyes (a fluorescein, tetramethylrhodamine, 4-chloro-7-nitrobenzo-2-oxa-1-diazole (NBD), and Texas red) that were attached to the 5' end of the sequencing primers. ABI has not disclosed the specific dyes that they use with their commercial system. Due to the significant differences in the excitation maxima of the dyes, this system uses two different wavelengths of light or laser lines (488 and 514nm) to excite the fluorophores. The fluorescent emission of the excited dyes is detected by a single photomultiplier tube (PMT). A set of four filters, mounted on a wheel, is used to discriminate the wavelengths of light that are being emitted. The laser focusing optics, filter wheel, and the PMT are mounted on an optical carriage that travels across the face of the gel (figure 1). The optics direct the laser beams at a point on the gel and the PMT detects the light that is emitted. Since the system uses a single PMT, the gel must be scanned four times for each "complete" data point. That is, one of four filters is placed over the PMT during each pass of the carriage in order to detect/discriminate between the various dyes.

Du Pont's Genesis 2000 DNA Analysis System also uses four fluorescently distinct labels (9). These fluorophores are members of a family of succinyl fluoresceins that differ from each other based on the extent and position of methyl substitutions in the fluorescein ring. The methyl substitutions shift the emission maxima of the dyes. The emission maximums range from 505 to 526nm. Since the Genesis 2000 uses four closely related dyes, the fluorescent DNA fragments can be efficiently excited by a single 488nm laser line. In this system, a mirror mounted on a high resolution stepper motor sweeps the laser beam across the face of the gel (figure 2). When a fluorophore is excited in the Genesis 2000 system, it is detected simultaneously by two stationary PMTs. A filter, mounted in front of each of the PMTs, limits and differentiates the wavelengths of light that are detected in each of the two channels. Since each of the dyes has a unique emission spectra, the relative amount of light detected in the PMTs will differ for each fluorophore.

The Hitachi SQ3000 DNA Sequencer (10) differs from the systems described above in that it uses a single fluorescein isothiocyanate (FITC) label, and the 488nm laser beam passes through the side of the gel

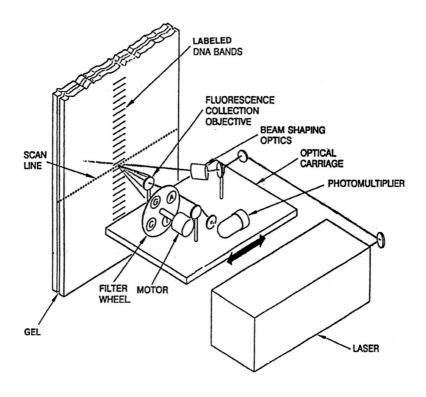

Figure 1. Schematic of the ABI Model 370A detection system. As the DNA sequencing fragments electrophorese through the detection zone, the fluorophores are excited by the laser and the emitted light is detected by the PMT. The gel is scanned by moving the optical carriage across the face of the gel. Since the system uses four filters with a single PMT, the gel must be scanned four times for each complete data point (i.e., one pass for each filter).

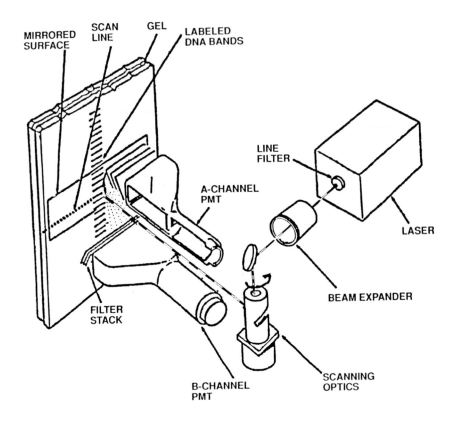

Figure 2.   Schematic of Du Pont's Genesis 2000 detection system.   This system uses a mirror mounted on a stepper motor to direct the laser beam.   As the motor rotates, the beam is swept across the face of the gel.   The fluorophores attached to the sequencing fragments are excited by the laser beam as they pass through the detection zone.   The emitted light is detected simultaneously by two PMTs that have differential band pass filters.   (Reproduced with permission from reference 9.)

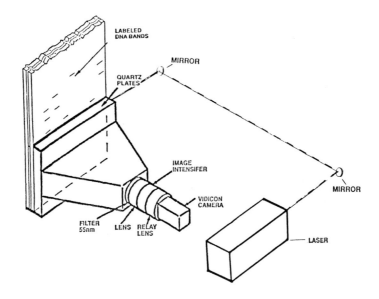

Figure 3. Schematic of the Hitachi SQ 3000 DNA Sequencer. In this system the laser beam enters through the side of the gel as opposed to sweeping across the face. As the sequencing fragments electrophorese past, the emitted light is detected by a camera and the position across the face of the gel is recorded.

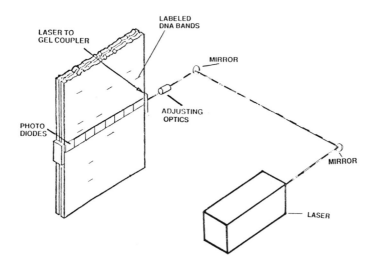

Figure 4. Schematic of the Pharmacia A.L.F. system. This system also uses a laser beam that enters through the side of the gel. With this system, the emitted light is detected by a linear array of photodiodes.

rather than across the face of the gel (figure 3). The emitted light in this system is detected with a vidicon camera system as opposed to the PMTs used in the ABI and Du Pont systems. Since each fragment has the same fluorophore attached, the reaction products from the A, C, G, and T reactions must be electrophoresed in four separate lanes. Therefore, in this system the nucleotide's identity is determined based on the lane position that it is detected in rather than as a function of its emission spectra. With the "side-on" laser detection systems, there are no moving parts, and a larger proportion of the band is excited compared with point excitation across the face of the gel. However, because four separate lanes are run, it is imperative that the rate of electrophoresis in each of the four lanes is identical. The rate of electrophoresis can vary from lane to lane due to differential rates of heat dissipation ("smiling"), differential amounts of salt in the sample (this could be in the form of residual reaction buffer, unincorporated nucleotide, etc), different extents of cross-linking within the gel matrix (or other irregularities within the matrix), and/or different masses of material applied to a lane.

The A.L.F. system developed at EMBL (11) and commercialized by Pharmacia also uses a side-on laser, a single-dye species, and therefore four lanes per sample (figure 4). This system also relies on equivalent rates of electrophoresis between the four lanes for accurate base assignments. In contrast to the other systems, the emitted light in this system is detected by a linear array of solid state photodiodes (one for each lane). To ensure alignment of the lanes with the photodiodes, the system uses a gel casting plate with lines etched in the glass. While this will effectively position the plate properly, it does not force the DNA to electrophorese in a straight path. As with the Hitachi SQ3000, there are no moving parts in the Pharmacia instrument.

GENERATION OF LABELED DNA SEQUENCING FRAGMENTS

The majority of this section will deal with generating fragments by dideoxy DNA sequencing. However, it should be noted that the ability to do chemical degradative sequencing has been demonstrated on several of the instruments (12,13,14,15). Unfortunately, signal processing software for this type of data analysis has not been released.

There are several important factors that make the generation of fluorescent DNA sequencing fragments different from the generation of radiolabeled fragments. These should be taken into consideration when developing protocols for the generation of fluorescent DNA sequencing fragments. For optimal detection, it is desirable to achieve an equivalent amount of label associated with each size class of fragment. Fluorescent DNA sequencing fragments are typically labeled with a single fluorophore per fragment, whereas radiolabeled fragments contain internal labels throughout their length. This means that the fragment length distribution for fluorescent sequencing needs to be considerably different than that for radiometric sequencing. Fragments labeled with radioactive deoxynucleotides have a constant specific activity; the amount of label associated with a fragment is proportional to the length of the fragment. As the fragments get longer, they contain increasing amounts of label. Therefore, in order to have equal intensity of two fragments of different length, there must be a larger quantity of the shorter fragment. In fluorescent sequencing, where every fragment has a single label, there must be an equal number of fragments of each length to achieve a uniform signal intensity. The consequence of this is that as the fragments get larger, there is an increase in the mass of DNA in a given band (decreasing density of label) as opposed to an equal mass in traditional sequencing (even density of label). This can have an adverse effect on the resolution of the larger DNA fragments. Another difference between the two general labeling methods involves the removal of unincorporated label. With radiometric sequencing the unincorporated nucleotide is generally electrophoresed off the bottom of the gel and therefore, not detected on the X ray film. With fluorescent sequencing, the unincorporated label is excited and detected as it electrophoreses through the detection zone. Since the detection systems are designed to be sensitive to very low light levels, if large amounts of fluorescent material (unincorporated nucleotide) are allowed to pass by the detectors, the detectors may become saturated (no longer sensitive to small changes in light level). If the detectors do become saturated, they may take several minutes to recover and stabilize. Therefore, post-reaction cleanup (ethanol precipitation or spin column chromatography) is usually employed. Reaction cleanup has an additional benefit in that it can also reduce the amounts of other contaminants (i.e., proteins, salts, and glycerol) that may be present in the sample. While these compounds may not be fluorescent, they may have an adverse effect on resolution and/or scatter light as they pass by the detectors and therefore, interfere with data collection.

Two general methods have been described for fluorescently labeling DNA fragments. In one method, fluorescent oligonucleotides are used as sequencing primers (8). Generally, the fluorophore is covalently attached to the 5' end of the oligonucleotide following synthesis. The labeled oligonucleotide is then annealed to a target sequence and elongated with a DNA polymerase in the presence of all four of the deoxynucleotides and one of the dideoxynucleotide analogues. With this approach, independent of whether a single dye or four distinguishable dyes are used, four separate sequencing reactions must be performed per template. If a single fluorescent label is used, the reactions are very similar to those used in traditional sequencing, that is, following annealing, the sample is divided into four reactions each containing a single dideoxynucleotide. However, if four different labels are used in addition to the four separate enzymatic reactions, four separate annealings are also required. While additional annealings may not have a significant impact on the amount of effort required to generate the sequencing fragments or the results obtained with simple single-stranded templates, it can affect the quality of the results obtained when double-stranded templates are used. With the more complex templates there is a greater likelihood of achieving differential annealing efficiencies between the four annealing reactions. Since the signal is proportional to the amount of primed template, poor or variable annealing efficiencies can result in an under-representation of a given set of fragments. Primer labeling, while still sensitive to secondary priming (nonspecific annealing of the primer to a site that has a related sequence), is not sensitive to false priming (generation of sequencing fragments from RNA or other short nucleic acid fragment contaminants acting as primers).

The alternative to 5' end labeling with a fluorescent primer is 3' end labeling with a fluorescent terminator (9). This approach offers several advantages over both the traditional and the labeled primer methods. In this procedure, one unlabeled sequencing primer is annealed to a template and elongated with a DNA polymerase in the presence of four deoxynucleotides and fluorescently labeled dideoxynucleotide(s). If the system uses a single dye and four separate lanes, four individual reactions are still required. However, if the detection system has the capability of distinguishing multiple fluorophores, then a single reaction can be run per template. In these systems, each dideoxynucleotide is labeled with a unique dye. In the case of the Genesis 2000, each terminator is labeled with a succinyl fluorescein which has a different emission maxima: the ddGTP has a maxima at 505nm, the ddATP at 512nm, the ddCTP at 519nm, and the ddTTP at 526nm. This not only simplifies the preparation of the sequencing fragments by reducing the number of reactions per template from four to one, it also minimizes the potential difficulties that can arise from differential priming/elongation efficiencies between reactions. A traditional (or single fluorescent label) sequencing reaction requires four separate nucleotide mixtures and in fact is really four separate reactions. Each mixture must contain an equivalent amount of the nucleotide corresponding to the labeled nucleotide in order to achieve equivalent specific activities of label in each of the reactions and a carefully balanced ratio of deoxy to dideoxynucleotide in order to generate equivalent fragment length distributions between the reactions. With terminator labeling systems that use four distinct fluorophores, a single reaction mixture containing all four dideoxynucleotides is used. This allows suppliers to offer the nucleotides as a single prepared mixture in their optimum relative ratios and therefore eliminates a point of potential error. Recently, Du Pont has taken this one step further and produced a "tableted" reaction mix that contains both the deoxy and dideoxynucleotides. With this system, the polymerase is simply added to the hydrated tablet and an aliquot of this mixture is dispensed into each of the annealed template solutions. Tableting of the reagents not only eliminates the need for mixing of individual components and increases the reproducibility of the reactions, it also simplifies the execution of the reaction by reducing the number of pipetting steps and increases the reagent shelf life.

An additional benefit to terminator labeling is the elimination of the detection of dideoxy independent stops. A dideoxy independent stop (strong or false stop) is an accumulation of a large number of DNA fragments of the same length which do not end with a dideoxynucleotide. In traditional sequencing, this is easily identified as a band appearing in two or more lanes at a single position on an autoradiograph. In automated systems that employ primer labeling, they appear as bands in all four lanes (similar to the autoradiograph) or as overlapping signals in single-lane systems. In fact, only a portion of the fragments in only one of these lanes is actually terminated with a dideoxynucleotide. The other signal is a result of the polymerase's failure to elongate those fragments. In some cases, the presence of the dideoxy independent stop appears to be polymerase dependent; in others it is more likely a function of secondary structures present in the template. In either case, fragments resulting from the polymerase's inability to elongate through a particular region are not labeled if terminator labeling is used. With terminator labeling, only fragments that have incorporated the fluorescent dideoxynucleotide are labeled. The other fragments are not labeled, and therefore do not contribute to background or confusion during data analysis. It should be recognized, however, that these fragments still exist and that they can still affect the quality or quantity of data obtained. In the case of an extremely strong stop, there may be a significant reduction in signal following the stop due to the lack of available template for synthesis/labeling. In addition to an overall loss of signal, there can also be effects on the resolution of DNA fragments immediately preceding or following a stop. This is due to a local overloading of the gel at this point. In systems that run four separate lanes, the effect on resolution should be reduced as there is 1/4 the amount of DNA in any given lane; however, there will still be a loss of available template.

While terminator labeling offers several benefits to the sequencer such as simplicity, flexibility in primer selection, elimination of strong stop detection, etc., it also imposes limitations. The most significant limitation is in the choice of polymerases available for carrying out the sequencing reaction. This is a function of the polymerase's ability to efficiently incorporate the dideoxynucleotide analogue. Several potential problems arise if the polymerase requires too high a concentration of terminator. These include difficulty in removing the unincorporated nucleotide, increased potential for mis-incorporation, and increased expense in performing the reaction. This has resorted in most of the resources being focused on developing procedures for polymerases that use lower amounts of terminator (e.g., T7 DNA polymerase and AMV Reverse Transcriptase).

Reaction protocols have been developed for all of the automated instruments for dealing with the sequencing of single-stranded, double-stranded and PCR-amplified DNAs. Some differences exist between the procedures due to the differences in the detection systems, but essentially, most of them are quite similar to the procedures used in traditional sequencing. There have been some exceptions, however, particularly with the systems that use the fluorescent terminators. An outgrowth of the PCR technology has led to the development of a unique approach to sequencing double-stranded DNA templates. This approach, applied initially to radioisotopic sequencing (16), can be used with either fluorescent primers or terminators. In this procedure, all of the reaction components including undenatured supercoiled plasmid, buffer, thermal-stable polymerase, primer, and nucleotides are mixed, and the reaction is performed in a thermal cycler. If fluorescent terminators are used, a single reaction mix is prepared. If labeled primers or radio-labeled nucleotides are used, four separate mixes need to be prepared. The reaction mix(s) typically goes through 20 cycles of a 98°C denaturation step, a 55°C annealing step, followed by a 65°C elongation step. This procedure is simpler, less tedious and more reliable than the traditional alkaline denaturation method (4) of preparing sequencing templates from supercoiled DNAs. Researchers at EMBL have also described an alternative method for doing double-stranded sequencing. In this system, rather than precipitating the alkaline denatured DNA, the solution is neutralized and used directly in the sequencing reaction (17).

In addition to eliminating detection of sequencing fragments originating from strong stops, labeling at the 3' end with a dideoxynucleotide has also made it possible to sequence templates that would otherwise have significant problems with background due to run-off fragments. Run-off fragments are fragments that are generated from nicked templates. The polymerase cannot extend the fragment beyond the nick, so it simply stops at that point leaving a fragment of fixed length that does not have a dideoxynucleotide at the 3' end. An example of how this property can be exploited is the procedure developed at Du Pont for generating and sequencing RNA transcripts (18). In this procedure, a mini-prep purified supercoiled plasmid, containing phage polymerase promoters (i.e., T7, T3, or SP6) is used by a phage RNA polymerase as a template for the synthesis of RNA transcripts. These transcripts are then primed with oligonucleotides and sequenced with fluorescent terminators using AMV Reverse Transcriptase. In addition to generating multiple copies of a single-stranded template from a double-stranded template, this procedure offers several benefits to the sequencer: it requires only small amounts of supercoiled plasmid (approximately 200ng), is rapid and is simple. Following a 5 minute transcription reaction, primer, dNTP, ddNTP, and AMV Reverse Transcriptase are added and the reaction is continued for an additional 30 minutes. This procedure has been further developed to allow the generation of sequence data from the plasmid extracted from a single isolated colony. This general approach is successful because cDNA run-off products, generated from partially degraded and prematurely truncated transcripts, are not detected in the system.

Terminator labeling has also permitted the development of an efficient and reliable method of sequencing PCR products (19). This procedure takes advantage of the naturally occurring truncated PCR fragments generated in the amplification reaction. These products can act as sequencing primers in the sequencing reaction. By adjusting the PCR primers to achieve a slight asymmetric ratio (e.g., 1/10), high yields of PCR product can be obtained and easily sequenced with minimal sample cleanup (G-50 spin column chromatography), low background, and high reliability. Neither primer labeling nor isotopic labeling can take full advantage of these fragments. Both of these systems work much more effectively on completely single-stranded templates. The labeled primer would have to compete against the truncated PCR product for the annealing site. Any fragments generated from the truncated products would not be labeled. Therefore, only a fraction of the potential template would be used in generating signal. Radiometric sequencing relies on increasing the specific activity of a fragment as the fragment gets larger. With truncated products acting as primers, the ratio of the deoxy to dideoxynucleotides would have to be altered, based on the truncated product distribution, to achieve the desired specific activities of the sequencing fragments.

DATA ANALYSIS / SIGNAL PROCESSING

Due to the differences in the detection systems and the computers used in data analysis, each system has developed its own specialized software for signal processing. Instruments that use a single label and electrophorese the sequencing reaction products in four separate lanes (Hitachi and Pharmacia), deduce the nucleotide sequence from the chronological order in which the fluorescence is detected, in each of the four lanes (figure 5). The use of four separate lanes that must be aligned places additional burden on the electrophoresis of the samples. With these systems, temperature control of the gel is critical because the temperature of the gel affects the rate of electrophoresis. If the gel's temperature is not carefully controlled, the lanes toward the outside of the gel will tend to be cooler, and therefore, samples in those lanes will electrophorese more slowly. Differential rates of electrophoresis, independent of whether they are a function of gel temperature or

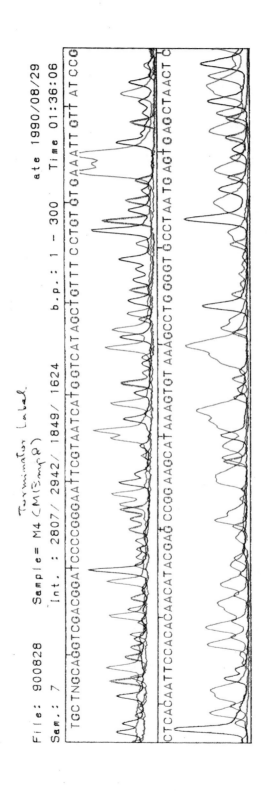

Figure 5. Data from four lanes per sample instrument following alignment. The tracing represents the first 150 bases of M13mp8 as detected on the Hitachi SQ 3000. The sequencing fragments were generated using fluorescent terminators with T7 DNA polymerase. Each color tracing represents the signal detected in an individual lane. The dashes above the base assignment reflect every tenth base.

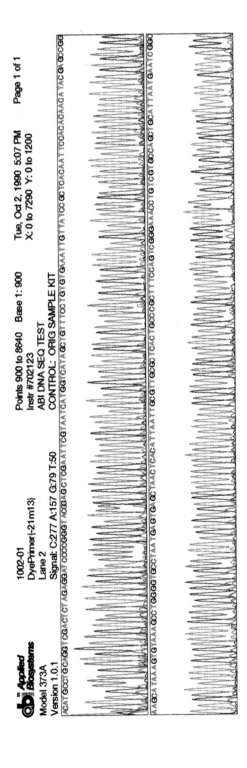

Figure 6. Data from the 373A using a labeled primer. M13 was sequenced with T7 DNA polymerase in a manganese-containing buffer.

contaminants (e.g., salts) present in the sample, can make the task of aligning the data from the four lanes difficult. The complications are quite different in the single-lane, four-dye instruments. Although detection is required for only one lane, the spectral characteristics of the fluorescence emission must be analyzed to determine which nucleotide has passed by the detector. The "purity" of the signal is crucial for spectral analysis. High resolution between adjacent fragments and minimal background from secondary priming, strong stops, etc., are important factors in achieving high quality signal processing. Uniformity of signal intensity between adjacent fragments is also important for accurate data analysis, especially as the resolution between fragments becomes reduced. If a very large signal is adjacent to a smaller signal, the larger signal has the potential for contaminating the emission of the smaller signal. This problem has been addressed with the substitution of manganese for magnesium in the sequencing reaction buffer (20). The manganese reduces the distinction between the deoxy and dideoxynucleotides and results in a more even signal distribution (figure 6). In addition to producing a more even signal distribution, this substitution has also reduced the concentration of terminator required in the reaction.

As mentioned in the System Overview, the ABI 373A uses a set of four filters to discriminate between four fluorescent labels. In the setup of the instrument, a matrix is established by electrophoresing four sets of fragments, one corresponding to each terminator, in individual lanes on the instrument. As the fragments pass through the detection zone, the relative amount of light detected with each filter for each fluorophore is determined and used to establish the matrix which is then used for sequence data analysis. For example, the blue channel may detect all of the fluorophores, the green channel detect A, T, and G, the red channel detect T and A, and the black detect G and A. The base assignment for any given signal is then made based on the relative detection of the fluorophore in the four channels and where this fits in the matrix (i.e., when there is signal in blue channel only, the base is a C, if red is larger than green with nothing in black, it is a T, if green is larger than red and black, it is an A, and if black is larger than green and red, it is a G - figure 7). An additional complication in the signal processing required for this system is a function of the fluorophores used with the system. In order to simplify the spectral analysis of the dyes, dyes with significantly different emission characteristics were

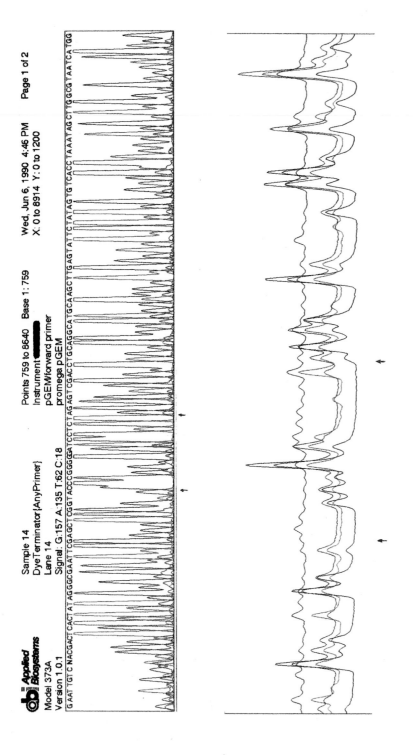

Figure 7. Data from ABI 373A using fluorescent terminators. The upper figure shows the processed data output from the system. The lower display provides a magnified view of the unprocessed raw data. The four colored tracings represent the light detected by the PMT with each of the four filters. The same portion of the sequence is represented in the area between the rows

used. Unfortunately, this also resulted in the dyes having significantly different physical properties (i.e., different charges and masses). This meant that they also had differential electrophoretic properties and therefore, differentially affected the mobility of the fragment to which they were attached. To further complicate the analysis, the contribution of the fluorophore to the fragment mobility was inversely proportional to the fragment length; as the fragment got larger, the contribution was reduced.

The Genesis 2000 also must discriminate between four fluorophores. However, in this case, the fluorophores used are much more similar both spectrally and physically. In this system, the relative amount of light detected by each of two PMTs indicates which base has passed through the detection zone (figure 8). The Du Pont system differs from the ABI system in that with the ABI system, the data from all of the lanes is collected through the same portion of the filter, the optics move from lane to lane. In the Du Pont system, the filters are stationary and there is the potential for slight variability across the face of the detection system. This means that the ratio of light detected in the two channels for the 505 fluorophore may be different in lane 1 vs lane 2. Therefore, a histogram, based on the relative amount of light detected in the two channels for each base detected, is generated for each lane of data (figure 8b). While the absolute ratio of light detected for 505 fluorophore may vary slightly from lane to lane, the relationship between the 505, 512, 519, and 526 dyes within a lane remains constant. Since the relationship does not vary, each region of the histogram can be assigned to a specific nucleotide. The Genesis then makes the nucleotide assignments based on where the ratio of the signals for any given peak falls within the histogram. Therefore, the accuracy of the base call is dependent on being able to accurately determine the ratio of the two signals. There are two methods used to determine this ratio: one is based on determining the area under the peak, the other is based on determining the peak shape. Since the emission spectra of the dyes are very similar, the accuracy of peak area, or peak shape determination, becomes critical. This places a greater emphasis on the resolution of the labeled fragments (purity of signal) in obtaining accurate base assignment.

While all of these systems are capable of achieving high accuracy in base determination, they are also sensitive to the quality of the data. In general, systems that use labeled primers are sensitive to dideoxy

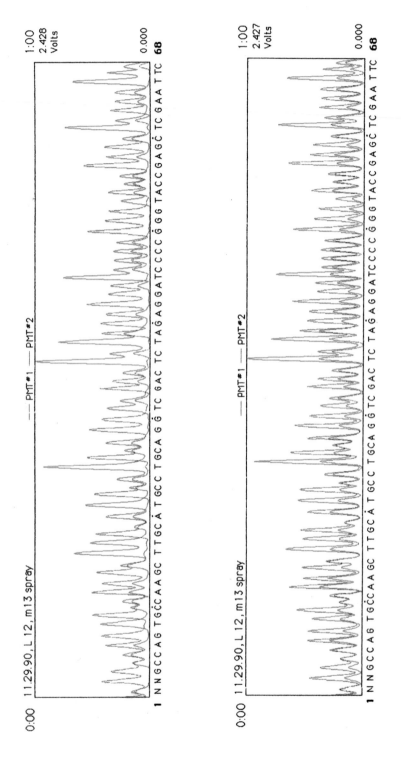

11.29.90, L 12, m13 spray

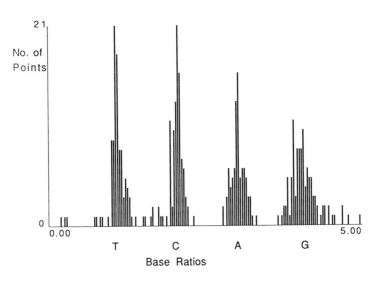

Base Ratio Histogram

Distribution		Count		Orthogonal ratios	
T:	1.02 ± 0.19	98		T:	0.3785
C:	1.98 ± 0.13	100		C:	0.8177
A:	2.98 ± 0.11	83		A:	1.6201
G	4.04 ± 0.23	109		G	2.7975
Ambiguous:		2			
N calls:		8			
Total bases :		400			
CV:		4.27%			

Figure 8. Data from the Genesis 2000 sequencing system. A) The colored lines represent the output from each of the PMTs. The upper portion is a display of the processed data in the "orthogonal" view. Based on the histogram, the signal is transformed to produce a display whereby, for example, the T-nucleotide, instead of having a peak in each of the channels, has a peak in the red channel only. The orthogonal view is the typical display used for editing. The lower portion is the same data in the raw data view. B) The ratio of the two signals for each of the peaks is used to generate a histogram. The ratios fall into four distinct groupings, each grouping representing a particular fluorophore. Base assignments are made based on where a particular peak's ratio falls within the histogram.

independent stops, and systems that use four fluorophores and do spectral analysis are sensitive to low intensity signal. As might be expected, most of the errors in signal processing are associated with these "sensitive" regions. While software can be designed to identify these potentially error prone regions and flag them for the user, the user may still not be able to make an accurate base assignment. Therefore, as is the case with traditional sequencing, the sequence of the complementary strand must also be determined with the fluorescent systems in order to generate a high degree of accuracy in the final data.

SUMMARY

The introduction of systems designed to generate and detect fluorescent DNAs has offered an alternative to the traditional radiometric approach to DNA sequencing. A number of different approaches have been taken towards automating DNA sequencing with fluorescence-based DNA detection systems. These systems automatically collect and process the data generated by the electrophoresis of fluorescent labeled dideoxy sequencing fragments through a standard DNA sequencing gel. In addition to saving the time typically required in generating, reading, and inputting the data from the autoradiographs used with radiometric sequencing, several of these systems can offer significant timesavings in the generation of the sequencing fragments as well. The introduction of terminator labeling has also opened the door to the design of new sequencing methods and strategies. Even though the technology is quite young, we have already begun to see some examples of new traditional methods being applied to the automated systems (e.g., cycle sequencing) and new methods developed specifically for these systems (e.g., Du Pont's RNA sequencing and PCR sequencing methods).

From an instrument perspective, there are several groups that are evaluating the coupling of capillary electrophoresis with fluorescence detection (21, 22, 23). This offers the potential for a faster rate of data collection and the use of longer gels. The increased number of theoretical plates obtained with longer gels could improve the resolution and ultimately increase the number of bases that can be determined per experiment. It seems unlikely that the faster rate of data collection will have a significant impact on the speed of the overall process. However, the potential gain from the acquisition of larger blocks of data may be significant as it relates to the computer time necessary for assembly and analysis.

With the existing systems, the generation and input of sequencing data into computers has been dramatically simplified. The time previously spent in performing these tasks now seems to be occupied with the generation of templates and the analysis of the data. Having recognized this, a number of approaches are being taken to simplify the process(es) of template preparation (PCR and RNA transcript sequencing). These approaches, coupled with robotic work stations, can satisfy this need in the short term. This now leaves data analysis as being one of the most time-consuming activities associated with DNA analysis. While a number of powerful software packages are available, they require user intervention and often "lock" researchers to the terminals during the analysis. Development of computer "pathways" that automatically shunt data through a variety of routines immediately after base determination could save significant time in a molecular biology laboratory.

ACKNOWLEDGMENTS

I would like to thank Hideki Kanbara of Hitachi's Central Research Lab, David Vissing of Nissei Sangyo, Carla Baehler of Pharmacia LKB Biotechnology and Steve Lombardi of Applied Biosystems for providing data. I would like to thank Robert Blakesley, Bethesda Research Labs, for his helpful comments and suggestions. I would also like to thank George Tice, Tina Hatfield, Jim Shultz, and Brenda Asbury of E. I. Du Pont's Medical Products for their demonstrated stamina in reviewing the various drafts and their invaluable assistance in the preparation of the manuscript.

REFERENCES

1   Maxam, A.M. and Gilbert, W., Sequencing End-Labeled DNA with Base-Specific Chemical Cleavages, Methods Enzymol. 65 (1980) 499-560.
2   Sanger, F., Nicklen, S. and Coulson, A.R., DNA Sequencing with Chain Terminating Inhibitors, Proc. Natl. Acad. Sci. USA 74 (1977) 5463-5467.
3   Messing, J., New M13 Vectors for Cloning, Methods Enzymol. 101 (1983) 20-78.

54

4   Chen, E.Y. and Seeburg, P.H., Supercoil Sequencing: A Fast and Simple Method for Sequencing Plasmid DNA, DNA 4 (2) (1985) 165-170.
5   Biggin, M.D., Gibson, T.J., and Hong, G.F., Buffer Gradient Gels and ^{35}S label as an aid to Rapid DNA Sequence Determination, Proc. Natl. Acad. Sci. USA 80 (1983) 3963-3965.
6   Olsson, A., Moks, T., Uhlen, M., and Gaal, A.B., Uniformly Spaced Banding Pattern in DNA Sequencing Gels by use of Field-Strength Gradient, J. Biochem. Biophys. Methods 10 (1984) 83-90.
7   Robertson, C.W., Automated DNA Sequencing Instrumentation, Biochemical Instrumentation for Protein and DNA Chemistry, Verglagsgesallschaft,(In Press).
8   Smith, L.M., Sanders, J.Z., Kaiser, R.J., Hughes, P., Dodd, C., Connell, C.R., Heiner, C., Kent, S.B.H. and Hood, L.E., Fluorescence Detection in Automated DNA Sequence Analysis, Nature 321 (1986) 674-679.
9   Prober, J.M., Trainor, G.L., Dam, R.J., Hobbs, F.W., Robertson, C.W., Zagursky, R.J., Cocuzza, A.J., Jensen, M.A., and Baumeister, K., A System for Rapid DNA Sequencing with Fluorescent Chain-Terminating Dideoxynucleotides, Science 238 (1987) 336-341.
10 Kambara, H., Nishikawa, T., Katayama, Y. and Yamaguchi, T., Optimization of Parameters in a DNA Sequenator using Fluorescence Detection, Bio/technology 6 (1988) 816-821.
11 Ansorge, W., Sproat, B.S., Stegemann, J. and Schwager, C., A Non-Radioactive Automated Method for DNA Sequence Determination, J. Biochem. Biophys. Meth. 13 (1986) 315.
12 Freeman, M., Baehler, C. and Spotts, S., Automated Laser-Fluorescence Sequencing, Bio/technology 8 (1990) 147-148.
13 Rosenthal, A. and Bankier, A., Fluorescent DNA Sequencing by Chemical Degradation: On-line Detection of One or Two DNA Clones in One Lane using Four Different Fluorophores, Genome Mapping and Sequencing, Cold Spring Harbor Laboratory, USA, May 2-6, 1990, Cold Spring Harbor, New York, 1990, 151.
14 Voss, H., Schwager, C., Wirkner, U., Sproat, B. Zimmermann, J., Rosenthal, A., Erfle, H., Stegemann, J. and Ansorge, W., Direct Genomic Fluorescent on-line Sequencing and Analysis using *in vitro* Amplification of DNA, Nucleic Acid Res. 17 (7) (1989) 2517-2527.
15 Cocuzza, A. Personnel Communication.
16 Murray, V., Improved double-stranded DNA sequencing using the linear polymerase chain reaction, Nucleic Acid Res. 17 (21) (1989) 8889.
17 Zimmermann, J., Voss, H., Schwager, C., Stegemann, J., Erfle, H., Stucky, K., Kristensen, T. and Ansorge, W., A Simplified Protocol for Fast Plasmid DNA Sequencing, Nucleic Acid Res. 18 (4) (1990) 1067.
18 Amorese, D. A., Fry,D. D., Tice, G., Coupled Transcription/RNA Sequencing; a New Sequencing Strategy (*in preparation*).

19 Shultz, J., Johnson, J., Norris, T., Hendrickson, E., and Amorese, D. A., A Model for Understanding the Sequencing of Asymmetric PCR Products (*in preparation*).

20 Tabor, S., and Richardson,C. C., DNA Sequence Analysis with a Modified Bacteriophage T7 DNA Polymerase, J. Biol. Chem. 265 (14) (1990) 8322-8328.

21 Zagursky, R. J. and McCormick, R. M., DNA Sequencing Separations in Capillary Gels on a Modified Commercial DNA Sequencing Instrument, BioTechniques 9 (1) (1990) 74-79.

22 Swerdlow, H. and Gesteland, R., Capillary Gel Electrophoresis for Rapid, High Resolution DNA Sequencing, Nucleic Acids Res. 18 (6) (1990) 1415-1419.

23 Luckey, J. A., Drossman, H., Kostichka, A. J., Mead, D. A., D'Cunha, J., Norris, T. B., and Smith, L. M., High Speed DNA Sequencing by Capillary Electrophoresis, Nucleic Acids Res. 18 (15) (1990) 4417-4421.

I. Karube (Ed.) *Automation in Biotechnology*
Proceedings of the 4th Toyota Conference, 21–24 October 1990

AN AUTOMATED SYSTEM FOR SEQUENCING OF THE HUMAN GENOME

E. SOEDA[1] ,Y. MURAKAMI[1], K. NISHI [2] AND I. ENDO[3]

[1]RIKEN Gene Bank, The Institute of Physical and Chemical Research, 3-1-1 Koyadai, Tsukuba Science City 305 (Japan)
[2]Division of Supersensor's Engineering, Instrumentation and Characterization Center, The Institute of Physical and Chemical Research, 2-1 Hirosawa, Wako 351-01 (Japan)
[3]Laboratory of Chemical Engineering, the Institute of Physical and Chemical Research, 2-1 Hirosawa, Wako 351-01 (Japan)

SUMMARY

Several machines and devices involved in DNA sequencing stages have been developed in order to organize them in an automated line system for massive sequencing. We took advantage of shotgun sequencing strategy and chose a cosmid clone as the maximum unit of the human genome. For evaluation of the machines and interpretation of sequencing, yeast chromosome VI was selected as a model organism.

INTRODUCTION

Shotgun is currently regarded as one of the rapidest sequencing methods, being most applicable to an automated line system equipped with machines where sequencing raw data would accumulate massively once the libraries are constructed. We took advantage of this approach, and developed machines and devices involved in sequencing stages since 1981 under the auspices of the Science and Technology Agency (STA) of Japanese Government. Our strategy was to replace hand-manipulation by automated machines. The machines were developed in collaboration with private companies, integrating their expertises in electronics, robotics, computers and materials science. Seiko Electronic and Instruments Co., Tosoh Co., Hitachi Co. and Mitsui Knowledge Industry Co. were participated in this project.

METHODS

The modified shotgun method (ref.1) developed by Deininger (ref2) was used. It consisted of seven successive stages; construction of the libraries, isolation and amplification of recombinant phage, extraction of DNA from phage particles, dideoxy-reaction (ref.3), sequencing gel

58

electrophoresis, gel reading and connecting numbers of short sequences to compile a long DNA sequence (Fig.1). The goal has been to increase the rate of DNA sequence output starting from the libraries, replacing hand-manipulation by automated machines involved in the stages.

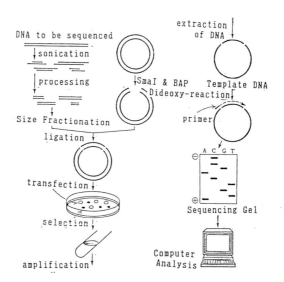

Fig. 1 Flow sheet of shotgun sequencing

RESULTS AND DISCUSSION
Background Statements

A study program entitled 'Extraction, Analysis and Synthesis of DNA' was started from 1981, supported by the Special Coordination Fund for the Promotion of Science and Technology under the auspices of the Science and Technology Council of Japan. The program aimed to encourage DNA study by organizing the resources in universities, private companies and government research institutes. In addition, the Council has asked them that the project should give interest and benefit to scientists working in DNA study.

Under these conditions, an analysis section of the program, led by Professor A. Wada of the University of Tokyo, aimed initially to reduce the burden of time demanded of researcher for DNA sequencing analysis by developing automated machinery, utilizing the knowledge and resources of companies with expertise in electronics, robotics, computers and materials science. Seiko Electronic and Instruments Co., Fuji Photo Film Co., Tosoh Co., Hitachi Co. and Mitsui Knowledge Industry Co. were participated in the project. It was annexed to the Research on Development of Common Basic Technologies for Cancer Research from 1984 and resulted in completion of several machines including their proto-types. They were DNA extractor, dideoxy-reactor, machines for mass-production of ready-made gels, fluorescent sequencer and sequencing gel reader. In 1987, these machines and gels were submitted to evaluation and integrated into RIKEN research project started from 1988 in order to organize a total sequencing system for the human genome .

Evaluation of machines and devices

(i) Dideoxy-reactor: This machine was made in collaboration with Seiko Co. Ltd, watch-maker and designed originally to achieve faithfully a complete set of Sanger's sequencing reaction from annealing to denaturation before loading gel. These reactions might be characterized by repeated cycles of pipetting at a sub-microliter level, dropping down reagents by centrifugation, mixing and heating. In addition, it was applicable to any enzymatic reactions at a sub-microliter level according to user's program which was input from the ten keys connected to a computer. The major functions achieved by the machine were quantitative addition and dispense of reagents, mixing, heating and cooling. Centrifugation, most complicated step in robotics, for dropping small amount of solution down to bottoms of tubes was avoided by inserting a pipette-tip into solution.

Any dideoxy-reactions were applicable, no matter how they were carried out with RI and fluorescent methods or with any DNA polymerases. The resulting sequence profiles obtained with this machine were excellent with reproducibility, corresponding to those produced by hand-manipulation of experts. Cross-contamination which might be caused by using a single pipette-tip at each step was negligible because of hydrophobicity of plastic one. Loss of water caused by evaporation during heating was prevented by putting a heat panel over the reaction tray in the working station of the machine. It took two hours to carry out the complete cycle of 96 sequencing reactions. The machine has been in Japanese market since April, 1988 and integrated into the Riken project .

(ii) Ready-made gel Polyacrylamide sequencing gel films, 20 x 40cm and 0.2mm thick, covered with this acetate films were prepared by mass-production with machine of Fuji Photo Film Co. For loading samples, two type of gel films were available, shark and well. The gel films with twenty wells were examined. Autoradiography could be performed without removal of urea and fixing after drying the film. Resolution between bands separated on ready-made gels were comparable to that with hand-made gels. Approximately 150 nucleotides could be read at the first set of 2 hour running and another 100 nucleotides at the second set of 5 hour running.

iii) <u>Gel autoReader:</u>

Gel autoReader was an instrument equipped with an optical scanner and computer program for pattern recognition, developed by Seiko Co. Evaluation of its machine was examined, including reproducibility, accuracy and the maximum readings by comparison with eye-power and the equivalents from other companies. The instrument could read sequences data up to 150 nucleotides from autoradiogram on film where more than 250 nucleotides could be read by eyes. Because unreadable or unrecognized regions were marked with red ink, it was easy to correct them quickly. It took twenty minutes for complete reading of one set of ladder pattern. An automated film-feeder equipped with the instrument was able to read out twenty gels, generating output of 18,000 nucleotides per day.

(iv) <u>Fluorescent sequencer:</u> The maximum output of sequencing data is restricted by the legal allowance for usage of radioisotope. Usage of fluorescent dyes instead of radioactive chemicals in sequencing extended its capability. We have introduced the sequencer made by Hitachi Co. Ltd. where technologies developed by Smith et. al (ref4) were modified. A laser beam coming from the side of polyacrylamide gel activated the dye and a fluorescence detector behind the gel read the sequences as the fragments passed there.. The quality, the maximum reading, accuracy, sensitivity and stability were examined, comparing with the radioisotope method. The primer labeled with a single fluorescent dye was added to 0.5 µg template DNA in wells of a multi-titer plate and the plate was put into the working station of dideoxy-reactor. After Sanger's reaction, the sample was loaded on gel mounted in the sequencer. After 3 hour run, readable sequences ran off from gels. Error was less than 1% up to 200 bases and 5% up to 300 bases. The maximum reading was 350 bases. The sequencer was applied to yeast sequencing system. The maximum output per day was 2,400 bases.

v) <u>DNA extractor and other interfaces:</u> The remaining steps to be mechanized were selection of plaques of M13 phage in shotgun library and extraction of DNA from phage particles after propagation. These were not serious rate limiting factors but they became new bottle-necks against massive sequencing.

For selection of the plaques from the libraries, a robot with a sensor and arms was applied. This machine was able to recognize the plaques in the libraries with a Video monitor, selected white plaques of recombinant phages from blue plaques of M13 phages and inoculated them to E. coli cells with disposable tubes. It took 30 min for recognition of the library in which less than 300 phage plaques grew and 10 sec. for inoculation of a plaque.

After cultivation, the recombinant phages were submitted to purification of DNA after extraction from the phage particles. Differential filtrations were applied to purify DNA from phage particles, using two sets of cassett filters, 0.45 μm filter to remove host cells and ultrafilter to collect phage particles and to concentrate DNA (Fig. 2). Thus, after removal of host cells, DNA was extracted the phage particles with SDS and concentrated on the filter after washing with TE. The yield of DNA was approximately 20 μg from 2-ml culture.

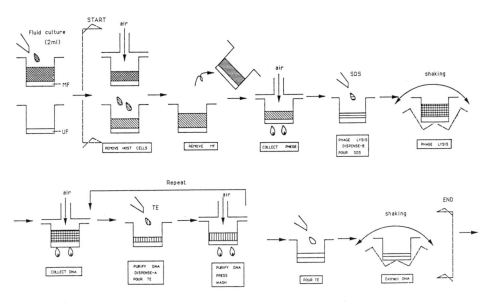

Fig. 2 Extraction and purification of the template DNA from M13 culture fluid , using differencial filtration method

It became one of the most laborious and time consuming stages in the present sequencing to load many samples to sequencing gels after dideoxy-reactions. We applied small volume of sample to a sequencing thin gel of 0.3 mm in width, using micro-injection with in 15 sec per one sample. Before loading sample, urea extracted from the gel was removed by blowing out with the buffer through a pipette device, forming a sharp band containing the sample. The robot was able to recognize more than 25 μm in diameter.

<u>A new dideoxy- reactor equipped with PCR function:</u>

Compositions of dideoxy-reaction mixture were compatible to those of PCR except that dideoxynucleotides and ^{32}P labelled nucleotides or primers tagged with fluorescent dyes were used in dideoxy reactions (Fig. 3). We have integrated PCR function into dideoxy-reactor so that several stages involved in template preparation could be eliminated. They were M13 propagation, sample transfer, concentration of phages, extraction and purification of DNA. Thus, template DNA was prepared in PCR reaction from a white plaque in the library using forward and reverse primers. The reaction mixture were submitted directly to dideoxy-reaction after dilution. By choosing primers, sequences from both strands of DNA could be read by either RI or fluorescent method.

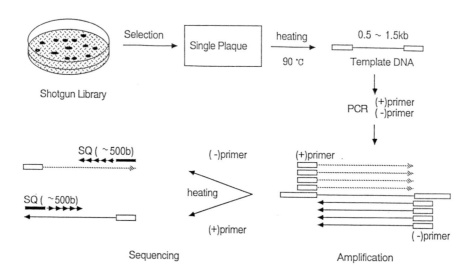

Fig. 3 Introduction of PCR function into Sanger's dideoxy-reaction

<u>Construction of massive sequencing system</u>

In shotgun approach, once succeeding in construction of the library, we have to employ the routine work without any strategy except how many

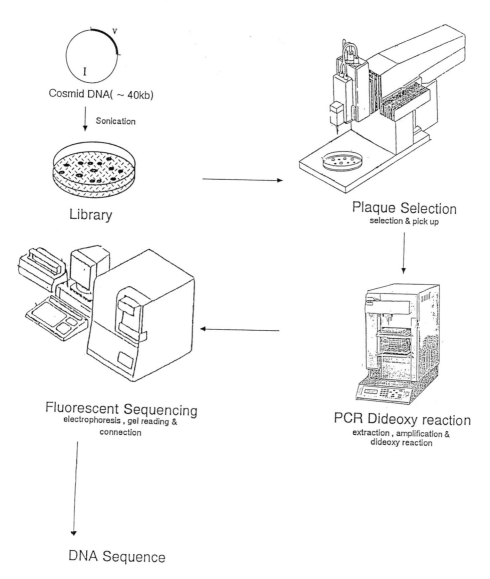

Cosmid DNA(~ 40kb)

Sonication

Library

Plaque Selection
selection & pick up

Fluorescent Sequencing
electrophoresis , gel reading &
connection

PCR Dideoxy reaction
extraction , amplification &
dideoxy reaction

DNA Sequence

Fig. 4 Possible organization of a sequencing line system

clones we can sequence and how fast we can do them. Therefore, we try to replace hand manipulation by automated machines. The method, combined with cloning and sequencing technologies, was virtually divided into seven steps, construction of a shotgun library, selection and amplification of recombinant phages, extraction and purification of the recombinant DNA, dideoxy reaction, electrophoresis on polyacrylamide gels, gel readings and finally compilation of accumulated sequence data in a computer .

Of these successive steps, the initial halves were not rate limiting and the latters were critical. Therefore, our effort has been made to remove the rate limiting processes by application of robotics. Thus, they were dideoxy-reactor, ready-made gels, gel autoReader, fluorescent sequencer and computer program for assembly of massive sequences as raw date. Most of them were commercially available.

Of these, dideoxy-reactor which was able to carry out automatically dideoxy-reaction from annealing to denaturation before loading to the gel was very useful. It could be applicable to either radioisotope or fluorescent method. Introduction of ready-made gel and gel autoReader made it possible to achieve rapid and massive sequence. However, shorter reading with more errors with the machine than with manuals led development of fluorescent sequencer for construction of improved system..

With present machines and technologies, sequencing are able to organize virtually as six stages, construction of library, plaque selection, preparation of template DNA, dideoxy-reaction, fluorescent sequencer and management of sequence raw data by computer. It should be note that interfaces such as gel making and sample loading become new limiting rates. Integration of PCR function into dideoxy-reaction and usage of capillary gel in fluorescent sequencer will make the stages shorter, eliminating interfaces and possibly into four stages (Fig.4).

A Strategy for human genome sequencing

In order to make sequencing of the human genome feasible, it is essential to construct a line system which is built up with automated machines and generate massive sequence data one after another once put DNA into the system. In this case, it is conceivable that DNA to be applied to the system should be cloned into cosmid vectors mapped on the chromosome because sufficient amount of DNA to sequence can be obtained from cosmid clone and the vector regions is negligible in size as compared with the inserts A cosmid clone involved in the human chromosome condensation was originally isolated from the revertant of the BHK mutant (ref.5) which showed premature chromosome condensation at non-permissive temperature. The revertants were obtained on transfection of the human DNA. The DNA was purified through the repeated cycles of transfection to the BHK mutant Alu sequence as a probe. Therefore, the insert must be contained Alu sequences.

The cosmid clone was sheared randomly to prepare the shotgun libraries. Six hundred fifty clones were picked up and sequenced. From each clone, about 250 nucleotides could be read from two sets of run in the ready-made gels. One hundred sixty thousand nucleotides were read and put into a computer. As the result, approximately 37 kb DNA was determined without constructing restriction map (Fig. 5).

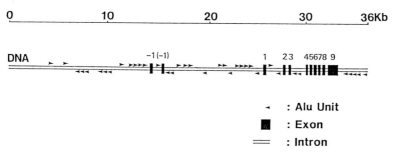

Fig. 5 Sequencing of the human cosmid clone by the shotgun method and distribution of Alu repeats

At first, we tried to compile raw sequence data with a personal computer using commercially available program and failed because the presence of the repeats in human DNA interrupted their connections. Thus, with the conventional method where 17 out of 20 base matches were sought for connection between two sequences, more than sixty clones appeared as candidates and even with perfect matches, several clones appeared. Most of them were derived from Alu-repeats characteristic of the human genome.

Improvement of computer program for data assembly and compilation

We prepared a computer program for sequence assembly and edition in collaboration with Mitsui Knowledge Co. Ltd. In assembly of random sequences, computer sought extensively around homologous regions once finding parts of them and aligned them, inserting gaps automatically to meet the maximum matching.

In shotgun approach, sequences in the same region are read approximately four times from the different positions on the same strand and the opposite strand of DNA and are aligned. The complete sequence is decided from the majority in the homologous sequences, removing the minor sequences. It should be noted that inevitable errors occurring in raw sequence data will be corrected or reduceed by overlapping sequences.

With integrating our experiences, the improved program for compilation of sequence is being prepared, applicable to any DEC computes.

Sequencing of yeast chromosomes as an model organism

Yeast, *Saccharomyces cerevisiae*, has already been identified as an important model organism for interpreting sequence data and understanding biology of the human because yeast is one of the smallest eukaryotes and its genomic organization is simpler than that of the human. In addition, extensive overlapping clone sets are useful for the development and testing of sequencing system needed for analysis of more complex genome.

TABLE I Compilation of sequences of yeast chromosome VI in shotgun approach

Clone Name		6341	6606	4361	4682	4584	3655	4900
Length (kb)		12	15	12	17	13	17	18
Tag		RI&F	F	F	F	RI	RI	RI
Gel Readings		391	145	138	139	177	150	378
Completion		95%	100%	99%	95%	66%	45%	50%
	>2 kb	5.33	6.41	8.10	4.38	2.42		
		2.87	5.77	2.29	3.73			
			2.50		3.55			
	1-2 kb	1.58		1.19	1.91	1.04	1.46	1.78
		1.16		1.09	1.19		1.37	1.25
Islands					1.13		1.34	1.19
							1.25	1.08
							1.12	1.05
							1.05	
	0.5-1 kb	x5		x1	x1	x16	x4	x13
	<0.5 kb	x20	x2	x8	x8	x9	x4	x9

*F and RI denote fluorescent tag and radioisotope methods, respectively.

Project to sequence yeast chromosome VI has been initiated in collaboration with Prof. M. Olson. The yeast DNA was cloned into lamba vectors as four sets of overlapping clones, extending from 10 to 20 kb in length. Shotgun strategy was totally applied to overlapping clones. As shown in Table I, sequences read with fluorescent sequencers were effectively contributed in linking and those in manual sequencing did not at all. The raw sequence data were almost randomly distributed along the fragment to be sequenced. Under the present system, approximately 400 nucleotides can be read from one template, allowing walking 100 nucleotides on yeast chromosome VI. The complete sequence in an editorial form will be obtained from a consensus sequence after assembly of raw sequence data

REFERENCE

1. S. Yasuda et. al, An improved method for shotgun DNA sequencing, Agr. Biol. Chem., 49 (1985) 1525-1526.
2 P.L. Deininger, Random subcloning of sonicated DNA. Application to shotgun DNA analysis. Anal. Biochem., 129 (1983) 216-218.
3 F. Sanger et. al, DNA sequencing with chain terminating. Proc. Natl. Acad. Sci. USA. 74 (1977) 5463-5467.
4 L.M. Smith et. al, Fluorescent detection in automated DNA sequence analysis, Nature, 321 (1986) 674-679.
5 R.Kai, M. Ohtsubo, M.Sekiguchi and T. Nishimoto, Molecular cloning of a human gene that regulates chromosome condensation and is essential for cell proliferation, Mol. Cell. Biol., 6 (1986) 2027-2032.
6 M. Olson et al, Random clone strategy for genomic restriction-mapping in yeast, Proc. Natl. Acad. Sci. USA 83 (1986) 7826-7830.

I. Karube (Ed.) *Automation in Biotechnology*
Proceedings of the 4th Toyota Conference, 21–24 October 1990
© 1991 Elsevier Science Publishers B.V. All rights reserved

AUTOMATED LABORATORY INSTRUMENTATION FOR THE PREPARATION AND ANALYSIS OF BIOTECHNOLOGY SAMPLES

James N. Little, Zymark Corporation, Zymark Center, Hopkinton, MA 01748 (U.S.A.)

SUMMARY

Increased emphasis on biotechnology research and new product development has lead to a rapid increase in the number of biotechnology samples to be analyzed. Many of these assays have required new bioanalytical procedures as well. Many methods were initially qualitative but new demands require them to be run routinely, efficiently and quantitatively. The biotechnology assays are often tedious and time-consuming and are ideal candidates for automation. Many methods have been developed in a research environment and are incompatible with automation techniques and require modifications.

The paper explores several biotechnology assays such as HPLC, ELISA, colorimetric, turbidometric, radiochemical, screening, etc. and how they have been automated using flexible, robotic automation systems. It will include what modifications to the method were required for transferring a manual method to an automated method and give comparisons between throughput, accuracy and precision between the manual and automated methods. Important considerations in automating biotechnology assays are that the automation system must be designed for automation as the examples will show. Assays requiring precise timing parameters are usually much improved through automation.

AUTOMATED SAMPLE PREPARATION AND HPLC ANALYSIS OF BIOSYNTHETIC HUMAN INSULIN (ref. 1)

The sample preparation and HPLC analysis of biosynthetic human insulin is a very labor intensive and technique dependent assay. The process involves 27 discrete manual steps, with 8 different incubation stages totaling 11 hours. A major portion of the sample preparation utilizes CNBr, which is toxic and requires containment of hazardous vapors. The HPLC run time is 45 minutes, with a restriction that any given lot of samples must be run on the same instrument. Sample loads run as high as 80 samples per day.

The procedure requires 12 man-hours during a day and evening shift. Samples are brought to the lab throughout the day, but the length of the procedure allows samples to be started only at the beginning of the day, thus some samples are held an additional day. The method was an excellent candidate for automation. Because the automated method is slower than the manual method, interfacing to two HPLC instruments is necessary to maintain timely delivery of results.

The manual cleavage and sulfitolysis of the polypeptide chain prior to HPLC analysis involves 27 discrete steps. Adaptation of this methodology to a robotic method resulted in 11 stages or preps (Table 1), each separated by an incubation step.

TABLE 1. Robotic Procedure for Biosynthetic Human Insulin.

Step	Action
PREP1	Transfer aliquot of fermentation broth from bar coded submission tube (reading bar code) to assay tube. Centrifuge tube for 2 robotic cycles (36 minutes).
PREP2	Decant supernatent, add stir bar, place on magnetic stir plate in hood and add first two cleavage reagents. Stir for 18 minutes.
PREP3	Add CNBr. Stir for 4 hours.
PREP4	Add two inactivating reagents. Stir for 18 minutes.
PREP5	Add acetone. Stir for 18 minutes.
PREP6	Centrifuge tube for 36 minutes.
PREP7	Aspirate acetone, add second aliquot of acetone while stirring. Centrifuge tube for 36 minutes.
PREP8	Aspirate acetone. Place tube in drying rack for 1 hour.
PREP9	Add sulfitolysis reagent, remove from hood and place on magnetic stir plate outside hood for 4 hours.
PREP10	Remove tube from stir plate. Centrifuge for 18 minutes.
PREP11	Pipette aliquot of supernatent, standard or control sample into HPLC vial(s) for appropriate instrument, based on communications with HP1000. Pipette additional sample into chilled vial for long term storage. Remove stir bar from tube, decant excess sample and discard tube.

The robotic system included several standard components but also used several custom components such as chill racks, evaporating station, test tube and stir bar dispensers and a fume containment hood.

Error detection is very important because of the use of CNBr. A wet chemical segmented flow instrument is used to monitor for stray CNBr gas. A positive response will cause the robotic system to stop and signal plant security. All error situations sound both audible and visual alarms and a phone dialer sequence alerts appropriate people.

Daily precision and accuracy of the robotic method is about the same as that produced by a very conscientious and methodical technician. The automated system frees significant manpower for other projects and offers much greater flexibility in the processing of these type samples.

AUTOMATION OF FERMENTATION, SAMPLE PREPARATION AND ASSAY OF RECOMBINANT MICROORGANISMS (ref. 2)

The screening of recombinant colonies for the production of new pharmacologically active components is a labor-intensive, integrated process that consists of fermentation, sample preparation and assay.

Transformed protoplasts diluted in hypertonic buffer are delivered into "Nunc" plates by a robotic system. After suitable time for cell growth and genetic expression, the robot delivers an antimicrobial agent that selectively kills non-transformed cells. These plates are incubated until colonies of transformed cells appear. The colonies are subsequently transferred by a robotic system into a module containing a vegetative growth medium. The module consists of 120 scintillation vials that are interconnected by a polyurethane matrix. This module is incubated on a rotary shaker until the microbial cells enter the stationary phase of growth. An aliquot of the cells and growth medium is removed from each vial and transferred by a robot to vials in another module that contains a fermentation medium. The modules are incubated until the fermentation product reaches a maximum titer. There are two types of fermentation products which can result from genetic recombination. One is a hybrid compound and the other is an enzyme with altered substrate specificity. Each type of product requires a different sample preparation procedure.

A robotic system prepares fermentation broths which might contain hybrid compounds for assay. This robot adds a predetermined amount of solvent to each vial, followed by homogenization and dilution. When the fermentation product is an enzyme, the robot is used to sonicate concentrated cell suspensions, add appropriate substrates and cofactors, and quench the reaction after incubation. Any new hybrid compound or product of an in-vitro enzymatic reaction is then assayed either by agar well diffusion, microtiter assay or HPLC.

The robotic work cells were designed to be highly reliable and still possess flexibility for handling a number of different protocols. The goals were to:

1.   increase sample throughput
2.   increase precision
3.   provide a system with sufficient flexibility to enable further expansion of operations.

The first robotic system delivers cell suspensions onto an agar surface, liquid transfer and sonication. The robot removes a plate filled with agar from a rack, takes off the lid and delivers a diluted cell suspension onto the agar. The lid is replaced and the plate returned to the rack. In the next application, a drinking straw is fed to the robot and then cut to the appropriate length. A pipette tip is molded onto the end of the straw. A syringe is used to alternately remove an aliquot of liquid and dispense it to another location. In the final application, a template is used to guide a sonic probe to the correct location in a microtiter tray. The contents of each well are cooled with chilled water.

The second robotic system receives plates containing colonies and refrigerated modules. The colonies are inoculated in modules by this robot. The modules are removed from the workstation and incubated until maximum yield of the fermentation product is obtained.

The third robotic system moves a bridge over the modules and turns on a peristaltic pump which delivers an equal volume of solvent into a row of vials in the module. The robot then lowers a bank of high speed grinders which mix the contents of each vial and is repeated until all vials in the module are processed. The samples are assayed by either agar diffusion, microtiter systems or HPLC. A bank of 6 autodiluters is used for the agar diffusion assays. The tips of the autodiluters are mounted on a bridge. Samples are removed from the module and then delivered along with diluent into wells on an agar plate.

The fourth robotic system removes a 8 ml aliquot of sample from each vial in a refrigerated rack. The sample is delivered into a centrifuge tube and the sample centrifuged. The samples are then filtered and injected into an HPLC system.

Recombinant colonies can be fermented and assayed by the workstations. These workstations can screen an unlimited number of colonies for the production of hybrid antibodies or new beta-lactams generated by the mutation of key biosynthetic enzymes. They also work relatively unattended. On a sustained basis they are 30% faster than a technician performing an equivalent task over a short time.

LABORATORY ROBOTICS APPLIED TO TURBIDIMETRIC ENDOTOXIN ANALYSIS OF RECOMBINANT DNA - DERIVED PHARMACEUTICALS (ref. 3)

The analysis of endotoxin samples from recombinant DNA-derived pharmaceuticals is one of the most commonly employed assays utilized by the biotechnology industry. A turbidimetric endotoxin analysis procedure has been successfully coupled to a robotic system to perform sample preparation and complete analysis of the endotoxin content of a number of biochemical samples. The results obtained using the automated method were comparable to the manual gel clot method, having an observed recovery of 98% with an RSD of 50%.

Endotoxins are lipopolysaccharide components of the cell wall of gram-negative bacteria. Because of their potency, stability and ubiquity, endotoxins are of major concern to the manufacturers of parental products. Nanograms quantities can induce a response in humans ranging from fever to shock or even death.

The traditional assay for pyrogens has been the USP Rabbit Pyrogen Test. The Limulus Amebocyte Lysate (LAL) assay was originally developed as a semiquantitative method for the detection of endotoxins and has been steadily replacing the Rabbit Pyrogen test.

There are four different LAL methodologies: gel clot, colorimetric, chromogenic and turbidimetric assays. The turbidimetric test measures the increase in turbidity that precedes clot formation. The test is performed by adding a small amount of LAL reagent to the sample, mix and the turbidity read.

The turbidimetric test was chosen over the other tests for automation based on the following:

1. cost savings - personnel and reagents
2. increased accuracy and precision
3. concern for employees - LAL testing is repetitive and monotonous
4. high number of samples for daily analysis
5. increased sensitivity
6. computerized data reduction and analysis
7. robotic sample preparation for consistency of results

The accuracy and precision were evaluated by spiked placebo recovery studies. The results are listed in Table 2.

TABLE 2. Results Obtained for the Spiked Placebo Recovery of Endotoxins in Water using the Robotic Analysis System.

Spike[a]	Percent Recovery	RSD (percent)	Replicates
0.031	104.8	46.9	20
0.062	97.3	33.2	17
0.125	112.3	52.6	20
0.250	93.7	56.9	20
0.500	90.4	74.1	21
1.000	89.2	47.9	20
	98.0	51.2	

[a] Values are in EU/mL

The average mean recovery (accuracy) was 98.0% with an average RSD (relative standard deviation, precision) of 51.2%. The error in the assay can probably be attributed to the variation in the kinetics of the LAL reaction. Because the gel clot test is a limit test with standards that are two fold increments of each other, the test has an inherent error of plus or minus twofold (i.e., plus 100% to minus 50%). The accuracy and precision of the turbidimetric test using the robotic system were, therefore, substantial improvements over those obtained using the gel clot test. Also, endotoxin could be determined at much lower levels by taking advantage of the turbidimetric capabilities of the robotic method than that possible using the gel clot method.

AUTOMATED SAMPLE PREPARATION FOR A RESTRICTION ENZYME ANALYSIS (ref. 4)

This section describes an automated procedure for the isolation of plasmid DNA to prepare enzyme digests prior to electrophoretic analysis. The manual preparation requires twenty-four discrete steps, and requires about four hours to complete twelve samples. The automated system will complete seven samples in four hours, with a continuous output thereafter of four samples/hour.

The manual procedure to be automated is shown in Table 3.

TABLE 3. Manual Procedure for Recombinant-DNA Sample Preparation.

1. Place 1.5 mL of culture in a microcentrifuge tube and centrifuge 1-2 minutes.

2. Remove the medium by aspiration, leaving the pellet as dry as possible.

3. Add 200 uL of an ice cold solution containing glucose, EDTA and Tris-HCl. Resuspend pellet by vortexing.

4. Incubate 5 minutes at room temperature.

5. Add 200 uL of a freshly prepared solution containing NaOH and SDS. Mix gently by inverting. DO NOT VORTEX.

6. Incubate on ice for 5 minutes.

7. Add 150 uL of an ice-cold solution of potassium acetate buffer. Mix gently by vortexing.

8. Incubate on ice for 5 minutes.

9. Centrifuge for 10 minutes.

10. Transfer supernatant to a fresh tube avoiding the white precipitate.

11. Add an equal volume of phenol saturated with Tris-HCl buffer. Mix by vigorous vortexing.

12. Centrifuge 5 minutes.

13. Transfer supernatant to a fresh tube.

14. Repeat phenol extraction and again transfer supernatant to a fresh tube.

15. Add 1 mL of ice-cold ethanol. Mix by gentle vortexing.

16. Incubate 5 minutes at room temperature.

17. Centrifuge 5 minutes.

18. Remove supernatant by aspiration.

19. Add 500 uL of ethanol. Gently vortex to wash pellet.

20. Centrifuge 2 minutes.

TABLE 3. continued.

21. Remove supernatant by aspiration.

22. Dry pellet under vacuum.

23. Add 50 uL of a Tris-HCl, EDTA mixture. Dissolve pellet by vortexing occasionally while heating at 70°C.

24. Store solution at 4°C.

While at first this procedure is rather overwhelming, it can be broken down into repeated simple cycles of adding reagent, mixing, incubating and centrifuging. This pattern appears throughout the procedure. One of the keys to automating this procedure is to modify these cycles so that each is a multiple of the full cycle. It was estimated that the total cycle time would be about fourteen minutes, so those incubation and centrifuge times that were less than seven minutes were lengthened to seven minutes and steps longer than seven minutes were lengthened to fourteen minutes. To decrease time, all reagents are added to the test tubes while in a vortex mixer. The automated procedure is shown in Table 4.

TABLE 4. Robotic Sample Preparation for Recombinant-DNA Products.

Action	Incubation Location	Prep	Tools
Add 6 mL broth to a 12 x 75 mm tube, centrifuge, decant waste.		Done by submitter	
Add 300 uL of cold reagent in vortex.	Incubate 7 min in vortex.	1	Vortex 1 Dispenser
Add 600 uL of reagent in vortex, gently invert.	Incubate 5°C, 7 min in chill rack.	2	Tube inverter Vortex 1 Disp. Chill rack
Add 450 uL of cold reagent in vortex.	Incubate 5°C, 7 min in chill rack.	3	Vortex 2 Disp. Chill rack
Switch tubes in centrifuge, transfer supernatant, add equal volume of reagent, vortex.	Centrifuge 14 min	4	Centrifuge Transfer device Vortex 2 Disp. Tube dispenser
Switch tubes in centrifuge, transfer supernatant, add equal volume of reagent, vortex.	Centrifuge 7 min	5	Centrifuge Transfer device Vortex 2 Disp. Tube disperser
Switch tubes in centrifuge, transfer supernatant, add 3 mL of cold ethanol in vortex, vortex.	Centrifuge 7 min Incubate 7 min in vortex.	6	Centrifuge Vortex 3 Disp.

TABLE 4. continued.

Action	Incubation Location	Prep	Tools
Switch tubes in centrifuge, robot decant, add 1.5 mL of 70% ethanol in vortex, vortex.	Centrifuge 7 min	7	Centrifuge Vortex 3 Disp.
Switch tubes in centrifuge, decant to waste, put tubes in dryer.	Air dryer 14 min	8	Air Dryer
Add 200 uL of reagent in vortex, vortex.	Incubate 70°C in vortex 14 min	9	Vortex 4 Disp. Heated 70°C
Put tube in chill rack for storage.	Chill rack	10	Chill rack

To achieve minimum lag time before the first sample is done, the preps are actually done in the order: 1, 3, 5, 7, 9, 10, 2, 4, 6, and 8. This takes advantage of the operations that have only a seven minute incubation time and can be taken to the next prep during the same robot cycle, rather than waiting for the next full cycle. With this approach the robot can complete approximately four samples/hour with the first sample being done after eight cycles or about two hours, rather than after twelve cycles (three hours) if the preps were followed sequentially one through ten.

The major benefits of automating this assay are time savings, long-term sample preparation consistency and flexibility of scheduling by allowing sample prep to be started or added to any time during the day.

ROBOTIC AUTOMATION OF A COLORIMETRIC ASSAY FOR PROTEOGLYCANS (ref. 5)

This section describes an automated colorimetric assay for proteoglycan in microtiter plates and after initial set-up, the robotic system carries out the entire assay without human intervention. The method prepares three standard dilutions of samples, aliquotes dilutions in triplicate, adds dye, and reads absorbances. The system is interfaced to a PC allowing on-line calculation of a standard curve and storage of assay data in files on a floppy disk. The software also directs the robot to prepare an additional dilution if the sample absorbances fall outside the limits of the standard curve.

With the automated system, reproducibility from sample to sample is improved over the manual method with reduction of the standard error of the mean to less than 1%. Accuracy of the method is improved 18%. The greatest advantage of the robotic method is unattended operation. Initial investment of time and money for the automated method are small when compared to increased accuracy and reproducibility, high throughput and independent operation.

This laboratory is interested in discovering drugs which block the erosion of cartilage in arthritis. A primary test involves evaluating the ability of potential drugs to block the stimulated breakdown of cartilage in organ culture. A colorimetric assay is used to monitor the loss of cartilage proteoglycan. The assay required dilution of culture media samples with water and addition of dye. The samples were then mixed and the absorbance read. Each dilution was assayed in triplicate. If absorbance of a sample was beyond the limits of the standard curve, an additional dilution had to be prepared and the sample reanalyzed.

Approximately forty samples could be assayed and analyzed in an eight hour workday. At first the assay was adapted to 96-well tissue plates and absorbances read using a microplate reader. With the automated procedure, once started, the entire assay is performed by the robot.

Cartilage is incubated in growth media in the presence of a stimulant that induces breakdown of its proteoglycan (PG). The degradation products are released into the media and can be quantitated by combining an aliquot of the diluted media with a dye that changes color according to the levels of PG present. A test group of cartilage is incubated in the presence of the stimulant, and additional test groups receive stimulation plus a potential inhibitor of cartilage breakdown. By monitoring the levels of PG in the culture media, one can determine if inhibition of cartilage breakdown has occurred and to quantitate the inhibition. Thus, a compound can be screened for the ability to inhibit cartilage degradation in response to the stimulant.

The assay was originally performed in test tubes and then adapted to 96-well tissue culture plates. This improved throughput but still a researcher was required to perform a very repetitive, tedious assay.

Calibration curves generated by the manual and robotic methods are superimposable indicating the assay method has no effect on the data obtained.

A comparison of the manual versus robotics method is shown in Table 5.

TABLE 5. Comparison Between Methods of Analysis.

Method	Precision[a] (% SEM)	% Difference[b] from Actual Value (+ SD)	% Human Participation	Sample/Day
Manual	8.1	23 ± 14	100	40
Robotic	1.3	7 ± 5	<5	216

[a] A known amount of PG (n = 11) was assayed using each analysis method. The amount calculated to be present was determined by comparing the sample absorbance to the standard curve prepared using the same method. The mean and standard error of the mean (SEM) were determined for each method and the % SEM calculated as:

TABLE 5. continued.

$$\% \text{ SEM} = \frac{\text{SEM}}{\text{mean}} \times 100$$

b   Using data obtained as described in a, we compared each calculated value with the actual amount present as:

$$\% \text{ Difference From} = \frac{|\text{Actual-Calculated}|}{\text{Actual Value}} \times 100 \atop \text{Actual}$$

    The table shows results from eleven replicates of a known level of PG assayed by both methods and the mean of the replicate absorbances was determined. The robotic method proved to be more reproducible than the manual method since individual absorbances obtained through this method varied less from the mean absorbances than those from the manual method.

    The robotic method also gave results closer to the actual level of PG in the sample. Table 6 compares human participation required by the manual and robotic method.

TABLE 6. Comparison of Human Participation Required by Manual and Robotic Method.

A. PROCEDURE		
Manual		**Robotic**
1. Pipet water into tube to dilute	1.	Load plates and tips
2. Pipet sample into test tube	2.	Start robot from PC
3. Vortex		
4. Add dye		
5. Vortex		
6. Pour sample into cuvette		
7. Read in spectrophotometer		Robotic
8. Compare absorbances to standard curve		Procedure
9. Decide if further dilution is necessary. If required, dilute sample and repeat steps 1-9		No Human participation required
10. Repeat steps 1-9 for each sample		
11. Type data into computer file		
12. Run program to analyze data		
13. Examine data	3.	Examine data

Current capability of the robotic system allows over five times more samples to be processed within a day (the robot does not stop after eight hours) and reduces human participation to less than 5%. The potential for discovering new drugs should increase using the robotic system because of greater productivity and throughput.

AUTOMATED MICROPLATE ASSAYS USING A ROBOTIC SYSTEM (ref. 6)

Many assays in biotechnology laboratories use standard 96-well microplate format. Enzyme immunoassay (EIA, ELISA) are routinely done in many labs and are now routinely performed using robotic automation systems. These assays are time-consuming and labor intensive but the 96-well format leads to low reagent and supply costs associated with assaying large numbers of samples. The technique is capable of high sensitively and specificity but requires a large number of precise, repetitive manipulations. This section will report on an automated system for screening assays for testing supernatants from fusion plates, assays for evaluating clone stability, assays for isotyping monoclonal antibodies, and assays for titering antibody activity in ascites fluid or antiserum.

The robotic system is capable of dispensing antigen (single or multiple) into microtiter plate wells for coating purposes, preparing dilutions of sample for analysis, preparing substrate, adding reagents, washing and blocking microtiter plates, and transferring completed plates to a plate reader interfaced to a PC for data acquisition and storage. The system operates unattended and frees laboratory personnel for other tasks resulting in improved productivity.

ELISA assays follow the same general sequence of steps, but many assays differ in the details of the incubation times, reagent volumes, plate washing protocols, and plate reading. A robotic ELISA system can achieve repeatability and consistency for a particular assay while maintaining flexibility to accommodate variations in procedure for additional assays as they are developed.

REFERENCES

1. S. D. Hamilton, "An Integrated Robotic Sample Preparation and HPLC Analysis of Biosynthetic Human Insulin", in Advances in Laboratory Automation Robotics Vol. 4, 1987. J. R. Strimaitis and G. L. Hawk, eds. (Zymark Corporation, Hopkinton, MA 1987) p. 195.

2. O. Godfrey, L. Ford, D. Berry, "Automation of Fermentation, Sample Preparation and Assay of Recombinant Microorganisms", in Advances in Laboratory Automation Robotics Vol. 5, 1988. J. R. Strimaitis and G. L. Hawk, eds. (Zymark Corporation, Hopkinton, MA 1988) p. 101.

3. R. H. Carlson, R. L. Garnick, M. M. Stephan, D. Sinicropi, C. P. du Mee, and C. Miller, "Laboratory Robotics Applied to Turbidimetric Endotoxin Analysis of Recombinant DNA-Derived Pharmaceuticals", in Advances in Laboratory Automation Robotics Vol. 4, 1987. J. R. Strimaitis and G. L. Hawk, eds. (Zymark Corporation, Hopkinton, MA 1987) p. 235.

4.	S. D. Hamilton, "Automated Sample Preparation for a Restriction Enzyme Analysis", in Advances in Laboratory Automation Robotics Vol. 6, 1989. J. R. Strimaitis and J. P. Helfrich, eds. (Zymark Corporation, Hopkinton, MA 1989) p. 13.

5.	M. A. Pratta, L. E. Sachau, P. W. DeYoe, E. C. Arner and D. C. Lapen, "Robotics Automation of a Colorimetric Assay for Proteoglycans" in Advances in Laboratory Automation Robotics Vol. 6, 1989. J. R. Strimaitis and J. P. Helfrich, eds. (Zymark Corporation, Hopkinton, MA 1989) p. 337.

6.	G. D. Hahn and B. Lightbody, "Automated EIA Microplate Management System Applications of a Monoclonal Antibody Development Laboratory: in Advances in Laboratory Automation Robotics Vol. 3, 1986. J. R. Strimaitis and G. L. Hawk, eds. (Zymark Corporation, Hopkinton, MA 1986) p. 167.

I. Karube (Ed.) *Automation in Biotechnology*
Proceedings of the 4th Toyota Conference, 21–24 October 1990
© 1991 Elsevier Science Publishers B.V. All rights reserved

POSSIBLE CONTRIBUTION OF STM TO BIOTECHNOLOGY

Masatoshi ONO

Electrotechnical Laboratory, Umezono, Tsukuba, Ibaraki 305 (Japan)

SUMMARY

A scanning tunneling microscope (or microscopy),STM , based on electron tunneling between the tip of a stylus and a sample, gives images of individual atoms on the surface of a solid in vacuum, air or liquid. New types of scanning probe microscopes, generically called SPM or SXM, utilizing the probe-sample interaction other than electron tunneling were derived from an STM. And these microscopes gave birth to technologies for handling atoms. This report describes the present statuses of these technologies and applications of STM related to biotechnology, together with discussion about their problems and possibilities.

INTRODUCTION

A Scanning Tunneling Microscope (ref.1), abbreviated as an STM, gives an image of individual surface atoms by visualizing the trajectory of a sharp metal stylus scanned over the surface of a sample with a given gap between the tip of the stylus and the surface. The gap distance of about one nanometer is kept constant by servo controls based on the fact that the tunneling current decays as a exponential function of the gap distance. The lateral resolution of the microscope is about 0.1 nm for the sharpest tip, and the probing depth is less than about 2 nm.

More precisely, the tunneling current depends also on local density of state of electrons in the surface layer of the tip and the sample. This was initially demonstrated by the spectroscopic image called STS, standing for Scanning Tunneling Spectroscopy, for 7 x 7 structure of Si (111) surface (ref.2), which was observed as the first STM image of atomic resolution by Binnig and Rohrer (ref.1) in 1982.

Rapid increase in number of papers reported at the annual international conferences on STM, firstly held in 1986 as "STM '86", clearly reflects the recent spread of the microscope over the various laboratories in different fields of science and technology. For example, the surfaces of semiconductors, metals and layered crystals, evaporated or adsorbed atoms and molecules on solid surfaces, and organic or biological samples have been studied by STM.

Those observations by STM have been conducted in different kind of environments, such as ultra-high vacuum (UHV), air, gas and liquid, at temperatures from

1.5 K to 600 K. And commercially available STM is now widely adopted as a surface roughness meter for industry as well as a scientific instrument for surface studies.

PRESENT STATUS OF STM RELATED TECHNOLOGIES
STM technology

An STM is composed of the stylus with a sharp tip, a piezoelectric actuator, called a scanner, for three dimensional displacement of the tip, a rough positionner to control the tip-sample gap down to a micron, a vibration isolator, electronic circuits for controlling the gap with an accuracy of 10 pm or better, a data processor and an image display. Fig.1 illustrates structure of an STM, excluding a rough positionner and a vibration isolator.

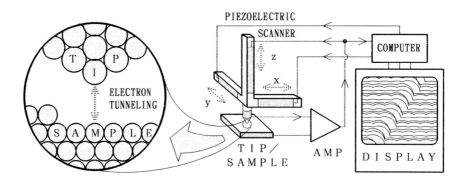

Fig.1. The principle of a scanning tunneling microscope. The left round inset gives an idea of geometrical relation of the tip and the sample surface with a gap of about 1 nm.

As it was initially called as a "scanning vacuum tunneling microscope", the microscope were designed and built for operation in UHV in the first few years. In 1984, successful operations of the microscope in air initiated the drastic increase in number of the microscopes including those put on the market. These microscopes for operation in air are applied to observe numbers of different samples including polymers and biological membranes. The most popular sample for STM observation in air seems to be a cleaved graphite crystal as shown in Fig.2. The surface is widely used as a substrate for STM observation of organic molecules, since it is flat and stable in air.

An STM with a vertical stylus over a level sample is applied to observe atoms and molecules at the interface of liquid and solid. In the case of electrolyte, a special stylus is developed to suppress ionic conduction, and special device was adopted for controlling potential of the tip to the electrolyte without affecting the tunneling.

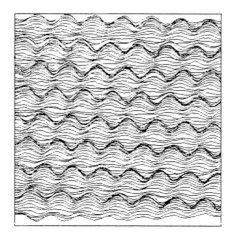

Fig.2. An STM image of surface atoms on cleaved surface of graphite, observed in air at room temperature. Constant current (gap) mode over an area of 1.7 nm x 1.7 nm (x,y plane). Trajectories of the tip in x,z plane is indicated as lines with corrugation up to 0.3 nm.

For surface studies, the combinations of an STM with conventional equipments, such as LEED, AES, and/or RHEED are adopted for the comparison with the results of previous studies and especially for the elementary analysis which can be done successfully by the tunneling spectroscopy only for limited cases where the electron energy spectra for several electron volts in the vicinity of Fermi level provide enough information for the analysis. Therefore, absorption or emission of photons has been studied to provide STM or SPM measurements with information to identify an element and a molecule.

SPM or SXM

Scanning Probe Microscope (SPM) or SXM (ref.3) stands for a microscope derived from the technology of STM. Each of them has a probe to trace the topography of the sample surface. The probe is mechanically driven to keep the strength of the interaction between the sample and tip of the probe, other than electron tunneling. Among them, Scanning Near-field Optical Microscopy (SNOM), Scanning Atomic Force Microscopy (SAFM), and Scanning Ion Conductance Microscopy (SICM) are expected to make important contribution to biotechnology by giving information about optical, mechanical and ion conductive properties, respectively.

Scanning Atomic Force Microscope (SAFM or AFM) (ref.4) images surface topography as a trajectory of the tip of a stylus controlled to keep constant force acting between the tip and the sample surface. The force is detected as deflection of a leaf spring with a stylus fixed on the end of it. The lateral resolution of the microscope is around 0.2 nm, almost same as that of an STM.

Scanning Near-field Optical Microscope (SNOM) (ref.5) relies on the exponential decay of light emitted through a hole much smaller than the wave length. The stylus made of transparent material is coated with a thin metal film which have a small hole at the tip of the stylus. This microscope is reported to visualize two dimensional distribution of optical properties of a sample surface with resolution of about 10 nm.

An SPM utilizing a very thin tube of insulator as a probe, is designed to operate in electrolyte, and called Scanning Ion Conductance Microscope (SICM) (ref.6). The probe is controlled to keep the ion current, through the hole at the end of the probe, constant. An SICM shows topography of an insulator and also distribution of ion conducting holes on the surface of the insulator, such as polymer film or cell membrane, and so on.

STUDIES BY STM & STS

The surface studies by scanning tunneling microscopy are reported on rearrangement of the surface atoms of clean semiconductor, those of metals, adsorbates on conductive crystals, initial stage of film growth on semiconductors and metals, and properties of irregular surface structures, such as steps and point defects (ref.7).

One of the major advantages of the scanning tunneling microscopy and spectroscopy is its capability of giving the real space image of the surface of a solid with very high spatial resolution which enables us to study about the behavior of individual atoms and molecules on the solid surfaces. The following examples of STM studies describe the capabilities and possibilities of these microscopy and spectroscopy to study samples related to biology.

Organic molecules

The most typical cases of successful measurements of adsorption of molecules seem to be benzene on Rh crystal surface (ref.8). The more larger or thicker molecules including organic and biological ones adsorbed on the solid surfaces have been studied by STM and STS. For example, copper phthalocyanine (ref.9), sorbic acid (ref.10) and so on are clearly imaged by the microscopy. Even the mechanism of electron tunneling and/or conduction has not been fully understood, molecules having thickness of more than ordinal tunneling distance of about 2 nm are observed by STM. For example, some of the double layers of Langmuir-Blodgett films of about 5 nm in thickness, gave STM images of two dimensional patterns having periodicity corresponding to the results of x-ray diffraction (ref.11).

Liquid-Solid Interface

Since the the microscopy enables us to observe atoms and molecules on the surface in liquid as well as in ultra-high vacuum, chemical reactions and electro

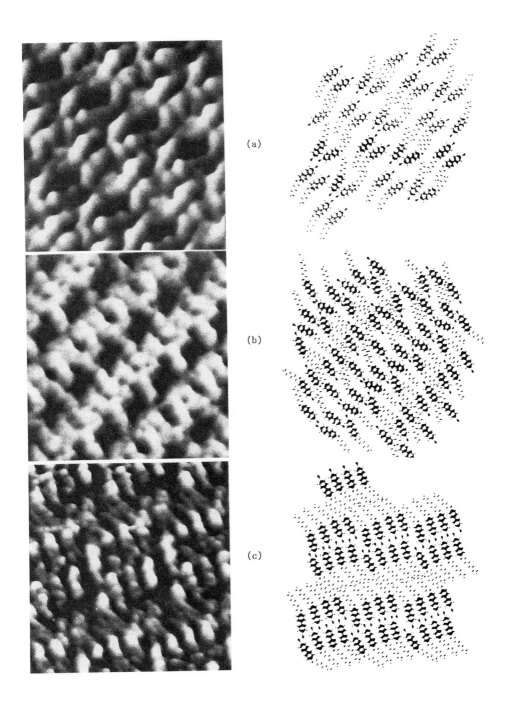

Fig.3. Variable current STM images of liquid crystal molecules on graphite at room temperature, with models of molecular arrangement (right).
(a) 6CB: 6 nm square, bias voltage; 1.8 V, average tunneling current; 70 pA,
(b) 7CB: 7 nm, 1.9 V, 100 pA, (c) 8CB: 8 nm, 0.8 V, 100 pA.

chemical processes at a liquid-solid interface are expected to be one of the most promising application fields of the microscopy.

Another important example of the observation in liquid was reported for liquid crystals. The molecules of liquid crystal material are found to be adsorbed on a graphite surface as one or two layers of periodic molecular arrangement, while the tip of STM is moving in the droplet of the material. Advances in the methods of sample preparation and in the condition of observation have been improving quality of observed images of the molecules (ref.12). In Fig. 3, STM images taken at room temperature are shown for liquid crystal of 4-n-hexyl-4'-cyano-biphenyl (termed 6CB), 4-n-heptyl-4'-cyano-biphenyl (7CB), and 4-n-octyl-4'-cyano-biphenyl (8CB), with models of molecular arrangement corresponding to calculated configuration of HOMO (Highest occupied molecular orbitals).

These cases of stable arrangement of molecules on graphite seem to be possible for STM observations of other organic molecules. For example, the similar technique seemed to be applied to the cases, such as imaging of single stranded DNA adsorbed on a graphite surface (ref.13).

Biological samples

Approaches to observe biological samples are inevitably different from the techniques described above for imaging organic molecules. The most successful method is supposed to be the STM observation of electroconductive replica prepared by the method similar to that for transmission electron microscope (TEM). This method was applied to observe morphology of two dimensional array of protein particles of 19 nm in diameter (ref.14) and topography of replica of a cell membrane (ref.15). The resolution of this kind of observation is limited to a few nanometers corresponding to fine structure of an evaporated film of Pt and carbon.

PROBLEMS & LIMITATIONS OF STM & SPM
Spatial Difficulties and Limitations

The sampling depth of an STM and SPMs is within a few nm. And the apex shape of the probe make it impossible to observe the area of the sample surface at a slope of more than about 1 in 1.

In the case of successful STM imaging of an insulating material, the sample on a flat conductive surface is generally less than 2 nm in thickness. . There are some exceptional cases, such as observation of Langmuir-Brodgedt films of double layers (ref.11) and that of paraffin (ref.16). In the former case, however, reproducibility of STM images are reported to be poor and the mechanism of electric conduction for the observation is not explained yet.

An SAFM gives an image of surface morphology even for insulators, however, sample deformation disturbs the observation of soft samples (ref.17), since the force on the sample applied by the stylus is about 10^{-9} N or greater.

Limitation on speed of observation

Another limitations of STM or SPM is speed of imaging. At the best condition, STM technology can image about 10^4atoms/sec, and it will take about 1 sec to image over area of 100 nm x 100 nm. This means that complete observation of 1 mm^2 will take more than 3 years. For this reason, combination with other microscope, such as optical or scanning electron microscope is needed for rapid selection of the imaging area of the sample surface.

HOW TO OVERCOME PROBLEMS OF STM AND SPM
What to be studied for STM

Among the means for surface studies, the scanning electron microscopy has the best spatial resolution which gives us new insight into surface phenomena. This fact, on the other hand, requires understanding of physical processes in the volume smaller than a few cubic nanometers, where the electron tunneling takes place, for appropriate interpretation of the images given by the microscopy.

For full comprehension of what we "see" on the display of an STM, theoretical studies on electron tunneling are required together with well controlled experimental studies. Since theoretical investigation must be based on the information about energy state and the molecular orbitals of the surface of the probe and the sample. For the understanding the mechanism of STM and STS measurements, technologies for the better control of the probe are indispensable.

The theoretical studies are now being done on rather simple models. The STM/STS measurements in UHV is suitable for supporting surface studies, however, present technology of STM is not able to give reproducible STS data. This suggest that the condition of the tip of the stylus is not well controlled during the measurement. Further researches on the theory and progress in experiments in UHV condition will provide the foundation for the real interpretation of STM images measured in the air, in gas and in liquid.

Variety of observation on biological samples

As well as direct observation of biological samples, STM and SPM are effective for biological applications as a tool to measure surface morphology after sample treatment.

Even in the case soft and nonconducting biological samples thicker than a few nm, the conductive replicas of the samples can be observed by STM and SPM to reveal detailed surface morphology, with spatial resolution better than a scanning electron microscope (SEM) or a TEM.

Another example is an application to readout of X-ray shadow micrograph. An AFM was used to measure surface topography of photosensitive plastics used as a

dry plate for X-ray shadow micrograph. The ultimate lateral resolution is limited
to x-ray wave length of water window, namely 2.5 to 4.4 nm, the technique can
give two-dimensionally projected flash image of three-dimensional distribution of
carbon atoms in live samples in water. The pictures was shown for spermatozoons
of a sea urchin (ref.18).

POSSIBILITIES OF NANOMETER EXPERIMENTATION

As stated above, the interactions between the probe and the sample surface,
including unintentional touching, cause displacement, deformation and destruction
of surface structure. However, if these interactions can be controlled, they can
be used as a means to handle the part of a sample by mechanical force, electric
field, and electrochemical reaction in nanometer scale. Thus the nanometer experimentation can be realized by STM and/or SPM.

The possibilities of the nanometer experimentations were demonstrated by the
following experimental results. Xe molecules adsorbed on Ni at liquid helium temperature were moved by an STM tip of W to display alphabetical characters
(ref.19). Dimethyl phthalate molecules were decomposed and fixed on graphite substrate by applying voltage pulse to an STM stylus (ref.20). Arrangement of liquid
crystal molecules adsorbed on graphite was gradually changed by varying tip voltage for STM observation (ref.21).

Now it can be stated that we are at the beginning of an era of nanometer scale
experimentation realized by handling and observing of samples by utilizing STM
and its related technologies.

ACKNOWLEDGMENTS

The author is very thankful to Drs. K. Kajimura, H. Tokumoto, W. Mizutani and
T. Tomie of Electrotechnical Laboratory, and Mr. M. Shigeno of Seiko Instruments
for their stimulating discussions and valuable experimental data.

REFERENCES

1 G. Binnig and H. Rohrer, Scanning tunneling microscopy, Helvetica Physica
   Acta. 55 (1982) 726-735.
2 R.J. Hamers, R.M. Tromp and J.E. Demuth, Surface electronic structure of
   Si(111)-(7x7) resolved in real space, Phys. Rev. Lett. 56 (1986) 1972-1975.
3 D.W. Pohl, U.Ch. Fischer, and U.T. Duerig, Scanning near-field optical microscopy (SNOM), J. Microscopy, 152 (1988) 853-861.
4 B. Drake, C.B. Prater, A.L. Weisenhorn, S.A.C. Gould, T.R. Albrecht,
   C.F. Quate, D.S. Cannell, H.G. Hansma, and P.K. Hansma, Imaging crystals,
   polymers, and processes in water with the atomic force microscope, Science,
   243 (1989) 1586-1589.
5 U. Duerig, D. Pohl, and F. Rohrer, Near-field optical scanning microscopy
   with tunneling-distance regulation, IBM J. Res.& Dev. 30 (1986) 478-483.
6 O. Marti, V. Elings, M. Haugan, C.E. Bracker, J. Schneir, B. Drake,

S.A.C. Gould, J. Gurley, L. Hellemans, K. Shaw, A.L. Weisenhorn, J. Zasadzinski, and P.K. Hansma, Scanning probe microscopy of biological samples and other surfaces, J. Microscopy, 152(3) (1988) 803-809.

7 H. Tokumoto, K. Miki, H. Murakami, H. Bando, M. Ono, and K. Kajimura, Real time observation of oxygen and hydrogen adsorption on silicon surfaces by scanning tunneling microscopy, J. Vac. Sci. Technol. A8 (1990) 255-258.

8 H. Ohtani, R.J. Wilson, S. Chiang, and C.M. Mate, Scanning tunneling microscopy observation of benzene molecules on the Rh(111)-(3x3) (C_6H_6+2CO) sur-surface, Phys. Rev. Lett. 60 (1988) 2398-2401.

9 P.H. Lippel, R.J. Wilson, M.D. Miller, Ch. Woell, and S. Chiang, High resolution imaging of copper-phtalocyanine by scanning tunneling microscopy, Phys. Rev. Lett. 62 (1989) 171-174.

10 D.P.E. Smith, M.D. Kirk, and C.F. Quate, Molecular images and vibrational spectroscopy of sorbic acid with the scanning tunneling microscope, J. Chem. Phys. 86(11) (1987) 6034-6038.

11 W. Mizutani, M. Shigeno, K. Saito, K. Watanabe, M. Sugi, M. Ono, and K. Kajimura, Observation of Langmuir-Blodgett films by scanning tunneling microscopy, Japn. J. Appl. Phys. 27(10) (1988) 1803-1807.

12 W. Mizutani, M. Shigeno, Y. Sakakibara, K. Kajimura and M. Ono, Scanning tun-tunneling spectroscopy study of adsorbed molecules, J. Vac. Sci. Technol. A8 (1990) 675-678.

13 R.J. Driscoll, M.G. Youngquist, and J.D. Baldeschwieler, Atomic-scale imaging of DNA using scanning tunneling microscopy, Nature, 346 (1990) 294-296.

14 R. Guckenberger, W. Wiegraebe, and W. Baumeister, Scanning tunneling microscopy of biomacromolecules, J. Microscopy, 152(3) (1988) 795-802.

15 J.A.N. Zasadzinski, J. Schneir, J. Gurley, V. Elings, and P.K. Hansma, Scanning tunneling microscopy of freeze-fracture replicas of biomembranes, Science, 239 (1988) 1013-1015.

16 B. Michel, G. Travaglini, H. Rohrer, C. Joachim, and M. Amrin, Z. Phyz. B 76 (1989) 99-105.

17 S.A.C. Gould, B. Drake, C.B. Prater, A.L. Weisenhorn, S. Manne, H.G. Hansma, P.K. Hansma, J. Massie, M. Longmire, V. Elings, B.D. Northern, B. Mukergee, C.M. Peterson, W. Stoeckenius, T.R. Albrecht and C.F. Quate, From atoms to integrated circuit chips, blood cells, and bacteria with atomic force microscope, J. Vac. Sci. Technol. A8 (1990) 369-373.

18 T. Tomie, H. Shimizu, T. Tajima, M. Yamada, T. Kanayama, H. Kondo, M. Yano, and M. Ono, Three-dimensional readout of flash X-ray images of living sperm in water by atomic-force microscopy, Science, 252 (1991) 691-693.

19 D.M. Eigler and E.K. Schweizer, Positioning single atoms with a scanning tunneling microscope, Nature, 344 (1990) 524-526.

20 J.S. Foster, J.E. Frommer, and P.C. Arnett, Molecular manipulation using a tunneling microscope, Nature, 331 (1988) 324-326.

21 W. Mizutani, M. Shigeno, M. Ohmi, M. Suginoya, K. Kajimura, and M. Ono, Observation and control of adsorbed molecules, J. Vac. Sci. Technol. B9 (1991) 1102-1106.

I. Karube (Ed.) *Automation in Biotechnology*
Proceedings of the 4th Toyota Conference, 21–24 October 1990
© 1991 Elsevier Science Publishers B.V. All rights reserved

STM INVESTIGATION OF MACROMOLECULES

B. Michel
IBM Research Division, Zurich Research Laboratory, 8803 Rüschlikon, Switzerland

SUMMARY

The scanning tunneling microscope (STM) facilitates novel accesses to matter on a molecular level: To observe and study individual objects or molecules in their natural environment, to follow and control molecular processes, and to modify molecular structures. Contact with established electron microscopy methods was made with STM investigations of biological samples coated with a conducting film. In view of the limited bulk conductivity, the first STM images of macromolecules came rather as a surprise. Meanwhile good imaging with STM has been achieved for all major classes of biological molecules: nucleic acids, proteins, carbohydrates and lipids. Not only individual molecules but also supramolecular structures like recA-DNA lipid bilayers, viruses, and even cell surfaces have been accessible to STM observation. STM experiments on paraffin structures with sizes up to 100 nm imply sufficient electron transfer many orders of magnitude larger than generally assumed. For STM investigations of macromolecules and for the design of biosensors it is essential to bind unmodified molecules strongly to an atomically flat reactive surface. Gold (111) surfaces have been functionalized with self-assembled monolayers and Langmuir-Blodgett films of mercaptanes and disulfides. The molecular recognition of biotin-functionalized self-assembled monolayers by streptavidin yielded protein covered surfaces with different crystallinity due to different lateral mobility.

INTRODUCTION

During the past decades, biochemical and biological research have undergone a transition from describing taxonomic and functional properties of living organisms to the investigation of the molecular basis of biological processes. Moreover, within these investigations they have to make use of molecular processes like site-specific proteolysis, in vitro DNA transcription, and site-directed mutagenesis. It is now widely accepted that all protein sequences in an organism are determined by the genome and that the primary structure of the protein controls the generation of the secondary structure and thereby the function of the protein. Due to the fact that with site-directed mutagenesis any protein can be produced in many variants, these techniques have found much wider application in industry. The marriage of biochemistry and biology with engineering for the purpose of industrial production of goods has given birth to biotechnology. The industrial application of biochemical analyses, in turn, has increased the demand for automated instruments. Meanwhile, for routine applications like DNA/protein sequencing, DNA/protein synthesis, cell sorting etc. have been automated. The automated instruments, however, cover only a few of the many analytic and synthetic methods in biochemistry and biology. Still most of the experimental work in these fields is done by hand. An important limitation to further automation in these fields is sensor technology.

Parallel to the progress in biological sciences chemistry has undergone a transition from molecular to supramolecular systems. In supramolecular chemistry the functional properties of molecules are not only determined by their chemical structure but to an even larger extent by their environment. A prerequisite for any of these molecules is the ability to recognize their neighbors and interact with them. The process where isolated molecules in solution recognize a partner molecule before they dock to the recognized site and form a complex is called self-assembling. The new goal is to synthesize molecules that self-organize to form complex assemblies where molecules interact in a purpose-oriented manner, like parts in a machine. This so-called "supramolecular engineering" has the potential to revolutionize technology in many fields, e.g. in microelectronics, integrated optics, and microsensors. The organization of monomolecular assemblies at solid surfaces provides a rational approach for fabricating interfaces with well-defined composition, structure, and thickness. Such assemblies provide a means to control the chemical and physical properties of interfaces for a variety of heterogeneous phenomena including catalysis, corrosion, lubrication and adhesion as well as for scanning tunneling microscope (STM) investigations of molecules. When self-assembling is used to build more complex structures, surfaces become very important, since the recognition of a new molecule always takes place at the surface of the preassembled complex. This and the many reactions at phase boundaries introduced in the past decade have shifted the focus of chemistry to reactions at surfaces.

When biological molecules are combined with structures built from molecular self-assembly on surfaces, a novel dimension can be brought to these systems: The reaction specificity of chemical sensors is upgraded with the substrate specificity typical for most biological processes and thereby becomes a biosensor. When such sensitive surfaces are combined on structures built with lithographic methods, sensors with many different specificities can be built.

The motivation to investigate biological systems from a general scientific standpoint is twofold: to discover biological concepts on molecular scale for later use of such concepts in human-built systems and to apply biological matter directly as tools in systems of contemporary concept.

Many varieties of molecular recognition are used in living organisms, the lowest level being the recognition sequences on proteins that guide the transport to the different compartments of the cell. Once in the correct compartment, the pre-sequence is cleaved off and the proteins self-organize to structures like cytoskeleton, microtubuly etc. In fact the genesis of any organism is guided by self-organization. The only input to that organization is the information stored in the genes. The best known example for this genetically determined self-assembly is the virus bacteriophage T4. The biological system of information processing is a good example for the mass storage of information. Transcription and translation process of DNA involves many selective molecular recognition processes: this causes the very low error rate of a few errors in one billion copied or translated bases.

The fault tolerance, the process of learning, and the efficient pattern recognition of our brain have led to the current interest in neural computing. The brain is a parallel distributed processing system that can adapt to new situations by modification of the interconnections. The basic unit of the brain, the neuron, uses principles that have been developed by millions of years of evolution: Our short-term and long-term memories are governed by the intracellular molecular organization of the neurons. The process of learning, the modification of the sensitivity of a neuron to external stimuli is based on molecular concepts. With instruments that can investigate processes at a molecular level this information can be made accessible for future use.

The STM has opened a new approach to molecules. They can be studied as individual objects in contrast to all other methods where properties of molecules can only be derived from the statistical behavior of many molecules. Since the STM has reduced the smallest size of observation to the atomic scale at a perfectly localized point in space, we are now able to make the connection between macroscopic structures and structures assembled from atoms. Once molecules are immobilized we can study their structure and watch functional molecules while they do their job. As a next step we can try to interfere with this function and control it so that it takes another direction. The tip can then be used to assemble molecular structures in an ordered manner so that they can perform a set of functions in controlled sequence.

RESULTS

A prerequisite for accurate studies on large biological molecules is a thorough understanding of the imaging mechanism on small adsorbed molecules. To that end several STM studies on adsorbed benzenes (Ohtani et al., 1988) and thiols (Hallmark et al., 1987) have been undertaken in UHV. It seems that the STM is able to visualize either lowest unoccupied or highest occupied orbitals of a molecule.

94

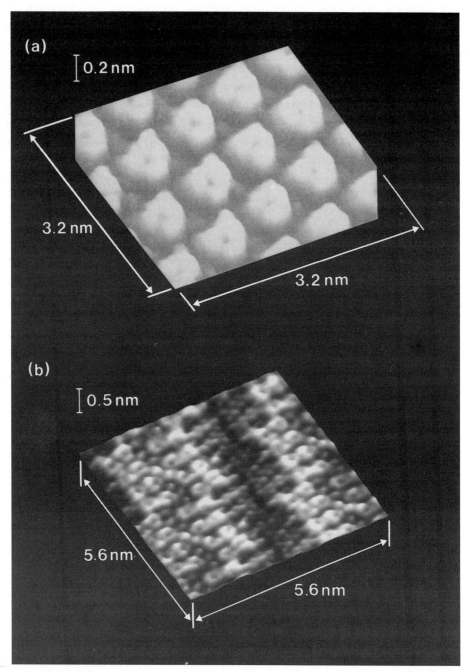

Fig. 1(a) Three-dimensional image of benzene and carbon monoxide coadsorbed as a 3 × 3 superlattice on Rh (111). Imaging conditions: voltage <0.5 V, current 4 nA. From Ohtani et al. (1988). (b) Three-dimensional images of octylcyanobiphenyl (8CB) liquid crystals on pyrolytic graphite. Imaging conditions: tungsten tip voltage 1.0 V, current 400 pA. From Smith et al. (1990).

The trifold symmetry found on a benzene ring in Fig. 1a is probably not due to the undisturbed orbitals but due to π or π^* orbitals of a benzene Rh π complex. The role of the coadsorbed CO is not clear, but it could contribute π orbitals to regularly saturate the Rh surface.

To achieve pure conditions and eliminate most unknown parameters due to contaminations, it is best to work with UHV. The easier interpretation in UHV has not prevented experiments under normal conditions whether in air, under protective gas, or under liquid. Good results have been obtained on smectic octylcyanobiphenyl liquid crystals adsorbed on highly oriented pyrolytic graphite (HOPG) (Foster & Frommer, 1987 and Smith et al., 1990).

In Fig. 1b orientation and packing of the liquid crystal molecules are clearly visible. Moreover, it is possible to resolve different functional groups: The aromatically bonded carbon atoms of the tilted benzene rings are visible as bright spots, the cyano group appears less bright, and the aliphatic carbons are faintly visible. By decreasing the tunnel gap resistance, the underlying graphite substrate is imaged, allowing the registry of the adsorbed molecules with the graphite to be deduced. The electronic nature of the STM contrast is confirmed by the good agreement of the images with results of ab initio Hartree-Fock orbital calculations. STM is sensitive to electron density at the Fermi level which is close to the level of lowest unoccupied molecular orbital (LUMO) or highest occupied molecular orbital (HOMO) of aromatic carbons or of the nitrogen.

STM studies of biological samples coated with a conducting film have been carried out and related to studies done by electron microscopy. The preparation method is very similar: The molecules are adsorbed on a freshly cleaved mica surface that had been pretreated with glow discharge. Then the sample is freeze-dried in high vacuum and covered with 1.5 nm of Pt-Ir-C (Scheme 1, Travaglini et al., 1987). The sample can be taken out of the evaporator and studied by STM in air with a resolution exceeding that of an electron microscope. This technique is quite reliable but the resolution is limited by the grain size of the evaporated film (about 10 Å). The sensitivity is good enough to resolve the single helix of recA-DNA (Amrein et al., 1988, Fig. 3a), a hexagonal lattice of proteins in cell membranes of bacteria (Michel & Travaglini, 1988, Fig. 2), and proteins that compose the head of the virus bacteriophage T4 (Amrein et al., 1989b, Fig. 3b). Figure 2 shows the typical zoom steps performed with an STM during an investigation: The left two images are from a series of images taken in search mode with image sizes of 3500 × 4800 nm. With the coarse positioning system each image has been shifted by 3000 nm. Once an interesting structure is found (box), the scan range of the piezo scanner is reduced and data points are acquired

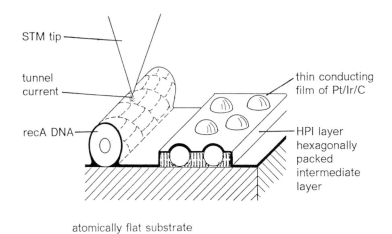

Scheme 1. Imaging of coated objects.

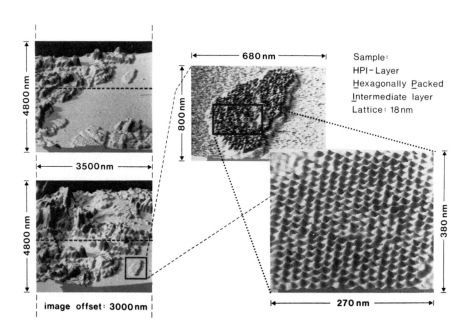

Fig. 2. Three-dimensional images with simulated shading of hexagonally packed intermediate layers of deinococcus radiodurans coated with 1 nm of Pt-Ir-C. Imaging conditions: gold tip voltage 300 mV (tip-sample), current 500 pA. Left side: Images with a size of 3500 × 4800 nm taken in search mode with an offset of 3000 nm. Right side: Enlarged images of membrane patch (box) with zoom factors of 4.25 and 17.

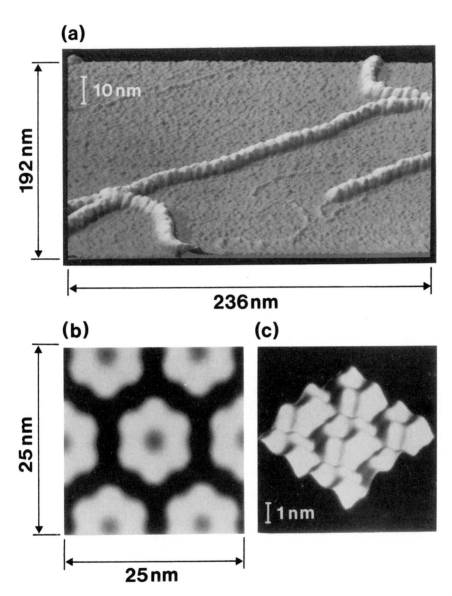

Fig. 3(a) Three-dimensional image with simulated shading of freeze-dried recA-DNA complexes coated with Pt-Ir-C film. Imaging conditions: gold tip, voltage 800 mV, current 500 pA. Large arrow points to recA-DNA complex, small arrow points to free DNA. From Amrein et al. (1988). (b) Unit cell of freeze-dried and Pt-Ir-C coated polyheads. (c) Three-dimensional image with simulated shading of (b). Imaging conditions: gold tips, voltage 0.1 - 1 V, current 0.1 - 1 nA.

at much closer spacing. The images on the right-hand side are zoom views of this structure. They show the regular lattice of the hexagonally packed intermediate layer (HPI-layer) of *Deinococcus radiodurans*. Such regular lattices of proteins that stabilize the cell membrane of bacteria are quite common. The proteins are hexagonally packed with a lattice size of 13 nm and the thickness of the structure is 6 nm. On the intact layer, the fragile membrane is not exposed at all; only at the edge of the membrane patch are some parts of the 4.5-nm-thick membrane visible. Figure 3a shows double-stranded DNA with a diameter of 2 nm partially complex with recA proteins. The complex appears as a right-handed single helix with a diameter of 10 nm and a pitch of 10 nm. The handedness of molecules was introduced by E. Fischer (1891). Ever since, the handedness of molecules was based on his rule with no relation to real space. With this investigation on recA-DNA the handedness of molecules could be related to real space and the rule of E. Fischer could be confirmed: D symmetry corresponds to right-handed and L symmetry to left-handed molecules. The head of bacteriophage T4 is composed of a hexagonal lattice of proteins with a lattice size of 13 nm and a vertical corrugation of 1 nm (Fig. 3b). The average capsomere morphology was determined by correlation averaging. When this is done, the sixfold symmetry and the central depression of 0.5 nm of the proteins becomes clearly visible (Amrein et al., 1989b). Owing to the high signal-to-noise ratio of the tunneling data, only a few unit cells were needed to reveal a stable average. With this highly reproducible sample, tip geometry could be deduced for individual experiments.

Already with the first set of experiments in air, several experimentalists noticed that the STM could image adsorbed molecules. With measures to prevent excessively high ionic currents from the tip it was even possible to detect biological structures adsorbed to surfaces and covered with a salt solution. With this method it was possible to image Langmuir-Blodgett (LB) films and membranes (Hörber et al., 1988) as well as proteins and DNA (Amrein et al., 1989a). As an example Fig. 4a shows an end of a purple membrane showing the thickness of this biological structure (6 nm). The purple membrane from *Halobacterium halobium* is densely packed with proteins that are used for the conversion of light energy into chemical energy. The dimensions of the recA-DNA complex imaged on a MgAc-treated Pt-C surface in Fig. 4b show that the topography with and without coating is essentially the same. From this we concluded that direct imaging of biological molecules reveals the true topography, even of large molecules.

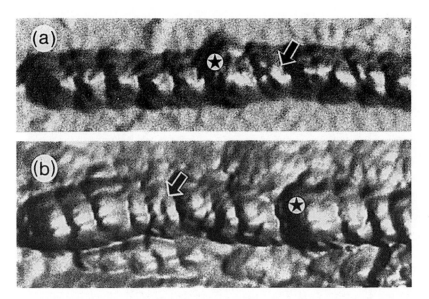

Fig. 4. Comparison of STM images of (a) a freeze-dried and metal-coated recA-DNA complex and (b) an uncoated recA-DNA complex adsorbed on a MgAc$_2$-treated Pt-C surface. The characteristic features appear to be in good agreement for the two methods. The right-handed helix partially shows two domains axially and 6 recA proteins per turn (arrows). In other parts two successive turns of the helix are in close contact (asterisks). Scale bar = 10 nm. From: Amrein et al. (1989b)

The study of proteins in their native environment is an important field of application for STM. Examples of structures that have not been accessible with high resolution, not even with diffraction methods, are surfaces of membranes. Figure 4c shows an STM image 60 × 60 nm in size of the interior surface of oocytes from the clawed toad *Xenopus laevis*. Following the dissection of the vitellin membrane, the oocytes have been adsorbed onto pyrolytic graphite. Upon removal of the oocytes, parts of the membrane remained tightly adsorbed to the graphite and were imaged using a current of 250 pA and a voltage of 0.6 to 1 V Ruppersberg et al. (1989). The observed filamentons structures with sizes of up to 10 nm are possibly due to vinculin and talin whose three-dimensional structure is still unknown. Figure 4d shows a 36 × 36 nm STM image phospholipid of molecules adsorbed to a graphite surface via an ester covalent bond. Single lipid molecules of about 2. to 2.5 nm in length and 1 nm in height ordered in stacks can be located. The dimensions observed for single lipid molecules are an indication for a lateral arrangement. Energy minimization calculations yield a length of about 2.2 nm for the lipid. This lateral orientation might be caused by the interaction of unreacted carboxyl groups on the surface and the lipid head group or by mixing of molecular states with the graphite wave functions (Heckl et al. 1989).

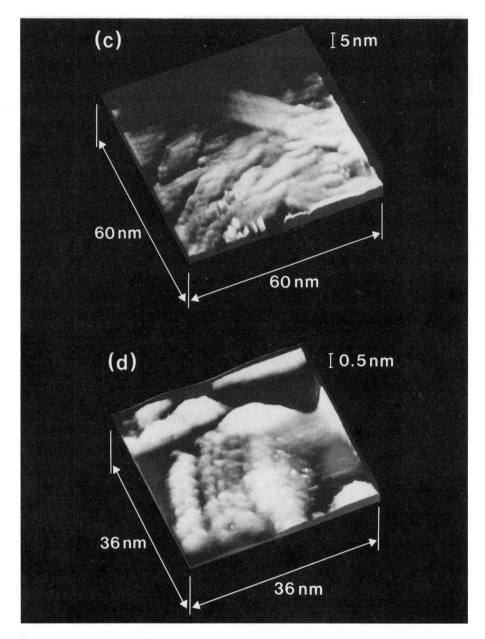

Fig. 4(c) Three-dimensional image of a *Xenopus laevis* oocyte interior surface. Imaging conditions: current 250 pA, voltage 500 mV. From: Ruppersberg et al. (1989). (d) Three-dimensional image of phospholipid molecules covalently adsorbed to pyrolytic graphite. Imaging conditions: current 250 pA, voltage 500 mV, the total height range in 2 nm. From: Heckl et al. (1989)

Direct imaging is not subject to artifacts due to sample dehydration, and the imaging conditions can be selected to simulate native conditions for the molecules under investigation. The method can be applied to physisorbed molecules when molecules are densely packed on a surface, or when the shape of molecules ensures good adhesion. In this case the resolution is only limited by the molecular vibrations and elastic tip-molecule interactions. For other cases molecules are often displaced by the scanning tip so that no high resolution can be obtained.

The different forms of DNA (alpha, beta, Z) have been demonstrated by X-ray structure analysis, but it is not known whether other structures exist. It is only possible to determine the majority structure of DNA. Recently, several investigators could verify the structure of alpha, beta and Z DNA by STM (Arscott et al., 1989, Bebee et al., 1989, Dunlap and Bustamante, 1989, Driscoll et al., 1990, Lindsay et al., 1989, Travaglini & Rohrer, 1987). It is well known that DNA is extremely compacted in bacteria, in cell nuclei, sperm heads and virus capsids. Livolant et al. (1989) have shown with electron microscopy and X-ray diffraction a columnar longitudinal and a hexagonal lateral order with intermolecular distances from 2.8 to 4 nm for the highly concentrated phase of DNA molecules. Figure 5a shows the surface of such a concentrated DNA phase with large arrays of ordered tubes having a lateral periodicity of 15 nm with a similar appearance but a different periodicity as observed by Livolant et al. (1989). The individual DNA double strands that are part of the supercoiled structure and some individual strands of DNA double helix on the surface can be detected in Fig. 5. The periodicity along the supercoil is approximately 3.8 nm, which is in accordance with the packing density of DNA found by Livolant.

The local deviations from the helical structure are supposed to play an important role in the binding of promotors and repressors. Such structures are not accessible for statistical methods. Since the STM is able to image unique molecular structures it is a good candidate to resolve unique structures of DNA and thereby to study problems of gene regulation. For STM investigations the bases are sufficiently exposed only in single-stranded DNA. Single-stranded DNA, however, is structurally less stable than double-stranded DNA and, therefore, more difficult to study by STM. Figures 5c and d show a preparation of single-stranded DNA of bacteriophage Φ X-174 RF I on a gold (111) surface. Under the selected conditions (10 mM Tris-HCl pH 7.4, 10 mM KCl), the single-stranded DNA seems not to be stable in the linear form. Figure 5c shows supercoiled structures with a lateral surface periodicity of 12 nm. The periodicity along the supercoil is approximately 4 nm. In some places a substructure with a periodicity of 1.2 nm is visible along an individual strand of the supercoil.

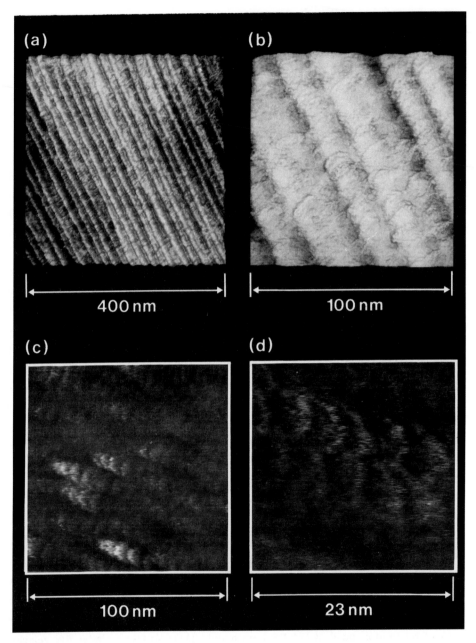

Fig. 5(a) Top view of the highly concentrated liquid-crystalline phase of sonicated salmon sperm double-stranded DNA on a Pt(111) surface. Imaging conditions: gold tip, voltage 300 mV, current 250 pA. (b) Close-up showing the 16 nm lateral and the 4 nm longitudinal spacing of the supercoiled structures. (c) Top view of a preparation of single-stranded DNA of bacteriophage Φ X-174 RF I on a Pt (111) surface. Imaging conditions as in (a). (d) Top view of one of the coils of (c).

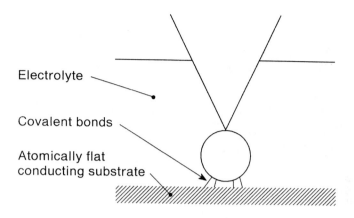

Scheme 2. Imaging of uncoated molecules.

For the investigation of untreated molecules, the dominant technical problems concern preparation of suitable substrates which remain atomically flat under various ambient conditions, immobilization of molecules in their active state preferably on a reactive surface (Scheme 2), and understanding the imaging mechanism.

The substrates used so far are cleaved pyrolytic graphite and flame annealed or epitaxially grown (111) surfaces of gold and platinum. The surface of these substrates is atomically flat on a lateral scale of hundreds of nanometers. Within the layers hexagonally packed gold atoms with a lattice size of 0.29 nm and a vertical corrugation of 0.01 nm can be seen.

From a macroscopic point of view, most biological substances are poor conductors or even insulators. In the case of adsorbates or adsorbate layers of a thickness of less than one nanometer, the imaging contrast may arise from adsorbate-induced changes in tunneling probability and/or changes of the adsorbate electronic structure due to adsorbate/substrate interactions. For large molecules or thick adsorbate layers, however, it appeared rather improbable to tunnel directly through them. Other electron transport processes from the surface of the adsorbate to the conducting substrate have therefore been proposed in order to allow STM imaging. They included band-type and hopping-type bulk conductivity and surface conductivity (Amrein et al., 1989a).

To explore STM imaging in the absence of polar and charged groups we investigated linear alkanes or paraffins adsorbed to gold (111) and HOPG. Linear

104

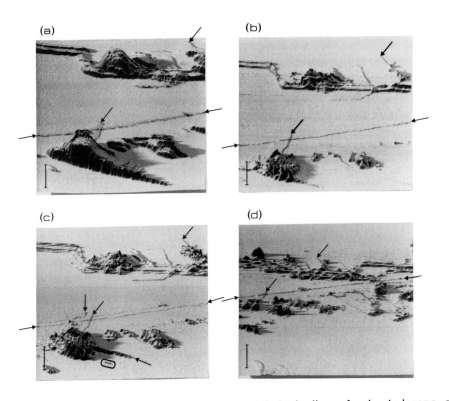

Fig. 6(a) Three-dimensional image with simulated shading of *n*-heptadecane on HOPG. Imaging conditions: Gold tip, current 150 pA, voltage 300 mV. (a) through (c) show the same area of 500×500 nm at temperatures of 18, 20 and 24 °C, respectively. (d) contains the area of (a) through (c) within its 1000×1000 nm at a temperature of 26 °C after a temperature pulse of approximately 36 °C for two minutes. The bar denotes a distance of 20 nm in the *z*-direction (the *z*-dimension is enlarged). The area can be recognized by the marked graphite step and two triangles marked on each picture (arrows). Large structures of *n*-heptadecane (20 nm) disappear as soon as the temperature is raised above 20 °C. Smaller structures (8 nm) and layers recrystallize even at a temperature of 26 °C. From: Michel & Travaglini (1989).

alkanes are the macromolecules with the most simple electronic structure; they are composed solely of C-C and C-H sigma bonds. They readily form orthorhombic and monoclinic prismatic layered crystals which have been extensively studied by electron diffraction methods (Dorset, 1978). The thickness of the layered crystals was found to be equal to the length of the chains (2.1 nm for heptadecane, 2.5 nm for octadecane, and 4.5 nm for hexatriacontane). Most of the crystallites were 2 to 10 nm high but in some places screw dislocation caused much larger structures. Figure 6 shows a sequence of four images of the same n-heptadecane covered area of graphite at increasing temperature. The area can be recognized by graphite structures (arrows) which remain stable during the melting. The crystallites shown in Fig. 6a could be reproducibly imaged at a temperature below 18 °C. As soon as the temperature exceeded 18 °C the large n-hepadecane objects melted away, but small structures remained (Fig. 6b). Above the melting temperature the small structures became quite mobile and changed shape constantly (compare Fig. 6b and c). Before taking image 6d the temperature was raised to 36 °C for 2 min. and lowered again. This caused a complete change of the adsorbed structures but the same graphite structures can still be recognized. In other experiments it was possible to observe pre-melting processes when the temperature was slightly below the melting temperature. In this temperature range the formerly stable shapes of the crystallites became much more mobile. It was possible to see layers of alkanes growing onto surfaces as well as structures disappearing. These melting experiments serve simultaneously to identify independently the observed structures as n-octadecane and n-heptadecane.

Initially, biological molecules were physisorbed to a surface or membranes were transferred with the LB technique prior to the STM investigations. These structures were seldom stable enough to facilitate an investigation with a local probe method. Due to difficulties in sample preparation the investigations were not reproducible, causing doubts about the feasibility of the method. This, however, changes when molecules are tightly immobilized by means of chemisorption. Chemisorption can be achieved by activating a surface via a bifunctional molecule. One functional group has to react with the solid surface and the other functional group has to react later on with the desired molecules. Different second functional groups can be used for a given first functional group. This facilitates a wide range of molecules to be reacted with the surface. Once the desired molecules are chemically crosslinked, they can be subjected to a prolonged and reproducible investigation with a local probe method.

Much of the interest in organic monolayers stems from their relationship to biological membranes and from their potential use as building elements of artificial

106

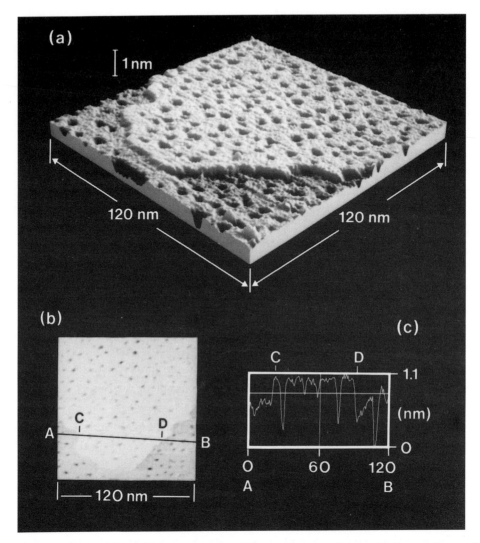

Fig. 7(a) Three-dimensional image with a chemisorbed self-assembled monolayer of mercaptoundecanol on a gold (111) surface. Imaging conditions: Tungsten tip, 200 mV, and 20 pA. (b) Top view of (a) with 1.5 nm from black to white. (c) Cross section along line A-B. From Häussling et al. (1990).

systems. Kuhn and co-workers have demonstrated that planned structures may be assembled by successive deposition of LB films (Kuhn, 1990).

Adsorption of amphiphatic molecules on solid surfaces has been known to lead to a formation of closely packed monomolecular films (Nuzzo & Allara, 1983, Sagiv, 1980, Bain & Whitesides, 1988). The experimental evidence points to a similarity between the structure of these self-assembled monolayers and of LB films. If the adsorption of such a film is followed by a chemisorption like that for mercaptanes on gold, the stability is increased by many orders of magnitude. Such films can be easily studied with local probe methods with no risk of displacing molecules (Häussling et al., 1990). In fact, the properties of the film can be altered by changing either the spacing of the head or tail groups. Figure 7 shows films of mercaptoundecanol on gold. The typical appearance of such layers is seen in Fig. 7a: a smooth surface with randomly distributed depressions. The smooth parts covering approximately 85% - 95% of the surface exhibit the same terraced structure as the uncovered gold films with the same height and comparable sharpness of the steps (Fig. 7a and bisection in Fig. 7c). The depressions have lateral dimensions of 3.5 to 5 nm and a maximum depth of 1.1 nm (Fig. 7a and bisection in Fig. 7c). The gold substrate is supposed to be covered by a densely packed monolayer of mercaptoundecanol except for the holes, which are supposed to be either empty or filled with loosely packed molecules. We interpret this to be due the van der Waals radii of the head groups differing from the van der Waals radii of the tail groups or from the lattice given by the underlying solid. For this reason the packing becomes irregular which causes randomly distributed "holes" in the film.

The shape and distribution of the depressions observed on the mercaptane films can vary slightly when a series of images is taken on the same spot. This variation is tip-induced since it occurs only when the current is set to a level larger than 1 nA. In this case the force exerted on the film is large enough to change the tilt of the chains. Monolayers of molecules are a well-known way to evaluate experimentally the electrical resistance of a single molecule. A classical example of such a film is the LB film. Here, the resistance associated with one molecule is 10^{17} ohms. With an STM the resistance of a molecule can be measured directly with I-V characteristics when the tip stops over one molecule. This (Tip - Insulator - Metal) situation is similar to a M-I-M junction. Still, the experimentally observed resistances are less than 10^9 ohms. Many explanations have been proposed to deal with this 10^8 difference. The I-V characteristics of the self-assembling monolayer shows a small gap at large tip - gold substrate distances. This gap, however, vanishes when the tip is closer to the substrate. We interpret this linear

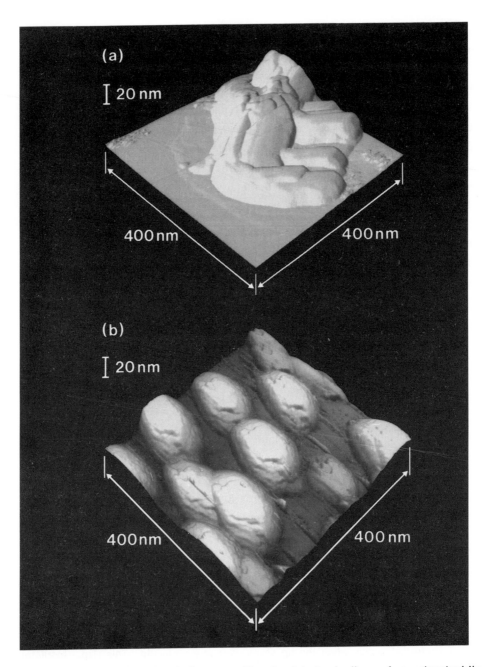

Fig. 8(a) Three-dimensional image with simulated shading of a streptavidin microcrystal grown in the liquid analogous phase at 1 mN/m at a layer of biotin disulfides spread on 10 mM NaCl, 10 mM Tris-HCl pH 7, and transferred to gold (111). Imaging conditions: gold tip, 200 mV, and 100 pA. (b) Three-dimensional image with simulated shading of a streptavidin microcrystal grown at the solid analogous phase at 4 mN/m at a layer of biotinylated lipid spread on buffer as in a and transferred to gold (111).

ohms regime with a valence band coming close to the Fermi level as the tip -
substrate distance is reduced (Dürig et al., in preparation).

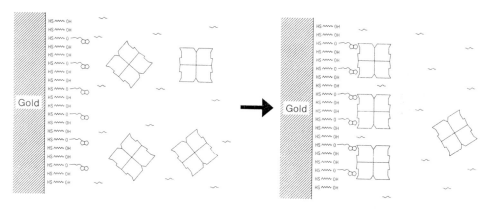

Scheme 3. Streptavidin adsorbed to self-assembled monolayers. From Häussling et
al. (1990).

If reactive groups are introduced on the surface, bigger molecules can be
adsorbed tightly to the self-assembled monolayers (Scheme 3). This has been done
to adsorb streptavidin on biotin functionalized self-assembled monolayers. Unlike
Langmuir monolayers on water, chemisorbed monolayers on solids exhibit only
limited lateral and no vertical mobility. The recognition of biotin functionalized gold
(111) surfaces by streptavidin is very slow and leads to randomly distributed
proteins on the surface (not shown). Since the recognition is very slow,
crystallization is sometimes favored over recognition. The molecular recognition of
biotinlipids spread on a buffer surface by streptavidin leads to the well-known
"H"-shaped structure of the streptavidin crystals (Ahlers et al., 1989). Such
structures were imaged with the STM after they had been transferred from the
water surface to a gold (111) surface with the Langmuir-Schäfer technique (Fig. 8b).

The stability of films was increased by spreading biotinylated disulfides which
chemisorbed upon transfer to the gold substrate. Fig. 8a shows crystallites of
streptavidin grown on a film of biotinylated disulfides in the liquid analogous phase
and transferred to the gold substrate by the Langmuir-Schäfer technique. Due to
the higher mobility during the process of molecular recognition, regular
arrangements of streptavidin either as a monolayer or a multilayer (Fig. 8a) were

found. These experiments demonstrate that vertical and lateral mobility are important parameters controlling the arrangement of chemisorbed molecules.

DISCUSSION

We have seen that the STM offers a wide range of novel investigations dealing with macromolecules.

STM imaging of objects coated with a conducting film is a reliable and quick method to obtain sub-nanometer resolution on any structure irrespective of the conductivity. The method is suitable for making contact to electron microscopy studies, since the coating technique is very similar. The low resistance of molecules adsorbed to surfaces to the tunneling current is still an open question. Within this context it was not possible to discuss all the aspects of conduction mechanisms through molecules. Most of these aspects as well as other currents problems of tunneling microscopy have been discussed elsewhere (Behm et al., 1990). In the investigation of paraffin molecules, however, we excluded surface conduction and ionic conduction. With these experiments we have shown the ability of the STM to study processes such as melting and recrystallization. Highly interesting was the reversible use of the STM tip for modification and investigation of a soft paraffin structure.

Some of the scientists in the field of STM are concentrating on imaging biological structures with the intention of being able to sequence DNA. To a non-expert the simple and straightforward design of the STM would seem to be a perfect solution to this problem. The STM would in fact read a single DNA molecule over its entire length with no need for restriction enzymes, cloning etc. The problems presented by this method start with the preparation of single-stranded DNA on a surface exposing all bases such that they are accessible to the tip. The properties of the four bases which are easily distinguished by a local probe method and thus allow an acceptable throughput have yet to be found. If parallel imaging of local probe instruments can be achieved, they might become a viable method for DNA sequencing. I think, however, that alternative applications are at least as promising: There are many possibilities to study unique biological objects. As mentioned above, the most interesting topics that are not accessible to other methods are the binding of promotors and repressors to DNA. In the future the STM could even help study the DNA/RNA transcription at a DNA/RNA polymerase or the RNA-protein translation at a ribosome.

Alternative methods for sequencing DNA which would read the DNA code directly at a DNA polymerase look equally promising. Moreover, novel procedures along such lines should be the basis for a wider application of local probe methods as an interface to molecular processes. In this sense, the local probe method would serve as an interface to the biological world rather than as a microscopic tool. In an ultimate step the local probe method for interfacing might be in a fixed array and the molecules would organize themselves accordingly.

The experiments with self-assembled monolayers have shown that methods like grazing angle infrared and contact angle microscopy are able to predict an average coverage of a surface. On a molecular basis, however, the structure of a given film can vary widely. This is even more pronounced when enzymes are immobilized on biosensor surfaces. Here, the enzymic activity is taken as a measure of determining coverage. The enzymes quite often assemble to larger complexes, leaving most of the surface uncovered, but the total activity can still be close to that of a monolayer. Self-assembly of structures on surfaces can make the connection between lithography and the self-assembly commonly found in biological systems.

The final goal, the use of molecules to produce artificial structures, can only be achieved by a combined effort of techniques from chemistry, biology, and physics. With these techniques we will be able to gain access to the experience and expertise of nature, a vast database which has been collected in the course of billions of years of molecular evolution and is stored in the genome of organisms. With some adaptations we can make use of this information to induce molecules to self-assemble to human-designed molecular devices.

REFERENCES

M. Ahlers, R. Blankenburg, D.W. Grainger, P. Meller, H. Ringsdorf & C. Salesse, Thin Solid Films 180 (1989) 93.

M. Amrein, R. Dürr, A. Stasiak, H. Gross & G. Travaglini, Science 243 (1989b) 1708.

M. Amrein, R. Dürr, H. Winkler, G. Travaglini, R. Wepf & H. Gross, J. Ultrastruct. Mol. Struct. Res. 102 (1989b) 170.

M. Amrein, A. Stasiak, H. Gross, E. Stoll & G. Travaglini, Science 240 (1988) 514.

P.G. Arscott, G. Lee, V.A. Bloomfield & D.F. Evans, Nature 339 (1989) 484.

C.D. Bain & G.M. Whitesides, Science 240 (1988) 62.

R.J. Behm, N. Garcia & H. Rohrer (Eds.), Proceedings of the NATO meeting "Basic Concepts and Applications of Scanning Tunneling Microscopy (STM) and Related Techniques", Erice Italy, April 17-29, 1989, Vol. 184 (1990), Kluwer Academic Publishers.

T.P. Beebe, T.E. Wilson, D.F. Ogletree, J.E. Katz, R. Balhorn, M.B. Salmeron & W.J. Siekhaus, Science 243 (1989) 370.

D.L. Dorset, Z. Naturforsch. 33a (1978) 64.

D.D. Dunlap, & C. Bustamante, Nature 342 (1989) 204.

U. Dürig, C. Joachim, L. Häussling & B. Michel, in preparation.

R.J. Driscoll, M.G. Youngquist & J.D. Baldeschwieler, Nature 346 (1990) 294.

E. Fischer, Ber. 24, (1891) 2683.

J.S. Foster & J.E. Frommer, Nature 333 (1987) 542.

V.M. Hallmark, S. Chiang, J.F. Rabolt, J.D. Swalen & R.J. Wilson, Phys. Rev. Lett. 59 (1987) 2879.

L. Häussling, B. Michel, H. Ringsdorf & H. Rohrer, Angewandte Chemie, in press.

112

W.M. Heckl, K.M.R. Kallury, M. Thompson, C. Gerber, H.J.K. Hörber & G. Binnig, J. Amer. Chem. Soc. 5 (1989) 1433.

H.J.K. Hörber, C.A. Lang, T.W. Hänsch, W.M. Heckl & H. Möhwald, Chem. Phys. Lett. 145 (1988) 151.

H. Kuhn, Molecular Electronics, F.T. Hong (Ed.), (1989) Plenum Publishing Co.

S.M. Lindsay, T. Thundat, L. Nagahara, U. Knipping & R.L. Rill, Science 244 (1989) 1063.

F. Livolant, A.M. Levelut, J. Doucet, & J.P. Benoit, Nature 339 (1989) 724.

B. Michel & G. Travaglini, J. Microscopy 152 (1988) 681.

B. Michel, G. Travaglini, H. Rohrer, C. Joachim & M. Amrein, Z. Phys. B 76 (1989) 99.

R.G. Nuzzo & D.L. Allara, J. Am. Chem. Soc. 105 (1983) 4481.

H. Ohtani, R.J. Wilson, S. Chiang & C.M. Mate, Phys. Rev. Lett. 60 (1988) 2398.

J.P. Ruppersberg, J.K.H. Hörber, Ch. Gerber & G. Binnig, FEBS Lett. 257 (1989) 460.

J. Sagiv, J. Am. Chem. Soc. 102 (1980) 92.

D.P.E. Smith, J.H.K. Hörber, G. Binnig & H. Nejoh, Nature 344 (1990) 641.

G. Travaglini & H. Rohrer, Spektrum d. Wissenschaft No. 8, (August 1987) 14; in English: Scientific American 257, (November 1987) 30.

G. Travaglini, H. Rohrer, M. Amrein & H. Gross, Surf. Sci. 181 (1987) 514.

I. Karube (Ed.) *Automation in Biotechnology*
Proceedings of the 4th Toyota Conference, 21–24 October 1990
© 1991 Elsevier Science Publishers B.V. All rights reserved

HANDLING OF BIOLOGICAL MOLECULES AND MEMBRANES IN MICROFABRICATED STRUCTURES

Masao Washizu

Department of Electrical Engineering, Seikei University,

3-3-1 Kichijojo-Kitamachi, Musashino-Shi, Tokyo 180 Japan

SUMMARY

Handling of biological materials, such as biological cells, organelles, molecules and membranes, is an important unit-operation in biotechnology and is considered to be one of the major targets for automatization. The author and his co-workers have developed the Fluid Integrated Circuit (FIC) technology for the handling of such small objects, where manipulations of the objects are made by the electrostatic field generated inmicro-fabricated electrode system. In this paper, several examples ofFIC-based molecular and membrane manipulations are presented: DNA molecules, being randomly coiled by thermal motion, are stretched to the full length in a high intensity electric field, and the molecular size can instantaneously be determined by measuring the stretched length. The transport and positioning of individual molecule is possible with the dielectrophoretic effect, so that 'molecular surgery', such as cutting of the molecule with an ultra-violet beam, can be applied to the precisely positioned molecules. The electrostatic stress not only stretches a flexible molecule but also can induce a reversible change in the geometry of a protein molecule. Such change may be used for the electrical control of the the biological activity of proteins. All these are made possible with the high frequency high intensity field created in micro-fabricataed electrode gaps. Another advantage of photo-fabrication processes, the ability to manufacture regularly arranged periodical structures, is utilized to fabricate uniform-sized liposomes. The growth rate of liposomes can also be controlled by the application of high intensity electric field.

INTRODUCTION

Handling of biological materials, such as biological cells, organelles,

molecules and membranes, is an important unit-operation in biotechnology and

is considered to be one of the major targets for automatization. Because the

dimension of these objects ranges from 10^2 μm of cells down to nanometer of

molecules, and they are often soft and fragile, their manipulations, especi

ally on one-by-one basis, require tools with comparable dimension and special

methods of actuation. For this purpose, the micro-machining technique based

on photo-lithographic processes provides ideal means to manifacture micro-

structured tools, whereas the electrostatic effects, such as dielectrophoresis

(DEP) and electrostatic orientation, are suitable for the soft actuation.

Hence, the combination of the microfabrication and electrostatic technique

can potentially solve the specific problems associated with the manipulation

of biological objects. In addition, the electrostatic actuation has an advantage in that it require no mechanical moving parts, and the difficulty with the fabrication and reliable operation of micrometer-sized mechanical components can be avoided.

Cell handling devices with the combination of micro-fabrication and electrostatic method have been developed by the authors, and are named Fluid Integrated Circuit (FIC) [1-3] because all the components for the manipulation are integrated on a substrate with the technique used for the production of semi-conductor IC's. It was demonstrated that the transport, sorting, fusion, etc. of cells are possible with the use of high frequency high intensity field created within micro-machined electrode gaps. It was also shown that the precise tailoring of field pattern is prerequisite for the actuation of cells, which is made possible only with the photo-lithographic technique that enables the fabricaton of arbitrary two-dimensional electrode patterns.

Handling of biological molecules, such as nucleic acids or proteins, are as important, but less developed area because of the difficulty to manipulate individual molecules. Its application is not limited to the analysis of molecular size and structure, but may in future be used for the construction of molecular devices or the control of molecular functions. In the latter applications, biological-membrane-like structure might play an important role as a reaction bed where systematically arranged molecules work cooperatively to realize their functions. The extension of FIC technology is considered to be one of the promising methods in these areas also. In fact, stretching of a DNA molecule, its fixation on a substrate, deformation of a protein molecule by the electrostatic stress, biological and artificial membrane manipulation, etc., are possible with the FIC method. Some of the results obtained throughout our experimental investigations are outlined in this paper.

PRINCIPLE OF ELECTROSTATIC MANIPULATION

The specific problem associated with the electrostatic manipulation of biological objects is that it must be made in aqueous solutions. Application of voltage on electrodes immersed in water often causes chemical reactions at electrode-solution interface, which result in generation of ionic species or bubble formation. This becomes a crucial problem in our case where objects are confined in micro-machined structures. The only way to avoid the chemical reactions is to use high frequency voltage. This in turn means that electrophoretic force $F_e = q E$, where E is the applied field and q is the charge the particle carries, cannot be used, because the time average of F_e vanishes for a.c. field. In contrast, we use kinetic effects based on the interaction of the field and the induced dipole moment $p = q L$, where q is the magnitude of the induced charge of both polarity, and L is the distance of separation.

One of such effects is the dielectrophoresis (DEP) [6], where the field non-uniformity yields the translational force to the polarized object. The magnitude of DEP force F_{dep} is given by

$$F_{dep} = (p \cdot \nabla) E = \alpha \nabla (E^2)$$

(1)

where α is the polarizability of the object. Eq. (1) states that, when $\alpha > 0$ the object is attracted towards where the field is stronger (positive DEP), and when $\alpha < 0$ it is pushed away to where the field is weaker (negative DEP). Experimentally, positive DEP is more often observed.

Another effect is the electrostatic orientation, where the torque T

$$T = p \times E$$

(2)

is exerted in non-spherical particle, regardless if the field is uniform or non-uniform. This torque directs the particle with its longest axis parallel (this case more often observed) or perpendicular to the field line [7, 8]. For a flexible object, this torque tries to align every portion of the object, say parallel to the field, so that it is stretched to a straight-line shape.

For non-spherical particle under non-uniform field, both above effects occur simultaneously, and for examle, the object can be positioned to where the field is strongest, with its longest axis aligned parallel to the field.

It should be noted that there is a constrains on the conductivity of the medium used. This is partly due to the general tendency that the available force becomes weaker in high conductivity medium, and partly because of the Joule heating under high intensity field necessary for the electrostatic actuation. The conductivity of the medium must therefore be kept low, typically below mS/cm, but at the same time, special precautions must be made for the stability of the object. For instance, osmotic pressure is an important factor when membrane structure of cells or liposomes is handled, while the existance of dication (Ca^{++} etc.) are crucial to maintain structure of some proteins, e.g. chromosomal proteins. Stability may be ensured with the addition of isotonic non-ionic agents or controlled amount of specific ions in these cases.

MANIPULATION OF DNA [4]

A DNA molecule, according to Watson-Crick model, is a linear molecule consisting of a pair of helical sugar-phosphate backbone and base-pairs stacked between them. The distance between adjacent base-pairs as obtained from x-ray diffraction is 0.34 nm, giving the information density of 2.9 kilo-base per μm, or 5.8 kbit/μm. A DNA molecule in water takes randomly coiled conformation due to thermal motion. If the molecule is stretched straight and its length measured, its information content can be determined.

As the DNA helix has the diameter of only 2 nm, it can hardly be grasped mechanically nor fully stretched without breaking. However, electrostatic method can realize such manipulations. Fig.1 a) shows l-phage DNA molecules (48 kb) stained with the DNA-binding fluorescent probe DAPI (4',6-diamidino-2-phenyl-indole) [9] and observed under a fluorescent microscope. Each white dot in the photo corresponds to single DNA momlecule. When the field of 1×10^6 [V/m] at 1 [MHz] is applied, the molecules are converted to linear conformation (fig.b)), and their length is measured to be 17μm, in accordance with the calculated value based on 48 kb and 2.9 kb/μm. The original random-coiled conformation is restored as soon as the field is removed. It should be emphasized that a high-frequency high-intensity field of this order is obtainable with a practical power supply only when microfabricated electrode gaps are used.

The positioning of the stretched molecules is possible with the aid of DEP effect, as shown in Fig.2. Here, yeast plasmid, 15 kb circular DNA, is used as the specimen. The thickness of vacuum evaporated parallel strip electrode used in this experiment has a thickness of less than 1 μm, and its edge is very sharp. The stretched DNA aligned parallel to the field, namely

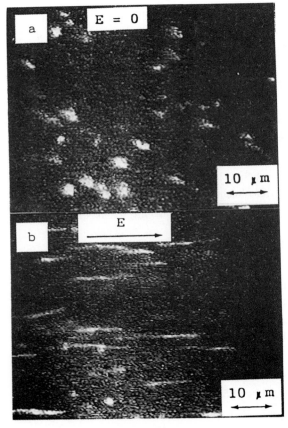

Fig.1
Stretching of DNA molecules.

a) l-phage DNA in the absense of electric field. Each white dot in the fig. corresponds to a DNA molecule.

b) l-phage DNA stretched by the electric field.

perpendicular to the electrode, are attracted by DEP force to the high-field region near this sharp edge, and anchored with one end on the electrode while the other end stretched towards the couter-electrode. The stretched length in the photo measures 2.5 μm, half of what is expected from 2.9 kb/μm. This is because the DNA is circular, and the circle is not broken up but just flattened straight.

Fig. 3 shows 'molecular surgery' performed to the positioned DNA. λ-phage DNA molecules aligned on the electrode are scanned by ultra-violet beam spot (shown by an arrow in the figure) so that the molecules are ampulated at this position, leaving about half of its length on the electrode while the rest diffused to the medium and lost.

Not only conductor electrodes are used for the manipulation. Fig. 4 depicts the bi-directional transport of DNA with the use of resistive elec-trodes[10]. In this case, a pair of parallel strip electrode made of resis-tive material is energized from one end, with the other end being terminated by a terminal resistor. Due to the voltage drop within the resistive strips, the voltage across the gap, and hence the field strength, becomes gradually

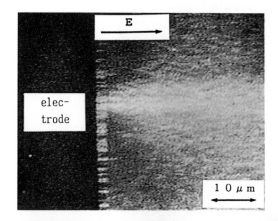

Fig. 2
Yeast plasmid DNA's positioned on the electrode.

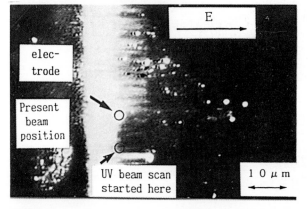

Fig. 3
Ampulation of DNA molecules with UV beam.

118

weaker towards the terminal end. Thus, the field gradient $\nabla(E^2)$ is created towards the power supply end, and DNA's being stretched by the field are dielectrophoretically transported to this direction. By exchanging the power supply end and the terminal end, the direction of transport can be reversed.

The resolution of the measurements and the accuracy of the molecular surgery presented above has the limitation due to the wavelength of light,

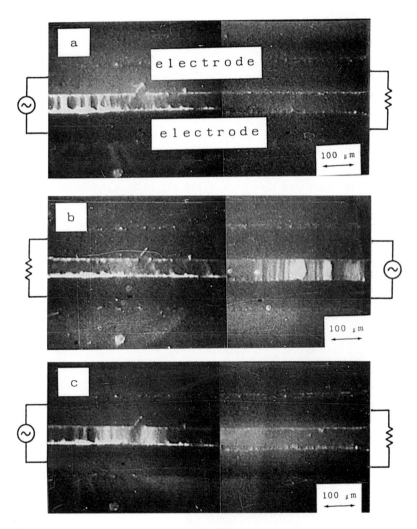

Fig.4 Bi-directional transport of DNA.
    a) The electrodes energized from the left.
       Transport of DNA's to the left occurs.
    b) Transport to the right.
    c) Transport back to the left.

which is of the order of a half micrometer. On the other hand, atomic resolution are obtainable with the use of a Scanning Tunneling Microscope (STM).
It may be possible to use DEP effect to position a DNA molecule under STM
tip, and scan longitudinally to directly readout the base sequence, and if
necessary, cut at the specific base sequence. Such technique has a potential
for the total replacement of the present day gene manipulations.

MANIPULATION OF PROTEIN MOLECULE [5]

The origin of physiological functions of bio-molecules, for example the
recognition and the processing of the ligand by an enzyme, are based on
contour fitting of the two molecules. Control on the reaction speed of some
enzymes are made through a slight deformation of molecular shape caused by
the binding of allosteric effectors. These facts suggest, that the activity
of bio-molecules may as well artificially be controlled through introducing a
strain by an external stress. From an engineering point of view, if such
effect is to be utilized, it is desirable that the stress can be applied
continuously and is controllable. The high intensity electrostatic field
generated in micro-fabricated gap can be a candidate for this pupose.

To investigate the field effect upon molecular shape, we chose bacterial
flagellum as the specimen. Flagella is a spiral organ extruded from the cell
wall of some bacterial species, for instance Salmonella, which is rotated
like a screw for locomotion. Flagella is made of non-covalently linked structural protain units called flagellin consisting of 494 amino-acid residues.
Depending upon the state of subunit association, flagella take saveral confor-
mations of different pitch and diameter besides the wild type 'normal' helix,
which are by appearance called 'straight', 'curly' or 'coiled'. A mutation
on the amino-acid sequence can be a cause of chronical abnormalities, but
environmental factors, such as temperature, pH, and externally applied force
can alter the association state also, by which drastic reversible transformations between these polymorphic states is induced [11]. The polymorphism of
flagella is observable with an intense illumination under a dark-field
microscope, and hence provides a good object for the electric field effect
studies.

In the follwing experiments, 0.3 mM Tris-HCl with the conductivity of 2
[mS/m] is used as the buffer. Flagellar filaments are observed to be stable
under this salt concentration. High intensity field is generated in a vacuum-
evaporated aluminum parallel strip electrode which is energized up to 2×10^6
[V/m] with a 1 MHz power supply.

When flagella are placed in an electric field, the electrostatic orientation first occurs. Under above condition of frequency and conductivity,

'straight' filaments align parallel to the field. However, helical filaments do not always align with its longest axis parallel to the field. Fig.4 shows the case for 'normal' and 'coil' at 0.5×10^6 [V/m], where 'normal' type filament is parallel to the field, while 'coil' type perpendicular. This is because every portion of the helix tries to be parallel to the field as in the case of 'straight' filament, and for 'coil' type where diameter/pitch ratio is large and projected filament length perpendicular to the helix axis is larger than parallel to it, this acts to align the helical axis perpendicular to the field. Combined with hydrodynamic considerations on viscous counter-torque exerted during the rotation, the magnitude of the polarization can be estimated from the speed of orientation[5].

The transformation between the polymorphic states are observed at the field strength of 1.0×10^6 [V/m] or more. Fig.5 shows one of the observed transformations of originally 'normal' type flagellum. First, the field of 0.5×10^6 [V/m] is applied (fig.5-a), where orientation parallel to field line occurs. Then the field is raised to 1.0×10^6 [V/m], transformation to 'coil' starts. Fig.5-b shows the instance when the transformation has propagated to the halfway, so that left half of the filament in the photo is still 'normal' while right half has changed to 'coil'. After several seconds (fig.5-c), the transformation is completed, and the filament is all 'coil'. At some instance which is not accurately predictable (after 15 second of voltage application in the case of fig.5), transition from 'coil' to 'curly' suddenly occurs. When the voltage is removed, 'curly' goes back directly to the original 'normal' state, not via 'coil'.

Fig.6 shows a more drastic case. Fig.6-a shows a 'coil' type filament under 0.5×10^6 [V/m] field, which is aligned perpendicular to the field. As the filament is attached to the substrate at one end (the uppper end in the photon) it is somewhat bent. When the field is raised to 1.3×10^6 [V/m], the coil first shrinks (fig.6-b), and then suddenly it uncoils to be 'normal' (fig.6-c), and at the same time, transformation from 'normal' to 'curly' is initiated from the attached end, finally to be all 'curly'.

Another interesting case is observed with 'normal' filaments cooled down to 5 C. Under the field of $1.0 \sim 1.3 \times 10^6$ [V/m], it shows a reciprocal change between 'normal' and 'curly'. This means that when the filament is 'normal' the field effect is to move the equilibrium to 'curly', and when it is 'curly' the field acts to make it 'normal'.

Unfortunately, the mechanism of the observed transformation remains to be investigated. The temperature rise due to Joule heating can cause similar transformations, but is measured to be small enough. In addition, the observed bi-stability between 'normal' and 'curly' clearly excludes this posiblity.

Electrohydrodynamic (EHD) flow induced by the electric field can be another cause of artifacts by exerting viscous force to the filament. But this too is excluded because the magnitude of this flow is not large enough to induce polymorphic changes, and because the filaments moving with the flow also show transformations. Therefore, we believe this to be a pure electric field effect. However, a simple tensile force, given by $T_f = q E = (p/L) E$ with the value of p evaluated from the measurement of orientation speed of the

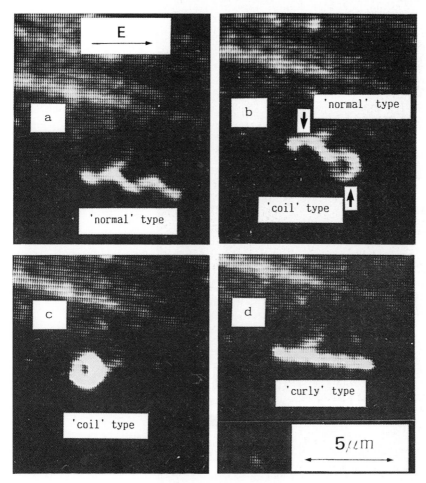

Fig.5 Transformation of 'normal' type flagellum.
  a) 0.5×10^6 [V/m] Orientation occurs.
  b) 1.0×10^6 [V/m], 4 sec. after.
    Transformation from normal to coil is proceeding.
  c) 1.0×10^6 [V/m], 6 sec. after. Transformation to coil completed.
  d) 1.0×10^6 [V/m], 15 sec. after. Sudden transformation to curly occurs.

straight type fibers, is more than an order of magnitude smaller than what is measured by viscous drag method[11]. How the electric field interacts with this protein molecule is still an open queston.

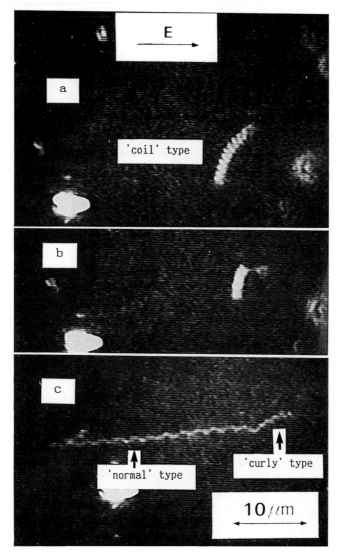

Fig.6 Transformation of 'coil' type flagellum.
   a) 0.5×10^6 [V/m] Orientation perpendicular to field occurs.
   b) 1.3×10^6 [V/m], 1 sec. after. The coil shrinks.
   c) 1.3×10^6 [V/m], 5 sec. after.
      The coil has transformed to normal, and is changing towards curly.

MANIPULATION OF MEMBRANE

In biological systems, membrane-like structures works not only as the barrier that separates inner and outer world, but also as the reaction bed where molecular devices such as receptors, ion pumps, and enzymes are fixed. An important application of electrostatic molecular manipulation is seen in the electrical cell fusion or the electroporation[12-14], where a pulsed voltage is used to induce perturbations and create reversible pores in the phospholipid bilayer of the cell membrane. On the other hand, artificial membranes, such as liposomes, Langmuir-Blodgett films and black membranes, are drawing more and more attentions recently, for both scientific researches and engineering applications. The use of micro-structure and electrostatic method, independently or in combination, has a potential of providing a new technology for the manipulation of these membranes. Its application to the cell handling and fusion is seen elsewhere[1-3], and the followings are the examples of other applications.

One of the advantages of photo-lithographic technique is its ability to fabricate regularly arranged periodical structures. This can be utilized to manifacture uniform-sized liposomes[15]. Its principle is to make use of the difference in hydrophilicity between Si and SiO$_2$. Regularly arranged circular pits are patterned on a surface oxidized Si wafer, and SiO$_2$ are etched away to yield a pattern of lipophilic (i.e. hydrophobic) Si exposed at the etched pit bottoms, which are surrounded by unetched hydrophilic SiO$_2$ surface. Lipid dissolved in organic solvent such as chloroform is poured here and allowed to evaporate very slowly. As the evaporation proceeds, the solvent tends to remain at hydrophobic Si pits, and when the evaporation complets, the lipid is left preferentially at the pits. The amount of lipid at each pit can be made equal by a careful control of the evaporation process. When the substrate is immersed in water, liposomes grow from each pit, and because all pits are identical, uniform growth of liposomes occurs. Fig.7 shows the

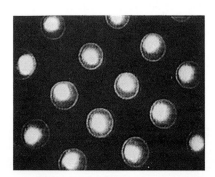

Fig.7 Fabrication of uniform-sized liposomes in the micro-machined structure.

5 0 μm

124

produced cardiolipin liposomes. The liposomes in the photo is still attached
to the lipid residues at the pit, but are released by a gentle heating of the
substrate.

The growth of liposomes can be enhanced with the use of high-frequency
high-intensity field. Fig.8 shows the result of such an experiment[16]. ITO
(Indium-Tin-Oxide) glass is used as the electrode, over which the phospholipid, phosphatidylcholine (PC), with 1 % fluolescent probe NBD-PC added to
facilitate observations, is uniformly spin-coated. The PC layer is estimated
to be about 30 lipid lamellar thick. A counter electrode is placed parallel
to the PC coated electrode at a distance of 0.5 mm, and 0.15 mM NaCl is introduced between them. Fig.8-a shows the PC layer observed at this instance
under a fluorescent microscope. No detectable growth of liposome is observed.
When a train of burst voltage, 1 [MHz] 0.4×10^6 [V/m] turned on for 40 [ms] in
every one second, is applied between the electrodes, the electrstatic force
lifts the layer towards the counter electrode, and inclusion of the medium
between the lamellae occurs. The layer then begins to swell at many locations
to form a mosaic-like pattern as shown in Fig.8-b. When the voltage is turned
off, these swollen parts become spherical to form liposomes. The reason why
the burst voltage is used instead of continuous a.c. voltage is that the
latter results in too much elongation of the swollen part towards the counter
electrode and is not suitable for liposome formations.

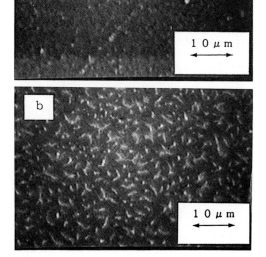

Fig.8
 Enhancement of liposome growth
by an electric field.

a) Phosphatidylcholine layer
   before voltage application.

b) With the application of the
   field, inclusion of the medium
   occurs, and the layer begins to
   swell.

CONCLUSIONS

Handlings of biological molecules and membranes in microfabricated structure with the use of electrostatic effects are presented. It is shown that

1) DNA molecules are stretched in 1 [MHz] 1×10^6 [V/m] electric field. The molecular size, and hence the information content, is instantaneously determined by a direct measurement of the length. Dielectrophoretic effect is used for the positioning of a molecule, and the ampulation of the fixed DNA molecule with UV beam is made.

2) It is shown by an direct observation under a microscope that an electric stress imposed on a molecule by a high frequency electric field in excess of 1×10^6 [V/m] induces transformation of flagellin protein.

3) A method for the production of uniform-sized liposomes using micromachined structure is developed. The growth rate of liposomes can be enhanced under high intensity electric field.

It has been demonstrated that such techniques provide a new and powerful method for the manipulaton of biological objects which are μm-sized and are often fragile. It is hoped that the molecular and membrane manipulations presented here may be coupled or used independently as the method for the molecular engineering in future, in such fields as the assembly of molecular devices and the activity control of functional molecules.

ACKNOWLEDGEMENTS

The author would like to thank Prof. Senichi Masuda of Fukui Institute of Technology, Dr. Shinichi Aizawa of ERATO JRDC, and Dr. Akira Mizuno of Toyohashi University of Technology and Science, Dr. Naoto Urano of Sapporo Breweries Ltd. for valuable discussions, Toshiyuki Nanba, Osamu Kurosawa, and Mayumi Toyoda of Advance Co., Mitsuhiro Shikida of Seikei University for help and discussions.

[REFERENCES]
[1] M.Washizu: Conf. Rec., The thrid Toyota Conf. (1989)
[2] S.Masuda, M.Washizu and T.Nanba: Trans. IEEE/IAS Vol.25, No.4, p.732(1989)
[3] M.Washizu, T.Nanba and S.Masuda: Trans. IEEE/IAS Vol.26, No.2, p.352(1988)
[4] M.Washizu and O.Kurosawa: Conf. Rec. IEEE/IAS '89 ann. meet., p.1978(1989)
[5] M.Washizu, M.Shikida, S.Aizawa and H.Hotani: ibid. '90 ann. meet.
[6] H.A.Pohl: "Dielectrophoresis", Cambridge University Press (1978)
[7] M.Saito, G.Schwartz and H.P.Schwan: Biophys. J. Vol.6, p.313 (1966)
[8] R.D.Miller and T.B.Jones: Conf. Rec. IEEE 9th Ann. EMBS Meet. p.710(1987)
[9] K.Morikawa and M.Yanagida: J.Biochem. 89, p.693 (1981)
[10] M.Washizu, O.Kurosawa and T.Hamayashiki: Conf. Rec. IEJ (The Institute of Electrostatics Japan) '90 ann. meet., (1990)
[11] H.Hotani: J. Mol. Biol. 156, p.791 (1982)
[12] U.Zimmermann, P.Scheurich, G.Pilwat and R.Benz: Angew. Chem. Int. Ed. Engl. 20, pp.325 (1981)
[13] U.Zimmermann, G.Pilwat and H.Pohl: Biol. Phys., Vol.10, pp.43 (1982)
[14] U.Zimmermann: Biochim. Biophys. Acta, Vol.694, p.227 (1982)
[15] M.Washizu and T.Hashimoto: Toyota Phys.Chem.Res.Inst., Vol.43, p.17 (1990)
[16] M.Washizu and M.Toyoda: Conf. Rec. IEJ '90 ann. meet., (1990)
[17] S.Takashima: "Electrical Properties of Biopolymers and Membranes", Adam Hilger, Bristol and Philadelphia. (1989)
[18] J.A.R.Price, J.P.H.Burt and R.Pethig: Biochim. Biophys. Acta, Vol.964, p.221 (1988)

I. Karube (Ed.) *Automation in Biotechnology*
Proceedings of the 4th Toyota Conference, 21–24 October 1990
© 1991 Elsevier Science Publishers B.V. All rights reserved

RECOGNITION AND COUNTING OF CELLS BY IMAGE PROCESSING

Toshio FUKUDA[1], Shigetoshi SHIOTANI[1], Fumihito ARAI[1],

Hajime ASAMA[2], Teruyuki NAGAMUNE[2] and Isao ENDO[2]

1 Nagoya University, Dept. of Mechanical Engineering,Furo-cho, Chikusa-ku, Nagoya, 464-01
Japan

2 Riken, 2-1, Hirosawa. Wako, Japan,351-01

ABSTRACT
This paper describes the recognition of small animal cells using image processing and fuzzy inference. The purpose is to count the number of the cells automatically. The cells (Interferon BETA) are cultured on "micro-carriers", which are beads of 150-200 mm diameter floating in culture solution. The cell images are obtained through a 200x microscope. Previously, cells were identified using image segmentation and an expert system. This method had some problems, such as taking too much time for the image processing and unrecognizing complicated cells. This study proposes a method for recognition of cells using fuzzy inference and a method of tracking the bead-contour..Experimental results shows, (i) improved recognition of cells using fuzzy inference, (ii) recognition of the micro-carriers and decrease of processing time using by bead-contour.

INTRODUCTION

Although recent advances in biotechnology have yielded great benefits both to medicine and industry, it seems that there will be an everpresent demand for in-vitro culture of a wide range of cell types. Cell culture typically involves optimizing both population growth rate and homegeneity. The growth rate, both of the desired population and of undesired ones, becomes an important optimization criterion. For this reason, cell-counting becomes an important task. But it is very difficult for human researchers to count them cultured under various conditions. To alleviate the problem, automated cell-counting, using computerized image-processing coupled to a microscope, seems a promising alternative.(ref.1 and ref.2) This paper reports on the second of two experiments in automated cell-counting. The first utilized image-segmentation, a data base of cell shape parameters, and an expert system. Several problems were experienced, including slow performance and difficulty with complex shapes. The present study describes a new approach in which many more parameters are extracted, fuzzy inference is used instead of the expert system, and the micro-carries, which are beads of 150-200 μm diameter, are also recognized by tracking bead-contour.

Experimental results demonstrate more accurate classification of cells, recognition of the micro-carriers, and decreased processing time.

PROCEDURE OF THE RECOGNITION

Figure 1 shows the flow chart of this study. This system is composed of three parts: (i) the bead-recognition, (ii) recognition of cells at the center of the bead, (iii) recognition of cells on the outskirts of the bead. In this paper, beads are assumed to be spheres and their appearance to be circular.

128

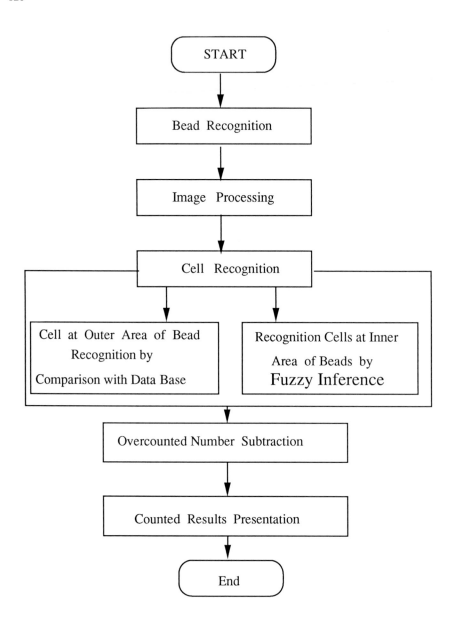

Fig.1. Flow Chart of Cell-Recognition

At first, a picture is taken through a video camera attached to the microscope. The location of beads is found by image processing. Generally, the center of a bead is bright, while the edge of the bead appears dark in this image. Because of this intensity difference, each bead must be

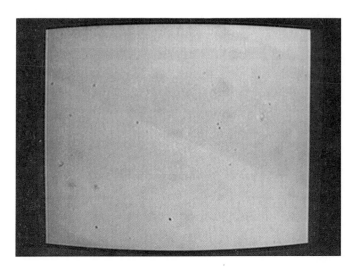

Fig. 2　Brightness　Picture

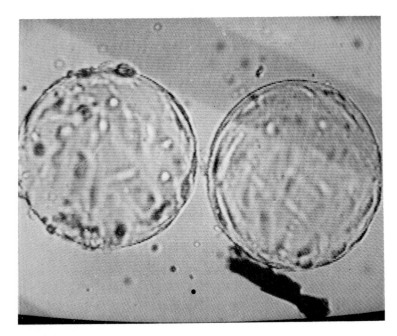

Fig. 3　Original　Picture

(*1) "Brightness Picture" - It is the background picture in which does not exist beads and is used to remove influences of brightness.

(*2)　"Original Picture" - It is picture in which exists beads.

segmented into two domains: inner and outer areas. Cells are taken from both domains with image processing and are counted by using fuzzy inference.

IMAGE PROCESSING SYSTEM

Usually pictures obtained through microscope are affected very much by the brightness of the lighting equipment and the filter in the microscope. These influences must be removed. In this study, the picture of a "Brightness Picture (*1)" (shown in fig.2) subtracted from an "Original Picture (*2)" (shown in fig. 3) is used for removing it. This picture is defined as "Subtraction Picture" after this.

There are many beads in culture solution. To count cells on a bead, the location of a bead must be found. Since the bead adheres to each other, it is impossible to separate the bead from the other by labeling (*3). The adhered beads can be separated into each distinct bead by the method described later. At first, the "Subtraction Picture" is digitized to the binary picture with the most suitable threshold. The threshold is determined by the next formulas.

Pixel value in a "Original Picture" $A(X,Y)$
Pixel value in a "Brightness Picture" $B(X,Y)$
Pixel value in a "Subtraction Picture" $C(X,Y)$
The error of the pixel value between a "Brightness Picture" and a "Subtraction Picture"
$$D(X,Y)$$
The maximum error of the pixel value between a "Brightness Picture" and a "Subtraction Picture"   MAX
The number of all pixels   N
The threshold of the binarization   T

$$C(X,Y)= A(X,Y)-b(X,Y)$$
$$D=\frac{1}{N}\sum_{X,Y} C(X,Y)$$

If $D = 0$ T=3
If $D = 1$ T=4
If $D = 2,3$ T=D+2
If $D > 3$ T=(D-MAX)/10

X is the horizontal direction and Y is the vertical direction.

To smooth the bead-contour, the binary picture is expanded (*4) as shown in fig. 5. Only the contour of the bead is extracted.

(*3) "Labeling" - It is popular image processing, which is used to classify objects in the image. But it can not divide objects adhere with some objects.

(*4) "Expanding" - It is used to smooth the picture, and is to turn four points around a black point, into black point as shown fig. 4.

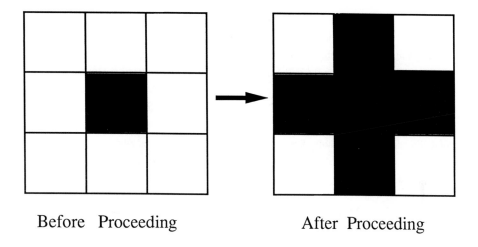

Before Proceeding   After Proceeding

Fig. 4   Expanding   Proceeding

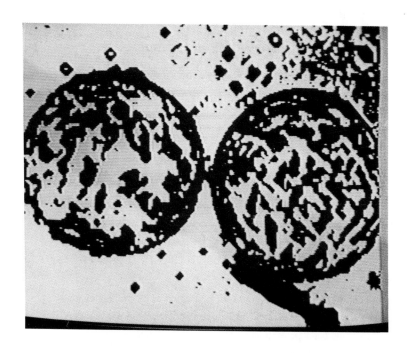

Fig. 5   Binary Picture (after Expanding Proceeding)

The location of the bead is found by contour-tracking method. Usually some dint particles adhere to the bead, so the bead-contour is corrupted by some spurious lines as shown in fig. 6. These lines are removed by the following method.

The contour-tracking is started from a random point on the bead-contour in the direction of clockwise. If it can not be carried out due to the discontinuity of the bead-contour, then it is done in the direction of the counterclockwise. Two kinds of lines are separated by using the gradient of the tangential line which is composed of 2 points on the bead-contour. If the contour branches off in some direction due to the noise lines as shown fig. 6, in case of the clockwise, the direction which gives the gradient a little increase is selected as the bead-contour. While in case of the counterclockwise, the direction which gives it a little decrease is selected.

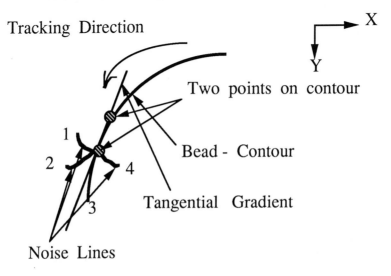

Tracking Direction

Two points on contour

Bead - Contour

Tangential Gradient

Noise Lines

1,2,3,4 are candidates of bead-contour.

In case of the counterclockwise the line (3) which gives the gradient a little decrease, is selected as bead-contour.

Fig. 6 Contour Tracking

This operation continues until the contour is disconnected or its gradient changes rapidly. Three points are extracted among the points on the tracked bead-contour and from three points the center point and the radius of the bead are calculated. Then the different three points are extracted and they are calculated as well as the above method. In this study, it is assumed that the standard radius be 130 pixels and that the measured circle be the bead-contour, if the error between the measured radius and the standard radius is within 15 percent, as shown in fig. 7.

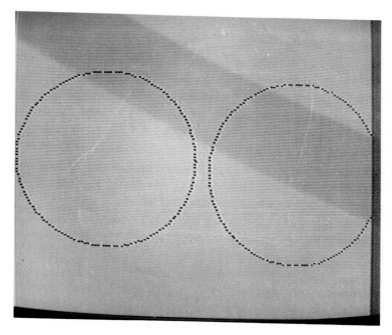

Fig. 7 Bead-Contour Recognition

The inner area and the outer area are shown in fig. 8.

In case of the inner area, the objects are extracted from "Binary Picture" shown in fig. 9. Then small objects are removed as noises in the image data shown in fig. 10. And rest objects are considered as candidates of cells.

In case of the outer area, the "Original Picture" is processed by the differential enhancement (*5) in 3*3 local domain and the binarization operations. The picture shown in fig. 11 is an example after these operation. And small objects is tied, if the distance between them is less than 3 pixels. Then small objects are removed as noise.

There may be some cells over-lapped between two domains. In this study, a rule is determined: If the distance between cells recognized in it is less than L, they are recognized as same cell. Where L is 3 pixels in this study.

(*5) "Differential enhancement" - It is used to extract object's edge, is to differentiate the change of brightness as next formula.

$B(X,Y)$ The pixel value

$D(X,Y)$ The differential value

$$D(X,Y)=\frac{\partial B(X,Y)}{\partial X} + \frac{\partial B(X,Y)}{\partial Y}=|B(X,Y)-B(X+1,Y)|+|B(X,Y)-B(X,Y+1)|$$

X: Horizontal direction,   Y: Vertical direction

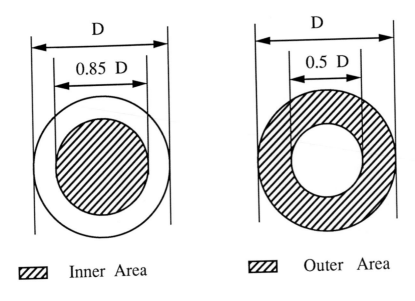

ZZZ Inner Area   **ZZZ** Outer Area

Fig. 8 Inner Area and Outer Area

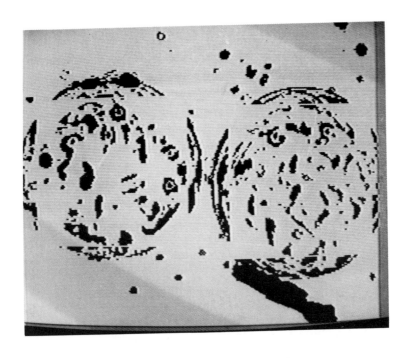

Fig. 9 Objects Extracted at Inner Area

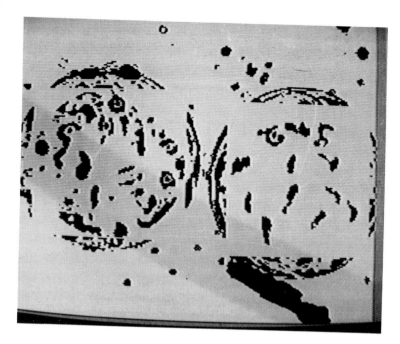

Fig. 10 Noise Reduction

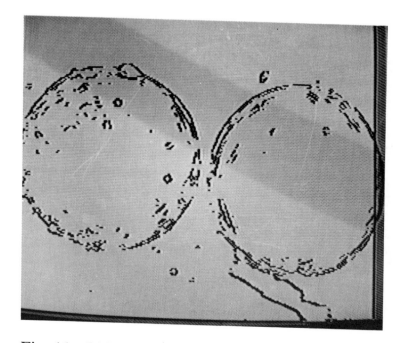

Fig. 11   Objects Extracted at Outer Area

METHOD OF CELL RECOGNITION

The standard shape of the cell is shown in fig. 12.

As shown in figs. 13 and 14, objects extracted at inner area have some noise. So taking account of the standard shape, the standard of the cell recognition must be determined.

Figure 13 shows the shape of objects extracted at inner area and the standard shape, and describes differences between cells and dusts. Four parameters are used to recognize cells:

(i) Width

(ii) Grade of the complexity

(iii) Ratio of length to width

(iv) Area

Fig.12. Standard Shape of the Cell

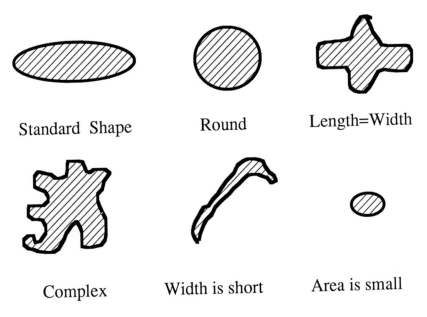

Standard Shape   Round   Length=Width

Complex   Width is short   Area is small

Fig. 13. Shapes of Cells Extracted at Inner Area

Figure 14 shows the shape of objects extracted at outer area, the standard shape and describes differences between cells and dusts. Thus, the following four parameters are used to recognize cells:

(i) Width

(ii) Hole area

(iii) Length

(iv) Area

These parameters are described in next chapter("Knowledge Expression for the recognition").

In case of the recognition of cells extracted at inner area, fuzzy inference is used, because shapes of cells are very complicated. The previous method using the data base, cannot recognize cells accurately, while the proposing method of using fuzzy inference for unclear elements, can recognize them more accurately.

In case of the recognition of cells extracted at outer area, they are compared with the recognition parameters, because they are recognized easily by comparison with the former.

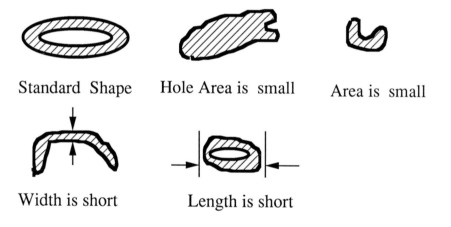

Fig. 14. Shapes of Cells Extracted at outer Area

KNOWLEDGE EXPRESSION FOR THE RECOGNITION

In this paper P is defined as the probability of the cell extracted at inner area. There are three elements to calculate P: they are the width, the grade of the complexity and the ratio length to width. Z is obtained by the fuzzy inference with them. This method are shown with Backs-Naur form:

< P >::=< Width, Grade of the complexity, (Length / Width) I (<Fuzzy Inference>) >

< Grade of the complexity >::=< Area, Perimeter I (<Perimeter*Perimeter/Area >) >

< Area >::=< The all pixel number occupied by the cell >

< Perimeter >::=< The perimeter of the cell >

< Width >::=< The Width of the cell >

< Length >::=< The Length of the cell >

The recognition of the cell extracted at outer area uses three element, which are the Width, the Hole area and the Length. Z is the condition for recognition:

< Z >::=< Width, Hole area, Length I (<Width ≥a>, <Hole Area ≥ b>, <Length ≥ c >) >

< a >::=< Threshold of the area obtained by the experiment >

< b >::=< Threshold of the blank obtained by the experiment >

< c >::=< Threshold of the Length by the experiment >

< Hole area >::=< The all pixel number occupied by blank portion in the center of the cell >

FUZZY INFERENCE

Four membership functions shown in fig. 15 are given based on the criterion of the human recognition. They are three membership functions on recognition parameters (width, the grade of the complexity and the ratio of length to Width) and one on the probability of the cell. Then the first two functions are divided three domains, and the rest two functions are divided two domains,

because the former influence the recognition result much more than the later do. But the area is out of all relation to the fuzzy inference, because it has been used at the noise reduction process already.

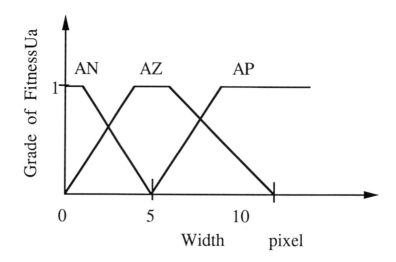

Fig. 15 -(1)   Fuzzy   Membership   Function

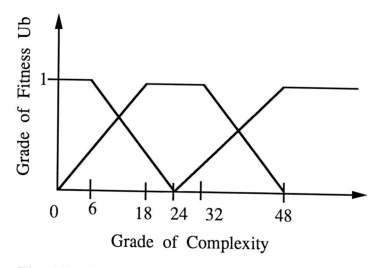

Fig. 15 -(2)　Fuzzy　Membership Function

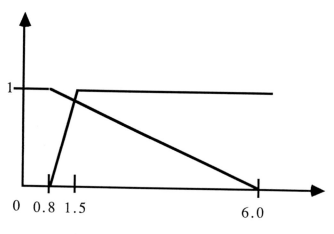

Fig. 15 -(3)　Fuzzy　Membership Function

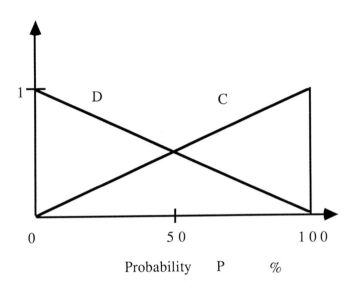

Fig. 15 -(4)   Fuzzy   Membership   Function

In the membership functions on the first two functions, AZ and BZ are the domains which are occupied by standard cells, AN and BN are ones which are done by cells with smaller volumes than standard cells, then AP and BP are ones which are done by cells with bigger volumes than standard cells.

In the membership functions on the ratio of length to width, CP is the domain which is occupied by objects, CN is one which is done by objects with smaller volumes than cells.

In the membership function on the probability of cells, D is the domain which assigned to the probability of being dusts, and C is one which assigned to the probability of being cells.

The fuzzy rule map shown in fig. 16 is given based on the experience. This rule map is the three-dimensional rule map, but is described with two dimensions for convenience' sake. Then it has 18 kinds of cases which are combinations of 8 domains in membership functions. And it shows the criterion, whether objects extracted in beads are dusts or cells in these cases.

In the fuzzy rule map, D assigns dusts and C does cells. For example, it is described in this map that it will be the dust in the case which combines with AN, BZ and CN.

The fuzzy inference is explained later. At first, only one case of 18 cases is explained. It is supposed that the grades of the adaptation for each domains of membership functions are Ua, Ub and Uc respectively, if the width of the cell is A, the grade of the complexity is B, and the ratio of

	AN	AZ	AP
BN	CN D CP C	CN D CP C	CN D CP D
BZ	CN D CP C	CN C CP C	CN D CP D
BP	CN D CP D	CN D CP D	CN D CP D

Fig. 16   Fuzzy   Rule   Map

length to width is C. And the whole grade of the adaptation, U, is calculated by using max-min operation.

$$U=min(Ua,Ub,Uc)$$

Then as shown in fig.17, taking account of the judge described in the rule map(C or D), the center of gravity of the shaded portion is calculated by using the following formulas.

The area $\qquad$ Sn $\qquad$ $(1 \le n \le 18)$

The center of gravity $Zg=Gn/Sn$

(The probability of being the cell in each case)

(1) If the judge in rule map is D

$Sn = $ U(z)dz $= 50(2-U)U$

$Gn = $ Z*U(z)dz $= 5000/3*[-(1-U)^3+1]$

(2) If the judge in rule map is C

$Sn = $ U(z)dz $= 50(2-U)U$

$Gn = $ z*U(z)dz $= 10000*[-U^3/6+1/2]$

And it is calculated by using same method in all of 18 cases. Finally the probability of being the cell, P , is the average of the probability of 18 cases, and is calculated by the following formulas.

$$P = \frac{\sum_{n=1}^{18} (\frac{Gn}{Sn})}{18}$$

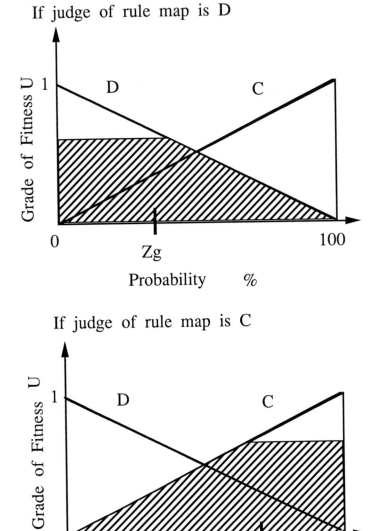

Fig. 17 Calculation of the Center of Gravity

The range of P is 33.3(%)< P <66.6(%), and the center of it is 50(%). So it is determined in this study that the cell is recognized if the probability(P) is over 50%.

EXPERIMENTAL METHOD AND RESULTS
    A drop of the cultured solution is put on the glass plate, the microscope is focused on the bead-contour, the brightness of the light is adjusted suitably. The "Original Picture" is taken through a 200x microscope, and "Brightness Picture" is also done in the same way.
    Then the number of cells on beads is counted by this system, and this is compared with the counted result by human.
    The outline of the experimental equipments is shown in fig. 18.
    Figure 19 shows the results compared with human performance. The average error between human and computer performance was 5%. This small error occurs due to the recognition error at outer area. Error became very small by using fuzzy inference.
    It takes much time to recognize beads with the method proposed before and with the new method, contour-tracking contributes to fast performance. And recognition time was reduced to 24 seconds per bead, which was 15 times faster than the method used before (ref. 1).

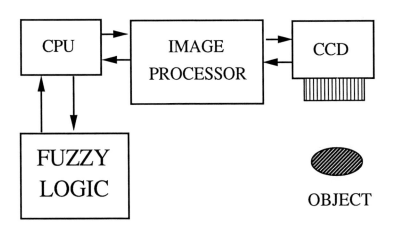

Fig. 18 Experimental Equipment

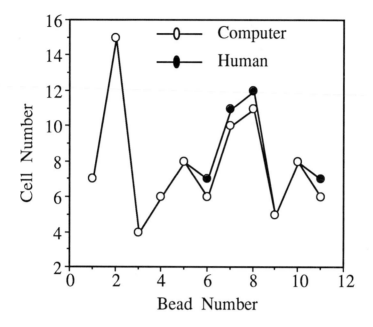

Fig. 19 Comparison Human with Computer

CONCLUSION

This study shows the validities of this proposed method.

(i) Improvement of recognition rate by using fuzzy inference

(ii) Accelerated image processing

REFERENCES

1 Fukuda, Hasegawa, Recognition and Counting Method of Biological Cells on Microcarrier Using Image Processing Based on Expert System, JSME, (1989), (Submitted)

2 Fukuda, Hasegawa, A study on control of a micromanipulator, JSME, 959, 55-512, (1989)

I. Karube (Ed.) *Automation in Biotechnology*
Proceedings of the 4th Toyota Conference, 21–24 October 1990
145

INNOVATIVE ACTUATORS FOR MICROMANIPULATION AND MICROINJECTION IN BIOTECHNOLOGY

T.Higuchi

Institute of Industrial Science, University of Tokyo, Roppongi 7-22-1, Minato-ku, Tokyo, 106 (Japan)

SUMMARY

New precise positioning mechanism which is suitable for micromanipulator and microinjectors is introduced. This mechanism utilizes frictional force and reactive force caused by rapid deformation of piezoelectric element. As it can move objects with resolution of from several nano meters to several micro meters, it can do precise manipulations like cell operation under a microscope and can move a piston of a syringe to inject very small drop. The developed micro manipulator can insert a thin capillary into elastic membrane of a cell with much less deformation of the cell than conventional hydraulic one. Since the developed positioning mechanism is controlled easily by computer, and can be used for both straight and rotational motion control by simple structures with small space, it is expected to contribute the automation in biotechnology.

INTRODUCTION

In the field of biotechnology, automated precise manipulations are necessary for laboratory research and industrialization. As an example, smart insertion and precise positioning of a sharp capillary into cell are necessary in cell operations. Precise positioning is also required in various advanced fields like semiconductor industries and optics. For these demand, in my laboratory and its cooperated groups, a new mechanism of precise positioning by using piezoelectric elements (impact drive mechanism) has been developed (Ref. 1,2) and studied for various kinds of applications such as precise positioning tables, micro robots, STM, stages for ultra high vacuum and micro electric discharge machining (Ref. 3). This new mechanism for precise movement utilizes frictional force and inertial force caused by rapid deformation of piezoelectric elements. It can make step movement of from several nano meters up to several micrometers. By repeating these step motions, movement of unlimited long distance can be obtained. To obtain small step movement and continuous movement by using piezo elements, inchworm mechanism is well-known. But the developed mechanism is quite different in the principle from that of inchworm where additional piezos are necessary for clamping and releasing moving parts against the guide. In this paper as an promising application of this new mechanism in biotechnology, micromanipulation for cell operations is treated. And a new device for microinjection by

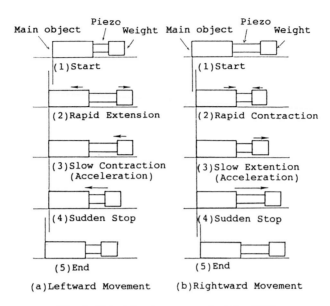

Fig. 1 Principle of movement of the IDM

using the impact drive method is also presented.

IMPACT DRIVE MECHANISM

A new driving method suitable for micro mechanism and micro motion control is introduced. It utilizes friction and inertial force caused by rapid deformations of piezoelectric elements. One dimensional linear positioner using this mechanism consists of one main object put on a guiding surface, a piezo and a weight as shown in Fig. 1. The weight is connected to one end of the main object via the piezo. By controlling rapid extension or contraction of the piezo, it can make step-like movements of several nano meters up to ten micrometers bi-directionally against friction. Thus, repeating this step movement, it can move for a long distance.

Principle of IDM

The mechanism of movement is explained in case of one dimensional linear positioner shown in Fig.1. The main object is put on a flat surface and held by the frictional force. A weight is connected to the main object attaching piezoelectric element in between. Controlling contraction or extension of the piezo by applying voltage waveforms to the piezo, the object can move step by step. Leftward (rightward) movements illustrated in Fig.1-(a) (Fig.1-(b)) are obtained in three ways as follows.

Type I : Movement by Rapid Extension (Contraction)

(1)-(2): Applied a steep rising (dropping) voltage shown in Fig. 2, the piezo extends (contracts) rapidly. While the weight moves rightward, the main object moves leftward by

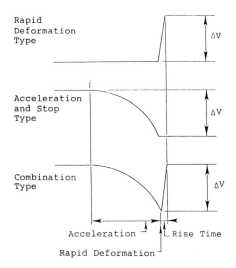

Rapid
Deformation
Type

ΔV

Acceleration
and Stop
Type

ΔV

Combination
Type

ΔV

Acceleration ⌐ ⌐ L Rise Time

Rapid Deformation ⌐

Fig. 2 Voltage patterns applied to the piezo

the reactionary force which is bigger than the frictional force.

(3):The piezo is made to contract (extend) to its initial stage softly enough not to move the main object.

Type II: Movement by Sudden Stop of Weight

(3)':While returning, the weight is accelerated by a constant acceleration to keep the inertial force less then the static friction force between the main object and guide surface. Otherwise, the main object makes reverse movement.

(4):At the time the piezo is contracted (extended) to the length at the beginning, a sudden stop of deformation of piezo is made. By this action, the main object gets momentum just like through the collision of the weight to the main object. Then the whole system starts to move leftward (rightward) against frictional force.

(5):The whole moving part runs against dynamic friction force until it loses its gained kinetic energy to produce another small displacement.

Completing the sequence from (1) to (3)' to (5), the main object is moved with two different step movements. Repeating the cycle from (1) to (5), the main object can be positioned for a long distance.

Type III: Movement Combining Type I and II

Combining the movements of Type I and Type II, the most convenient driving method can be obtained. To connect the stage (2) just after the stage (4), the movement of which displacement is almost equal to the sum of those of type I and type II can be produced at a time. In Fig.2 the three voltage patterns applied to the piezo corresponding to three type of movement are shown. From the voltage patterns, how to combine the two types is

148

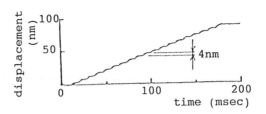

Fig. 3 Nano meter movement by the IDM

clearly presented. In this type, the movement occurs once in each cycle. Since the frequency of the repetitive cycle can be made quicker and larger movement can be obtained than the other two types, the drive pattern of type III is mainly used.

The rightward movement is acquired just in the same way only by exchanging contraction and extension of the piezo to each other as shown in Fig.1-(b) and described in parentheses in the above explanation.

Characteristics of IDM

Piezoelectric elements are remarked to be or make promising actuators for micro motion control. Ultrasonic motors and "inchworm"s are well known as mechanisms using piezoelectric elements. Ultrasonic motors, which utilize resonance to amplify the vibration of piezos to generate surface wave, are suitable for continuous traveling and not good for precise positioning with microscopic resolution. Inchworm mechanism, that consists of two clamps and a piezoelectric element in between, can make stepping movements for long distance and precise positioning with as same resolution as that of piezo itself. However, its construction is rather complicated, especially the clamp mechanism needs precise machining.

Comparing with the other mechanisms using piezos, the developed mechanism IDM has advantages as follows:

(i)Step movement of several nano meters can be obtained by the simple construction as shown in Fig.3.

(ii)As the position is kept by frictional force, it does not consume any power while holding its position.

(iii)In addition to the linear motion devices, apparatus which can control rotary motion and composite motions of multi-degrees of freedom can be fabricated easily in the same manner.

(iv)This mechanism consists of only weight and a piezo, so it is easy to fabricate devices in micro size.

MANIPULATOR FOR CELL OPERATION

Egg cell manipulation

As the first attempt of the application of the IDM to the field of biotechnology, a unique equipment for cell manipulation with a micro pipette or a capillary was developed. The specific task of the manipulation is insertion of the capillary into a egg cell under a microscope. This task is necessary to do the experimental researches related to embryology, especially artificial fertilization. Since the diameter of ova of animals is roughly 100 μm, the capillary should be moved with the resolution of at least one micro meter. Conventional equipments for positioning the capillary use hydraulic system, or precise gear mechanism in order to reduce the joy stick motion given by human hand. And the positioner of 3 degrees of freedom have been developed by using linear DC motors. Most of these devices can move the capillary rather smoothly and position its tip precisely. However, to do penetration of the membrane of the egg cell, which is the most important operation for the egg cell manipulation, most of them can not do well because of the elasticity of the membrane of the egg. Usually, tip penetration happens after deforming the egg to some amount and in worse case, the egg might be damaged by the operation.

Therefore the cell operation needs skills like that of a craftsman. And the experimental results of success rate of birth for example depends on not the fundamental method or idea itself but the experience and ability of the operators. In order to give generality to the experimental results, a standard manipulator by which any researcher can get the common results is demanded. Moreover, for the industrialization of artificial fertilization, it is necessary to provide easy tools for operators including even laymen. And of course, the manipulator should be controlled easily by computer for automation.

For the requirements, our research group began to develop a manipulator for insertion of micro pipette to ova by applying the IDM (Ref. 4).

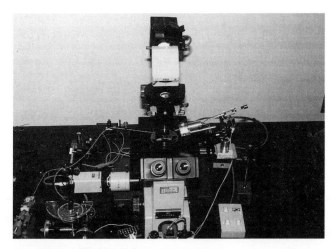

Fig. 4 Micromanipulation system for cell operation

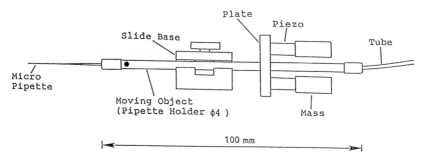

Fig. 5 Prototype I of micromanipulator

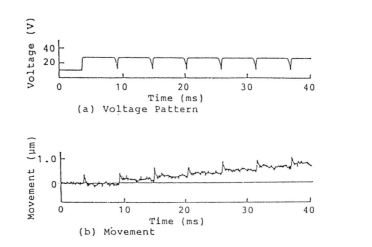

(a) Voltage Pattern

(b) Movement

Fig. 6 Movement of micro pipette and the applied voltage (Prototype I)

Prototype I

   The experimental system for micromanipulation of egg cell operation under a microscope is shown in Fig. 4. The developed equipment to position the micro pipette is shown in Fig. 5. It consists of a pipette holder, a guide, a pair of piezos and masses (weights). The diameter of pipette holder is 4 mm. The guide which is attached to a stand clamps the pipette holder to give the proper friction for IMD. To feed the tube for liquid to the micro pipette, two piezos are used to make room for the tube in the axis. The type of piezo is laminated type of PMN ceramic made by NEC. The cross section is 5 mm x 5 mm square and the length is 10 mm. The limit supply voltage is 150 V where the piezo produce 8 μm expansion. The mass for the weight is 8 g each.

   The example of movement of the pipette is shown in Fig. 6 where the step motion

151

(a) IDM manipulator

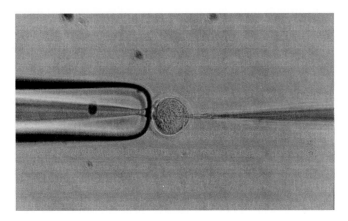

(b) Hydraulic manipulator

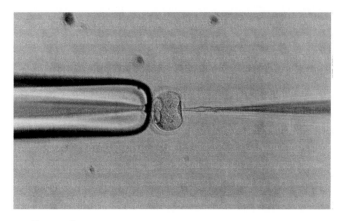

Fig. 7 Insertion of micro pipette into ovum of mouse

of 0.18 µm for each pulse of voltage amplitude of 20 V can be obtained.

The movement can be controlled by only supplying the voltage pattern to the piezos which is generated by computer. So the manipulator is essentially suitable for numerical control. For the direct human operation, a switch box is prepared to select the direction and velocity modes (high for rough positioning , medium and low for precise positioning).

The manipulators have been already used in several universities to do research such as artificial fertilization. So far it gains high reputations as to it abilities for precise positioning and its characteristics. And the number of users have been increasing.

The cell operation for an ovum of mouse is shown in Fig. 7. The diameter of the egg is about 70 μm. The developed manipulator can make the pipette considerably smoothly insert through the membrane of the egg with a negligible small deformation as shown in Fig. 7 (a). The hydraulic manipulator makes huge deformation as shown in Fig. 7 (b). This advantage of the IDM manipulator comes from the principle of the movement. The micro motion of the IDM shown in Fig. 6 contains rapid motion which is good for pipette to penetrate into elastic membrane. This property of the IDM leads to the expectation of the other applications like micro surgery of brains and eye operation where elastic membrane and flexible material should be cut or penetrated.

Improved type (prototype II)

For more precise operation like injection of DNA into the cell, the vibration of the tip of the micro pipette affects the performance of the operation. Since the prototype I has such a construction as loads a pair of driving mechanism as shown in Fig. 6, vibration of the tip may occur being influenced by the imbalance of the property of the piezos and the inaccuracy of their placement.

To reduce the vibration of the tip during movement and to make the manipulator smaller in diameter, improved type was developed by using one piezo with a through hole in center for the path of the tube as shown in Fig. 8. The sizes of the piezo are 10 mm square and 18 mm in length. The maximum elongation is 18 μm at 150 V. The movement of the pipette and the applied voltage are shown in Fig. 9. Comparing the prototype II to I, the movement is more faithful to the original linear motion of IDM. And to obtain the same displacement by a pulse, improved type needs much lower voltage than prototype I as presented in Fig. 10. For example 1.27 μm step can be obtained by 15 V amplitude in improved type, while 1.14 μm step by 60 V in the old type I. The vertical vibrations of the pipette holders for the above conditions are shown in Fig. 11 and Fig. 12. For the same displacement step, the vibration can be reduced considerably by the improvement. And even for the fast motion, the manipulator is allowed to use low voltage power source which is safe for life.

MICROINJECTOR

In usual cases of egg cell operations, a very tiny amount of liquid or a very small particle floating in liquid like a sperm is necessary to feed in or suck out through the thin hole of the micro pipette. So the microinjector is another important tool to do the cell manipulation. For the microinjector, some kind of mechanism for micro movement or rotation is necessary to move the piston of the syringe. For only an attempt to check the possibility of the application of IDM to the microinjector, a piezo and a weight are attached to the end of the piston of the conventional microsyringe as shown in Fig. 13.

Though the reconstruction is very easy and simple, the syringe driven by the IDM can control the amount of the injected liquid with very small resolution and precision as presented in Fig. 14. By designing the micro injector using the IDM from the beginning,

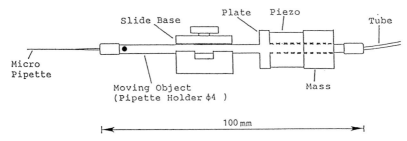

Fig. 8 Construction of improved IDM manipulator (Prototype II)

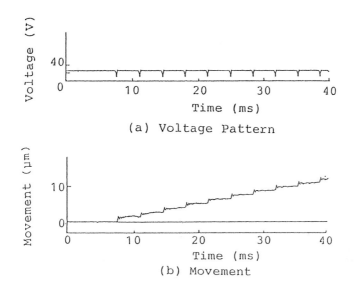

(a) Voltage Pattern

(b) Movement

Fig. 9 Movement of the micro pipette and applied voltage (Improved manipulator)

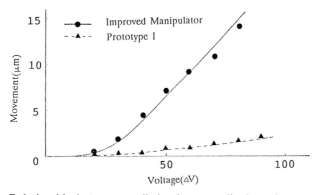

Fig. 10 Relationship between applied voltage amplitude and movement for one step

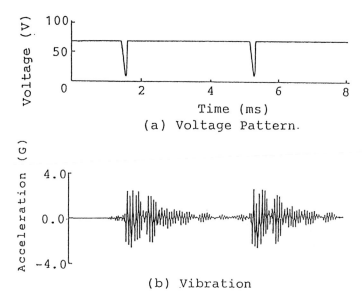

(a) Voltage Pattern.

(b) Vibration

Fig. 11 Vibration of pipette holder of Prototype I

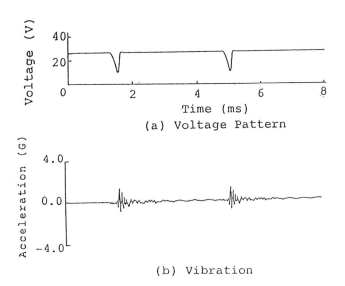

(a) Voltage Pattern

(b) Vibration

Fig. 12 Vibration of pipette holder of improvement manipulator

Fig. 13 Microinjector driven by the IDM

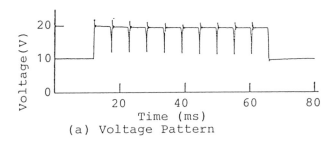

(a) Voltage Pattern

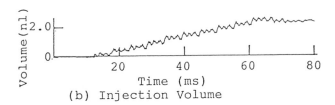

(b) Injection Volume

Fig. 14 Example of injet volume by the IDM microinjector

more precise and more convenient micro injector can be expected to be produced.

MECHANISM FOR ROTATIONS

The IDM can be also applied to rotational motion. Examples of the structures about the rotational movements are shown in Fig. 15, 16, 17. One-dimensional rotating joint made

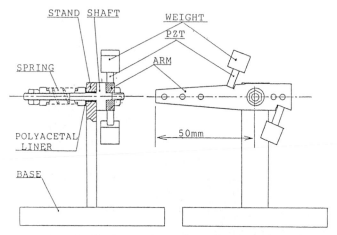

Fig. 15 Structure of one dimensional rotation joint

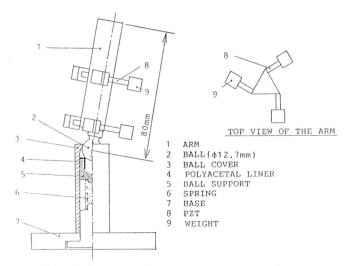

1 ARM
2 BALL(φ12.7mm)
3 BALL COVER
4 POLYACETAL LINER
5 BALL SUPPORT
6 SPRING
7 BASE
8 PZT
9 WEIGHT

Fig. 16 Structure of the three dimensional IDM joint

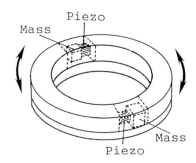

Fig. 17 Concept of ring mover by the IDM

for experiments is shown in Fig.15. It consists of an arm with a shaft, a spring and a stand. A pair of piezos with weights which are connected to the arm generate a couple of forces to cause the torque. At maximum, a movement of 5 μm can be obtained at the end of the arm (50 mm from the center) by one step of movement. The positioning resolution at the end of the arm is considered to be smaller than 0.1 μm. The rotating velocity of 0.048 rad/sec is obtained under the condition of pulse rate of 1.2 kHz and amplitude of the applied voltage of 80 V. The velocity of the tip of the arm is 2.4 mm/sec.

Three-degree of freedom joint was also developed as shown in Fig.16. The arm is supported by ball-joint, and has three degrees of rotating freedom. Six piezos which are driven simultaneously generate torques to rotate the axis of the arm for desired direction and angles based on the principle of the IDM.

The concept of the moving ring mechanism is shown in Fig. 17. By combining the rings and the joints with the linear mechanism, various kinds of apparatus can be designed for precise manipulations in biotechnology and micro surgery.

CONCLUSIONS

The Impact Drive Mechanism is a unique driving method using piezoelectric elements for precise positioning. It has been already applied to many industrial and scientific positioning devices. The attempt to apply the IDM to the automation in biotechnology makes it sure that the mechanism will be the inevitable technology for insertion operation of insertion of pipette to elastic membrane. The laboratory of the author and its research group have the plan to apply the IDM to various kinds of tools for advanced automation in biotechnology by combing the vision system in order to complete the whole automated system.

ACKNOWLEDGEMENTS

The author would like to express thanks to Mr. K. Kudou and Mr. Y. Yamagata for carrying out the research works. And the development of the micromanipulator has been done in cooperation with Prof. K. Sato of Nihon University. The author wishes to especially acknowledge him for the key contribution from the standpoint of the user of the manipulator.

REFERENCES
1 T.Higuchi,Y.Hojjat and M.Watanabe : Micro Actuators Using Recoils of an Ejected Mass; IEEE Micro Robot and Teleoperators Workshop Proceedings, Nov. 1987
2 T.Higuchi,M.Watanabe and K.Kudoh : Precise Positioner Utilizing Rapid Deformations of a Piezo Electric Element; Journal of the Japan Society of Precision Engineering Vol.54 No.11 Nov.1988
3 T.Higuchi, Y.Yamagata, K.Furutani, and K.Kudou: Precise Positioning Mechanism Utilizing Rapid Deformations of Piezoelectric Elements; Proc. IEEE Micro Electro Mechanical Systems Workshop, Napa Valley, CA, Feb. 1990, 222-226
4 K.Kudou,T.Gotoh,K.Sato,Y.Yamagata,K.Furutani, and T.Higuchi: Development of Piezo Micromanipulator for Cell Micromanipulation;J. Mamm. Ova .Res. Vol.7 No.1 7-12, 1990

I. Karube (Ed.) *Automation in Biotechnology*
Proceedings of the 4th Toyota Conference, 21–24 October 1990
© 1991 Elsevier Science Publishers B.V. All rights reserved

159

APPLICATION OF A.C. ELECTRICAL FIELDS TO THE MANIPULATION AND CHARACTERISATION OF CELLS

R. Pethig

Institute of Molecular and Biomolecular Electronics, University of Wales, Dean Street, Bangor, Gwynedd, LL57 1UT, U.K.

SUMMARY

The theoretical framework underlying the dielectric properties of cells is outlined to serve as the basis for understanding the phenomena of cell dielectrophoresis and cell electrorotation. Measurements of these effects can be used to characterise and monitor subtle changes in the physico-chemical properties of cell membranes, and in conjunction with modern microfabrication technologies they can be used to induce controllable translational and rotational forces on cells as a function of their physiological state and viability. Knowledge of the underlying principles that control the way in which cells respond to A.C. fields can also be applied with advantage to the technology of cell electroporation and electrofusion.

INTRODUCTION

The use of steady-state (D.C.) electrical fields in the électrophoretic characterisation of cells, in the Coulter method for cell counting and sizing, as well as in electroporation and electrofusion of cells, are well known and widely employed. Not so well appreciated, perhaps, are the phenomena and possible applications arising from the interactions of alternating (A.C.) fields with cells.

The history of A.C. studies of cells is in fact old and illustrious. For example, between 1910 and 1913 Höber conducted an important series of experiments in an effort to measure the internal electrical conductivity of red blood cells [ref.1 and works cited therein]. Finding that the resistance of the cell suspensions decreased with increasing frequency up to 10 MHz, Höber concluded that the cells consisted of a poorly conducting envelope (eine dielektrische Hülle) enclosing a conducting electrolyte. Amongst the first indications of the ultra-thin nature of this membrane were those obtained by Fricke (ref.2), who derived the (still acceptable) value of 0.81 $\mu F\ cm^{-2}$ for the membrane capacitance and a value of 3.3nm for the membrane thickness. This last result was still under debate ten years later (ref.3). The observations by Cole and Baker (ref.4) of an inductive (negative capacitance) property of squid axons led directly to the concept of voltage-gated membrane pores and to the Nobel prize for Hodgkin and Huxley.

More recently the phenomena of dielectrophoresis and electro-rotation, as applied to the characterisation and manipulation of cells, have gained prominence in the subject. The purpose of this contribution is to outline the basic physical principles involved and their possible biotechnological applications.

EXPERIMENTAL BACKGROUND

A suitable point to start is to refer to the work of Teixeira-Pinto et al [5], in which a variety of bacteria, protozoa and other particles were exposed to electric fields having frequencies between 100kHz and 100 MHz. At the lower frequencies the elongated organisms, such as Amoeba, Chlorella and Euglena, orientated in a direction parallel with the electric field. However, as the frequency was increased this orientation changed to being perpendicular to the field. Non-living particles, such as polystyrene spheres and potato starch grains, always aligned parallel with the field and formed long (pearl) chains. Other interesting effects were also observed, namely:

i) If an Ameaba was aligned with the field at 5MHz and the frequency rapidly changed to 27MHz, the cell would immediately reorient to lie across the field direction but the cytoplasmic inclusions would remain oriented with the field.

ii) In mixtures of Amoeba and Euglena exposed to strong fields, the Euglena were drawn towards the Amoeba and when nearly in contact would spin rapidly.

iii) Depending on the field strength, Amoeba could either be reversibly distorted in shape or ruptured, with extrusion and subsequent chain formation of the cytoplasmic contents.

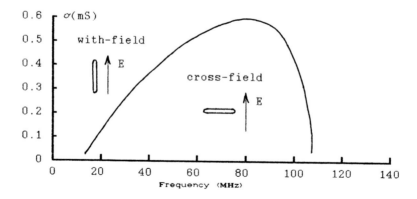

Fig. 1. Orientation of the phytoflagellate _Euglena gracilis_ as a function of the frequency of the applied electric field and the conductivity of the suspending medium. Orientation occurs across the field direction within the bounds of the solid line. Based on reference 6.

Griffin and Stowell (ref.6) repeated some of this work over a wider frequency range (10Hz-200MHz) and found that, depending on the conductivity of the aqueous suspending medium, a higher frequency range existed at which the orientation changed from across-the-field to align with the field, as for the lowest frequencies. This behaviour for _Euglena gracilis_ is shown graphically in figure 1.

In these early works we find examples of how A.C. fields may be used to orientate, distort and rupture cells, as well as to induce translational and rotational motions. Before such effects can be used in a controlled manner to characterise and manipulate cells, the underlying physical principles need to be understood. What follows is an outline description of these principles, with a sketch of the mathematical details.

THEORETICAL BACKGROUND

Particles in Fields act like Dipoles:

A well known problem in electrostatics concerns the effect of suspending a spherical particle of radius a and absolute permittivity ϵ_p into a medium of absolute permittivity ϵ_s in which a homogeneous electric field already exists (see figure 2.a). The result, as described in many textbooks on electrostatics, is that electrical charges are induced to appear on the surface of the particle so that is acts like a dipole (figure 2.b). The value of the induced dipole moment $\bar{m}$ is given by

$$\bar{m} = 4\pi \ \epsilon_s \left(\frac{\epsilon_p^* - \epsilon_s^*}{\epsilon_p^* + 2\epsilon_s^*} \right) a^3 \ \bar{E} \tag{1}$$

where $\bar{E}$ is the original field. In equation (1) we have modified the usual situation considered in textbooks to include the general case where the field varies sinusoidally in time and where both the particle and suspending medium possess a finite conductivity (σ_p and σ_s, respectively). The absolute permittivity is then a complex quantity, having real and imaginary components given by

$$\epsilon^* = \epsilon - j\sigma/\omega$$

where $j = \sqrt{-1}$ and ω is the angular frequency. We should note that the permittivity factor ϵ_s outside the brackets of equations (1) derives from Gauss' law and as such is not considered to include the effect of conductivity losses. The alternative problem is to consider a particle of conductivity σ_p

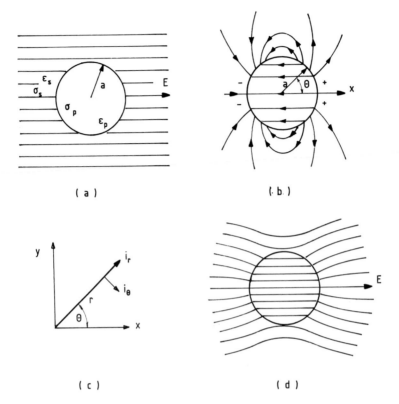

Fig.2. (a) A spherical particle placed in an electric field. (b) The electric field produced by the electric dipole induced in the particle. (c) Geometry of the coordinate system employed in equations 4 and 5. (d) Resultant field pattern within and around the particle ($\epsilon_p > \epsilon_s$).

suspended in a medium of conductivity σ_s. The induced dipole is then given by

$$\bar{m} = 4\pi\epsilon_s \left(\frac{\sigma_p^* - \sigma_s^*}{\sigma_p^* + 2\sigma_s^*} \right) a^3 \bar{E} \tag{2}$$

where now we have introduced the complex conductivity

$$\sigma^* = \sigma + j\omega\epsilon$$

Equations (1) and (2) can readily be shown to be equivalent

The surface charge that appears at the interface between the particle and the surrounding medium is not created immediately, but is established with a characteristic time constant τ given by

$$\tau = \left(\frac{\epsilon_p + 2\epsilon_s}{\sigma_p + 2\sigma_s} \right) \tag{3}$$

As a typical and relevant example we can consider a red blood cell suspended in an aqueous electrolyte, where the value for τ is around 8×10^{-8} seconds. If a d.c. or low frequency electric field is applied, then the surface charge is fully established. However, as the frequency is increased beyond the value $f = (2\pi\tau)^{-1}$, the total dipole moment $\bar{m}$ given by equation (1) cannot be established quickly enough. The frequency dependence for $\bar{m}$ is given by

$$\bar{m}(\omega) = \frac{\bar{m}}{(1 + \omega^2 \tau^2)^{\frac{1}{2}}}$$

and is shown schematically in figure 3.a. Likewise, the potential energy of the particle is given by

$$W = -\bar{m}.\bar{E}$$

and its frequency dependence is shown schematically in figure 3.b. The lines of equal potential (voltage) generated by the induced interfacial charge distribution are shown in figure 2.b, and the associated field $\bar{E}_s$ produced in the surrounding medium is given (for r>a) by:

$$\bar{E}_s = \frac{a^3 E}{r^3} \left(\frac{\epsilon_p^* - \epsilon_s^*}{\epsilon_p^* + 2\epsilon_s^*} \right) (i_r \ 2 \ \mathrm{Cos} \ \theta + i_\theta \ \mathrm{Sin} \ \theta) \qquad (4)$$

where r is the radial distance from the centre of the sphere and the geometrical coordinates of the unit vectors i_r and i_θ are as shown in figure 2.c. The resultant field outside the particle is given by the vector sum of $\bar{E}$ and $\bar{E}_s$. Within the spherical particle there is a resultant uniform field $\bar{E}_i$ given for (r < a) by

$$\bar{E}_i = \bar{E} - \bar{E} \left(\frac{\epsilon_p^* - \epsilon_s^*}{\epsilon_p^* + 2\epsilon_s^*} \right) (i_r \ \mathrm{Cos} \ \theta - i_\theta \ \mathrm{Sin} \ \theta). \qquad (5)$$

The resulting field patterns within and outside the particle are shown schematically in figure 2.d. From equations (1) and (4) we see that the polarity of the induced dipole moment depends on the relative magnitudes of ϵ_p^* and ϵ_s^*. The situation shown in figure 2.d is for $\epsilon_p^* > \epsilon_s^*$. For the case $\epsilon_s^* > \epsilon_p^*$ then the field lines would be "pushed" away from the particle rather than concentrated towards it.

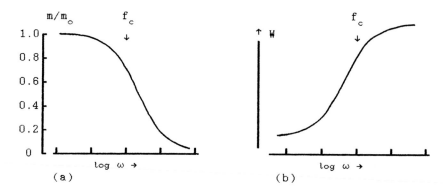

Fig. 3. (a) The variation of the induced dipole moment given by equation 1 as a function of frequency of the field E. The critical frequency is given by $f_c = (2\pi\tau)^{-1}$. (b) Frequency variation of the potential energy of the polarised particle of figure 2.

It is of interest to estimate the charge density ρ produced at the particle surface, and this is given by

$$\rho = \frac{3\,\bar{m}}{4\pi\,a^3}\ \mathrm{Cos}\ \theta \tag{6}$$

For a particle of $\epsilon_p^* = 10\epsilon_s^*$, subjected to a field of 10^4 V/m in an aqueous electrolyte, the induced surface charge density at the surfaces normal to the field is 1.6×10^{-5} Coul. m^{-2}. This charge density is smaller by some three decade orders of magnitude than the uniformly distributed net surface charge that exists (typically around $-1\ \mu C.\ \mathrm{cm}^{-2}$) on the surface of cells and micro-organisms. The important factor is that the field-induced charge distribution is not uniformly distributed about the surface.

From the symmetry of the situation shown in figure 2.d no net translational or rotational force is exerted on the particle. However, the particle does distort the field in its immediate vicinity. How will this affect a neighbouring particle? We can consider the general case of the second particle having point charges q_1, q_2, ... q_n situated at points $(x_1,\ y_1,\ z_1)$, ... $(x_n,\ y_n,\ z_n)$ over its surface, where the origin of the coordinates lies at the centre of the first particle. The external field can be written as

$$\bar{E}_x = \bar{E}_o + \left(\frac{d\bar{E}_x}{dx}\right) x + \left(\frac{d\bar{E}_x}{dy}\right) y + \left(\frac{d\bar{E}_x}{dz}\right) z \tag{7}$$

with similar expressions for $\bar{E}_y$ and $\bar{E}_z$. More succinctly we can write

$$\bar{E}\,(r) = \bar{E}_o + (r\ \mathrm{grad})\ \bar{E} \tag{8}$$

The force $\bar{F}$ acting on the second particle is given by

$$\bar{F} = \sum_{i=1}^{n} q_i \bar{E}(r_i)$$

$$= \bar{E}_o \sum_{i=1}^{n} q_i \quad + \left(\sum q_i r_i \, grad\right) \bar{E}. \tag{9}$$

The dipole moment of the second particle is given by

$$\bar{m}_2 = \sum_{i=1}^{n} q_i \, r_i$$

so that if the second particle carries no net charge (i.e. $\sum q_i = 0$) there is an attractive force given by

$$\bar{F} = (\bar{m}_2 \, grad) \, \bar{E}$$

acting on the second particle. The field $\bar{E}$ has the quality of being "irrotational", in which case we have that

$$(\bar{E} \, grad) \, \bar{E} = \frac{1}{2} \, grad \, \bar{E}^{\,2}.$$

It is also the convention to write "grad" as the del vector operator ∇ so that the force $\bar{F}$ can be written as

$$\bar{F} = \frac{\bar{m}_2}{2\bar{E}} \, \nabla\bar{E}^{\,2} = 2\pi a^3 \epsilon_s \, f(\epsilon_p^*, \epsilon_s^*) \, \nabla\bar{E}^{\,2} \tag{10}$$

The dependency on $\bar{E}^2$ indicates that the force is independent of the field polarity and thus exists for both D.C. and A.C. fields.

The translational motion imparted on particles when placed in a non-uniform electrical field is known as dielectrophoresis, and is a field of study brought into prominence by Pohl (ref.7). Another way of deriving this force of attraction between the two particles is by considering the interaction energy W_{12} of the dipoles $\bar{m}_1$ and $\bar{m}_2$ at separation r (see figure 4), where

$$W_{12} = \frac{\bar{m}_1 \, \bar{m}_2}{r^3} \, (1 - 3 \, Cos^2 \, \theta) \tag{11}$$

166

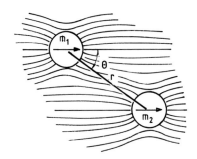

Fig.4. Attraction of two polarised particles by mutual dielectrophoresis, according to equation 11.

The force between them is

$$\bar{F} = - \text{grad } W_{12}$$

and for $\theta < 54.7°$ the particles are attracted towards each other. Also, now that the particles are located in a non-uniform field they will experience a rotational torque whose time-averaged value is given by

$$\langle \bar{T} \rangle = \frac{1}{2} (\bar{m} \times \bar{E})$$ (12)

These various effects provide a basis for understanding the observations by Teixeira-Pinto et al (ref. 5) of cells forming pearl chains and of Euglena cells spinning rapidly alongside Amoeba. Can the observations of changes in the orientation of ellipsoidal particles also be explained? The answer is "yes".

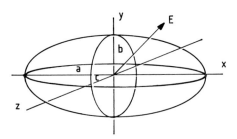

Fig.5. Ellipsoidal particle, showing coordinate axes and arbitrary field direction, used for equation 13.

For an ellipsoidal particle, with major axes of length 2a, 2b, 2c along the x, y, z coordinates, respectively, and experiencing an electric field of arbitrary direction (see figure 5), the principal dipole moments are given by

$$\bar{m}_x = \frac{4\pi\ abc\ \epsilon_s}{3} \left[\frac{\epsilon_p^* - \epsilon_s^*}{\epsilon_s^* + (\epsilon_p^* - \epsilon_s^*)\ L_x} \right] \bar{E}_x \qquad (13)$$

with similar expressions for $\bar{m}_y$ and $\bar{m}_z$, where L_x is an elliptic integral such that for $a \geq b \geq c$ then $0 < L_x \leq L_y \leq L_z < 1$. Saito et al (ref. 8) found that at sufficiently high frequencies the energy of stabilisation of the ellipsoid (for orientation along one of the axes) is inversely proportional to the associated elliptic integral (eg $W_x/W_y \propto L_y/L_x$). This means that the stable position at high frequencies is with the longest axis along the direction of the applied field. However, as shown in figure 6.a, the stabilisation energies for the different axial orientations were found to have different

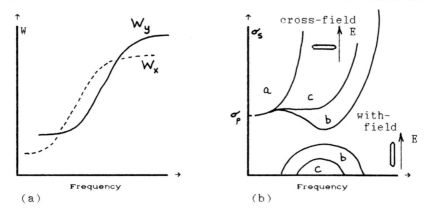

(a)   (b)

Fig.6. (a) Frequency dependencies of the stabilisation energies for different axial orientations of an ellipsoidal particle in an A.C. field. (b) Areas of stable orientation as a function of frequency and conductivity of suspending medium for an ellipsoid. Letters within areas indicate axis in stable alignment. Based on ref.9.

frequency dependencies. At the points where the stabilisation energy curves for W_x and W_y cross each other, the ellipsoid will make a 90° flip so as to minimise its energy. The theory adopted by Saito et al (ref. 8) does not allow for the fact that for systems where dielectric losses occur the field energies are not conserved. Miller and Jones (ref. 9) have addressed this problem by computing the various torque components using the basic form of equation (12). The conclusions are essentially the same as those derived by Saito et al (ref. 8), but the relative dependencies on the values of ϵ_p, ϵ_s,

σ_p and σ_s are more clearly evident. Some of the results obtained by Miller and Jones (ref. 9) are shown schematically in figure 6.b in the same style as figure 1.

CELLS IN ELECTRIC FIELDS

Cells are not homogeneous in their electrical and physical properties, but it turns out that all of the basic concepts described for homogeneous particles in electric fields are of relevance to the case of cells in electric fields.

To a good approximation, as first shown (ref. 1) by Höber in 1910, blood cells can be represented as a conducting sphere (the cytoplasm) surrounded by a resistive membrane (figure 7.a). Maxwell (ref. 10) demonstrated that such a concentric system can be replaced by a homogeneous sphere of the same outer radius having an effective resistance r_p (see figure 7.b and legend). When placed in an electrical field the "smeared-out" sphere can be substituted for the heterogeneous sphere without altering the field. Furthermore Maxwell, and later Wagner (ref. 11), extended this philosophy to derive the effective complex permittivity of a system composed of particles dispersed in a dielectric medium. This means that the interfacial charging effect that occurs for a homogeneous particle is reflected in the overall dielectric property of a heterogeneous mixture.

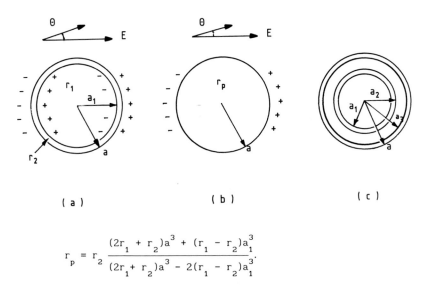

$$r_p = r_2 \frac{(2r_1 + r_2)a^3 + (r_1 - r_2)a_1^3}{(2r_1 + r_2)a^3 - 2(r_1 - r_2)a_1^3}.$$

Fig.7. (a) Single-shell model of a cell, showing distribution of induced interfacial charges. (b) Homogeneous sphere equivalent to single-shell model, having an equivalent resistance r_p. (c) Double-shell model of a cell.

The corresponding dielectric theories of relevance to biological particles were developed principally by Fricke (ref.12), Cole (ref. 13) and Dänzer (ref.14), and led to Fricke's (ref.2) major achievement in deducing the ultrathin, resistive, nature of the red blood cell membrane. Further details of these various theories and of the experimental works on biological systems have been provided elsewhere (refs. 15-18).

The process of calculating the effective complex permittivity of two concentric spheres in terms of a single homogeneous sphere can be repeated endlessly. So, by placing such a smeared-out sphere inside another sphere we can, for example, model a yeast cell having a cell wall. Repeating this once more provides a model of a mitochondria or a cell with a large nucleus, as shown in figure 7.c. Such a procedure has been extensively developed by Asami, Irimajiri, Hanai and coworkers (refs. 19-21), and Hanai (ref.22) in particular has extended the theory of heterogeneous dielectrics of relevance to biological systems beyond that originally developed by Maxwell and Wagner.

So, to summarise, the basic dielectric and field-induced phenomena that are understood for homogeneous spherical and ellipsoidal particles can be extended to describe the dielectric properties of biological particles. In principle "all" that is required is to calculate the effective value for ϵ_p^* in equations (1), (4) and (13). Irimajiri et al (ref. 20) show that

$$\epsilon_p^* = \epsilon_\infty + \sum_{k=1}^{n+1} \frac{\Delta\epsilon_k}{1 + j\omega\tau_k} + \frac{\sigma_o}{j\omega} \qquad (14)$$

where ϵ_∞ is the limiting high frequency absolute permittivity, σ_o is the limiting low frequency conductivity and $\Delta\epsilon_k$ is the dielectric dispersion associated with relaxation time τ_k. This function is shown sketched out in figure 8. A dilute suspension of "multi-shelled" spherical particles in general gives rise to a composite dielectric dispersion, with the maximum number of individual dispersions $\Delta\epsilon_k$ corresponding to the number of interfaces, each of which demarcates the dielectrically distinguishable subphases within the suspended particles. Whether or not the maximum number of dispersions actually occur depends on certain conditions being met, and Irimajiri et al (ref. 20) show that the single-shell model of figure 7(a) can only give rise to <u>one</u> relaxation time and hence one dispersion. This dispersion is centred typically around 1MHz, whilst $\Delta\epsilon_2$ etc appear at higher frequencies than this.

Mention has already been made of the fact that cells and micro-organisms usually carry a net negative surface charge of around $1\mu C \ cm^{-2}$. So far we have not considered how this might influence the dielectric properties of a

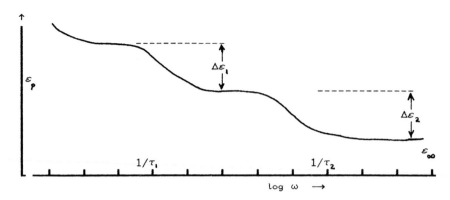

Fig.8. Frequency variation of ϵ_p^* as given by equation 14.

cell suspension. Counter-ions will be attracted to the cell so as to electrically neutralise it, and creates an electrical double-layer around the cell, which can be polarized by an external field. Field-induced relaxations of electrical double-layers around charged particles produce large, low-frequency, dispersions (ref. 23) and a theoretical basis for understanding this was developed by Schwarz (ref. 24). For particles suspended in dilute electrolytes the width of the double-layer (the Debye - Hückel screening length) can be large and of the order of a cell membrane thickness, for example. The diffusion of ions within the double-layer can then be a significant effect, and Garcia et al (ref. 25) have shown how this can modify the magnitude, shape and relaxation time of the interfacial charging polarisations. The interfacial dispersion appearing around 1MHz or so for cell suspensions and biological tissues was termed the β-dispersion by Schwan (refs. 15, 17), whilst the dispersion appearing in the kHz range of frequency and associated with counter-ion relaxations and ion diffusion effects at the outer membrane surface was termed the α-dispersion.

We are now in a position to consider in more detail the phenomena of cell dielectrophoresis and cell electrotation.

CELL DIELECTROPHORESIS

From equations (1) and (10) the dielectrophoretic force $\bar{F}$ acting on a particle in a non-uniform field is

$$\bar{F} = 2\pi\, a^3\, \epsilon_s\, \text{Re} \left(\frac{\epsilon_p^* - \epsilon_s^*}{\epsilon_p^* + 2\epsilon_s^*} \right)\, \nabla E^2 \qquad (15)$$

where Re means "the real part of". Taking into account the frequency dependent form of ϵ_p^* and ϵ_s^*, then for a particle of the form shown in figure 7.a it can be shown (refs. 26, 27) that

$$\bar{F} = 2\pi\, a^3\, \epsilon_s \left[\frac{\epsilon_p - \epsilon_s}{\epsilon_p + 2\epsilon_s} + \frac{3(\epsilon_s\sigma_p - \epsilon_p\sigma_s)}{\tau(\sigma_p + 2\sigma_s)^2(1 + \omega^2\tau^2)} \right] \nabla E^2 \qquad (16)$$

where ω is the angular frequency and τ is the characteristic (interfacial charging) time constant given by equation (3). This time constant is related (through ϵ_p and σ_p) to the cell membrane resistivity and capacitance, or for a more complicated structure such as that of a bacteria to the cell wall resistivity and capacitance. Also, at any one frequency, depending on the relative values of ϵ_p^* and ϵ_s^*, the dielectrophoretic force will be either positive or negative.

Taking values ϵ_p = 330 relative to vacuum and σ_p = 0.04 S m^{-1}, which are typical for <u>Micrococcus lysodeikticus</u> (ref. 28), then from equation (16) the relative dielectrophoretic force should vary with frequency as shown in fig. 9.

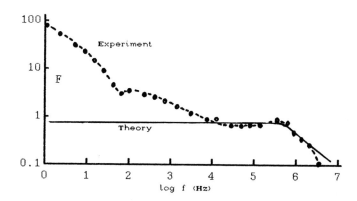

Fig. 9. Comparison of theoretical dielectrophoretic response for <u>Micrococcus lysodeikticus,</u> based on equation 16, against experimental data. See text for details.

The value of τ from equation (3) is 1.1 x 10^{-7} sec, corresponding to a characteristic frequency of 1.5 MHz, and this is reflected in the way that the dielectrophoretic force drops in magnitude in the MHz range. It is at this point that the electric field is beginning to penetrate more deeply into the cell. In electrical terms we can envisage that the cell membrane (wall) capacitive reactance is beginning to short-out the membrane (wall) resistance.

A common way to study cellular dielectrophoresis has been to observe the motion of the cells, using conventional microscopy and time-lapse photography, in non-uniform fields generated by well-defined electrode

172

geometries. Several such geometries are shown in figure 10, together with the mathematical forms of ∇E^2. A more convenient and rapid method is that of monitoring the optical density of a cell suspension as the cells are either attracted (positive dielectrophoresis) or repelled (negative dielectrophoresis) from micro-electrodes fabricated using conventional photolithography (refs. 27, 30). Also, Kaler et al (ref. 31) have developed a technique for obtaining dielectrophoretic spectra using quasi-elastic light scattering from cells suspended in an isomotive electrode system.

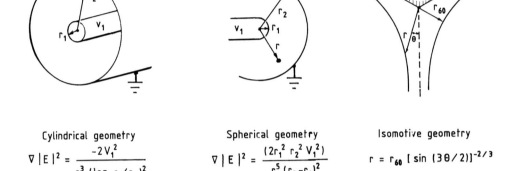

Cylindrical geometry

$$\nabla |E|^2 = \frac{-2V_1^2}{r^3 (\log_e r_1/r_2)^2}$$

Spherical geometry

$$\nabla |E|^2 = \frac{(2r_1^2 r_2^2 V_1^2)}{r^5 (r_2 - r_1)^2}$$

Isomotive geometry

$$r = r_{60} [\sin (3\theta/2)]^{-2/3}$$

Fig. 10. (a) Spherical and (b) Cylindrical electrode geometries used in dielectrophoresis studies. (c) Isomotive geometry where dielectrophoretic force is independent of distance from electrode (ref. 29).

The dielectrophoretic force spectrum observed (ref. 27) for Micrococcus lysodeikticus is shown in figure 9 alongside the theoretical prediction. The marked deviation from theory at the lower frequencies reflects the contributions from polarisations and ion diffusion effects associated with the electrical double-layer around the surface of the bacterium. The largest dielectrophoretic effects are usually observed below 10MHz, and so are related to the surface and bulk properties of the cell membrane or wall. (The electrical field is largely shielded from the cell interior for frequencies below 1MHz). From measurements in our laboratory and with collaborators at the University of Texas (refs. 32, 33) we have concluded that for frequencies below around 200Hz the overall dielectrophoretic response is dominated by membrane surface charge effects. For frequencies extending from around 200 Hz to 2kHz the effective conductivity of the cell is the dominating factor, whilst for frequencies higher than this the effective dielectric permittivity of the cell becomes more important.

The question of how to interpret the term "effective conductivity of cells" has been under debate for some time (refs. 15, 18) and is still not thoroughly resolved. We can imagine at low frequencies, where as shown in figure 2.d the electric field "skirts around" the cell, that the total current comprises a "surface" current around the cell plus a bulk membrane current, both having radial and tangential components. The overall conductivity of the cell can be written as (ref. 23)

$$\sigma = \sigma_b + \frac{2\,K_s}{a} \tag{17}$$

where σ_b is the effective homogeneous bulk conductivity of the cell, K_s is the surface conductance and "a" is the cell radius. No definite values appear to have been derived for the surface conductance of cells, but glass, ceramic and latex particles typically have values in the range 1 ~ 10 nS. (From equation (17) σ_p and σ_b have units of Sm^{-1}. Membrane resistances are typically quoted in $Ohmcm^2$, which for a membrane of thickness 10nm gives the corresponding membrane conductivity in units of $\mu S\ cm^{-1}$.)

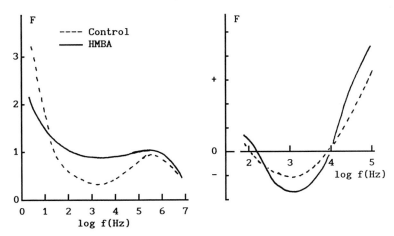

Fig. 11. (a) Dielectrophoretic response of erythroleukaemic cells before (---) and after (——) treatment with the differentiating agent HMBA (ref.32) (b) Dielectrophoresis data for same cells as (a) with measurements to emphasise negative dielectrophoretic response (ref.33).

Dielectrophoretic spectra vary significantly from cell type to cell type, and especially between viable and non-viable cells (refs. 7, 16, 33). The dielectrophoretic spectra exhibited (refs. 32, 33) by Friend murine erythroleukaemic cells (clone DS19) are shown in figure 11, before and after treatment with a chemical which induces differentiation. Subtle changes that have occurred to the cell surface charge and to the membranes of the DS19 cells are reflected by significant changes of their dielectrophoretic behaviour.

In the dielectrophoresis results of figures 9 and 11 the effect is seen to fall of rapidly at a critical frequency f_c in the MHz frequency range, and from this an estimate can be made of the characteristic time constant τ. From equation (3) we have, for the critical frequency

$$f_c = (2\pi\,\tau)^{-1} = \frac{\sigma_p}{2\pi\,(\epsilon_p + 2\epsilon_s)} + \frac{\sigma_s}{\pi\,(\epsilon_p + 2\epsilon_s)} \qquad (18)$$

By measuring f_c as a function of the conductivity σ_s of the suspending solution, then a plot of f_c versus σ_s should be a straight line of slope $[\pi(\epsilon_p + 2\epsilon_s)]^{-1}$ and intercept $\sigma_s = -\sigma_p/2$, from which both σ_p and ϵ_p may be determined.

For a spherical cell with a resistive membrane (i.e. the model of figure 7.a) we can make the substitutions (ref. 15).

$$\sigma_p = a\,G_m \quad \text{and} \quad \epsilon_p = a\,C_m \qquad (19)$$

where G_m and C_m are the conductance and capacitance of the membrane (per m^2) and "a" is the cell radius. These are strictly speaking low frequency approximations, and so their use at higher frequencies probably leads to an overestimation of G_m and an underestimate of C_m.

CELL ELECTROROTATION

Although theoretical and experimental studies of electric field-induced rotation of solids go back at least to the work of Hertz in 1880 (see ref. 34 for a good historical review), the first report of cell electrorotation appears to be that of Teixeira-Pinto et al [5]. Electrorotations of red blood cells and yeast cells were reported later (refs. 35, 36) but the first studies using rotating electric fields (rather than A.C. fields) were those of Arnold and Zimmermann (ref. 37) and Mischel et al (ref. 38). A comprehensive review of the subject has been given by Arnold and Zimmermann (ref. 34) and Hölzel (ref. 39) has provided details of the circuitry that can be used to generate rotating fields of frequency up to 120 MHz.

The basic format of the apparatus that may be used for studies of cell electrorotation is shown in figure 12. Basically, the test cell (or cells) experiences sinusoidally varying fields produced by four electrodes that are driven with equal voltages having four phases spaced $90°$ apart.

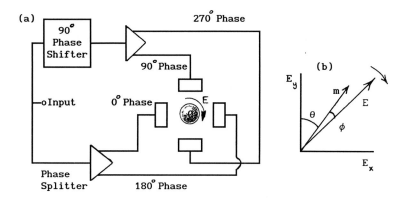

Fig. 12. (a) Schematic diagram of apparatus used to produce rotating electric field (see text and references 36-38 for details). (b) Superposition of two orthogonal linear field vectors to produce resultant field E at any instant and time.   The dipole moment m induced in the particle lags E by ϕ degrees, giving rise to a torque T = m x E = mE sinϕ.

In figure 12.b is shown the field E resulting from the superposition at a given instant in time of two orthogonal linear field vectors:

$$E_y = E_o \cos \omega t$$

and $E_x = E_o \cos (\omega t - 90°) = E_o \sin \omega t$.

The amplitude of the resultant field E is given by

$$E^2 = E_o^2 (\cos^2 \omega t + \sin^2 \omega t)$$

giving E = E_o at any time.   The angle θ in figure 12.b is given by

$$\tan \theta = \frac{E_o \sin \omega t}{E_o \cos \omega t} = \tan \omega t$$

so that $\theta = \omega t$.   The resultant field E thus rotates with angular frequency ω. This rotating field can be produced, as shown in figure 12.a, by four electrodes with $90°$ phase shifts, or by n electrodes phase shifted $360°/n$ apart.   As already discussed, the dipole moment induced in a particle reaches its maximal value with a finite time constant (equation 3), so that in a rotating field the direction of the induced moment does not coincide with the field direction.   A rotational torque T of magnitude

$$T = E_o m \sin \phi$$

176

is created, where ϕ is the angle between the dipole moment and field.

A good basis for understanding cellular electrorotation has been provided by Holzapfel et al (ref. 40), and greatly extended by Sauer and Schlögl (ref. 41). The theory shows that at any one frequency ω the torque exerted on a spherical particle, in a rotating field of strength E, is given by

$$T = 12\pi a^3 \epsilon_s \left[\frac{\epsilon_p \sigma_s - \epsilon_s \sigma_p}{(\sigma_p + 2\sigma_s) + \omega^2 (\epsilon_p + 2\epsilon_s)^2} \right] \omega E^2 \tag{20}$$

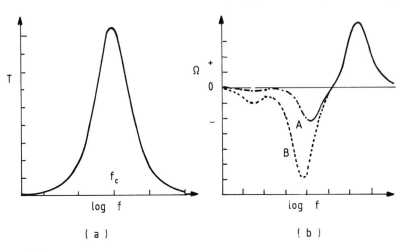

(a)   (b)

Fig. 13. (a) Frequency dependence of torque exerted on a spherical particle in a rotating field, according to equation 20. (b) Rotation spectra for a model cell (fig. 7a) having surface charges; (A) without and (B) with a significant membrane conductivity (after ref. 41).

This expression describes a curve of the form shown in figure 13.a, and rotational "resonance" occurs when $\omega \tau = 1$, where τ is the characteristic time constant given by equation (3). The sense of particle rotation (i.e. cofield or contrafield rotation) is given by the sign of the term $(\epsilon_p \sigma_s - \epsilon_s \sigma_p)$ in equation (20), and it is assumed that permittivity and conductivity parameters remain constant with changing frequency. Taking into account the complex permittivities, then the formal expression for equation (20) is (ref. 41)

$$T = 4\pi a^3 \epsilon_s \, \text{Im} \left[\frac{\epsilon_p^* - \epsilon_s^*}{\epsilon_p^* + 2\epsilon_s^*} \right] E^2 \tag{21}$$

where I_m means the "imaginary part of". For a multi-shelled spherical particle ϵ_p^* will take the form of equation (14) and now the torque T will show peaks (equivalent to dielectric loss peaks of ϵ'' versus frequency) centred at each dispersion $\Delta\epsilon_k$ shown in figure 8. For the single-shell model of a cell shown in figure 7.a then, according to the theory of Irimajiri et al (ref. 20), there will be just one cofield rotation peak centred around a frequency ω given by $\omega\tau = 1$. In the early studies the frequency of the rotating field was restricted to the range from around 100 Hz to 2 MHz, which does not extend quite high enough to observe in full detail the predicted cofield rotation. However, for many cells contrafield rotations (i.e. cells rotating in opposite sense to that of the rotating field) were observed in the KHz frequency range (ref. 34). This can be explained (refs. 41, 42) in terms of the effect of surface charge and ion diffusion conductivity at the cell membrane or cell wall surface. In fact, the angular frequency of rotation in the kHz range depends in a sensitive manner on the effective conductivity of the cell, whereas the effective permittivity plays a subordinate role (ref. 42). This mirrors to some extent the situation for dielectrophoresis (ref. 27). In their work, Sauer and Schlögl (ref. 41) considered the case of a sphere surrounded by a thin membrane, with and without surface charge and membrane conduction, and the corresponding theoretical rotation spectra are shown in figure 13.b. Hölzel and Lamprecht (ref. 43) were able to extend their range of measurements up to 120 MHz, and an example of how their results quite closely match theory is shown in figure 14 for viable and non-viable yeast cells.

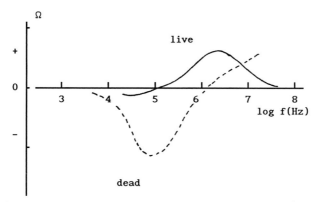

Fig. 14. Rotation spectra for living and dead yeast cells (after ref. 43).

Returning to equation (20) and the fact that the frequency giving maximum torque occurs when $\omega\tau = 1$, we see that use can also be made of equations (18) and (19), where now the frequency of the applied field at which maximum cell rotation occurs can be plotted against the suspending solution conductivity σ_s to give values for σ_p and ϵ_p, or G_m and C_m. Such plots are shown in figure 15 for the DS19 cells whose dielectrophoretic spectra are shown in figure 11. Use has been made of the equation favoured by Arnold and Zimmermann (ref. 34)

$$f_c\, a = \frac{\sigma_p}{2\pi C_m} + \frac{\sigma_s}{\pi C_m} \qquad (22)$$

which can be derived from equations (18) and (19), assuming $\epsilon_p \gg \epsilon_s$. ($C_m$ typically has a value of the order 1 $\mu F/cm^2$, giving $\epsilon_p/\epsilon_s \simeq 70$ for a cell of radius 5μm). The results shown in figure 15 indicate that when DS19 erythroleukaemic cells are induced to differentiate there is a decrease in the effective membrane capacitance (1.14 to 0.86 μFcm^{-2}) and membrane conductance (53.8 to 24 $mScm^{-2}$). This response effectively mirrors the increase in membrane capacitance (0.76 to 1.3 μFcm^{-2}) and membrane conductance (5 to 21 $mScm^{-2}$) induced by mitogenic stimulation of T and B lymphocytes (ref. 44).

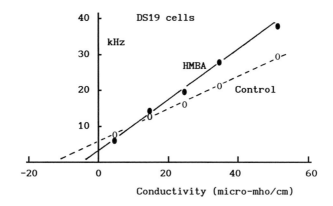

Fig. 15. Plot of frequency of applied field giving maximum cell rotation against suspending medium conductivity σ_s for the cells of figure 11.

CELL ELECTROPORATION AND ELECTROFUSION

The application of electric fields in biotechnology and genetic engineering is now widely employed, with electric cell fusion and electric gene transfer being at present the most prominent applications. Electroinjection of foreign molecules into freely suspended cells and electrofusion of adhered cells are both based on the controlled and reversible electrical breakdown of the cell membrane (refs. 45, 46). Furthermore, after electrofusion of the plasma membranes of two attached cells, nuclear membrane fusion must also be achieved for full recombination of the genetic material of the fused partners. Electroporation (also called electropermeabilisation) is typically achieved by subjecting the cells to a voltage pulse, where the field can range from around 40 kV cm^{-1} for 5μsec pulses and down to 1kV cm^{-1} for millisecond pulses, depending on the cell type. In electrofusion the cells are first brought into alignment and contact by applying an A.C. field and making use of dielectrophoresis (ref. 47, 48). Although some progress has been made, the basic procedures of electroporation and fusion are still largely an art rather than a science. Referring to some of the relevant factors of dielectrophoresis and electrorotation may therefore be of value.

An important factor that has been established is that the electric pulse treatment renders the cell permeable and fuseable by the formation of aqueous pores in the cell membrane, and that the direct cause of this is the high electric field stress created across the membrane, which can be calculated by subtracting equations of the form of (4) and (5) from each other. Referring to figure 7.a as a simple model of a cell of outer radius "a" exposed to an electric field E, then following Farkas et al (ref. 49) the fully established field acting across the membrane is given (neglecting the electrical double-layer) by

$$E_m(\theta) = \frac{9\sigma_s \sigma_i E \cos\theta}{(2\sigma_s + \sigma_m)(2\sigma_m + \sigma_i) - (2\sigma_s - \sigma_m)(\sigma_i - \sigma_m)(a_1/a)^3} \tag{23}$$

where σ_m and σ_i are the conductivities of the membrane and innerphase, respectively. Usually, σ_s and σ_i are much greater than σ_m, so that equation (23) simplifies to

$$E_m(\theta) = \frac{9E}{2[1 - (a_1/a)^3]} \cos\theta \tag{24}$$

The membrane thickness d is given by $d = (a - a_1)$ and usually $a \gg d$, so that in expanding the denominator of equation (24) it reduces to

$$E_m(\theta) = 1.5 \ (a/d) \ E \cos\theta \tag{25}$$

which is the usual expression quoted (refs. 45, 46). Slight correction of this is needed to account for the electrical double-layer around the cell (ref. 25).

Equation (25) shows that the greatest field stress is created across the membrane region that lies in a radial direction parallel with the field, and that there is an "amplification" of the field by a factor 1.5 (a/d), which can readily achieve a value of the order 10^3. The average field stress required across a membrane to achieve electroporation or electrofusion appears to be around $1 \sim 2 \times 10^6$ V cm^{-1}, and so E in equation (25) needs to be of the order

180

of kV cm^{-1} for a typical cell. The "amplification" effect can be appreciated by referring to figures 7(a & b), where the voltage drop across the whole sphere of figure 7.b becomes, because of the distribution of counter-ions, divided across the left and right halves of the vesicle of figure 7.a.

The full transmembrane stress will not occur instantaneously, but will follow a time course given by

$$E_m(\theta) = 1.5(a/d) \ E \cos\theta \ [1 - \exp(-t/\tau)] \tag{26}$$

where t is elapsed time after application of the voltage pulse and τ is the characteristic time constant of equation (3), also given from equation (22) by $\tau = (2\pi f_c)^{-1}$. From equation (22) we see that τ depends on the cell radius and the conductivities of the suspending medium and cell membrane (C_m is about 1 μF cm^{-2} for most cells). For a cell of radius 5μm and σ_s, σ_m in the 10μs/cm range, then τ is around 0.2 μsec. The variation of E_m in the frequency domain is given by

$$E_m(f) \ \frac{1.5(a/d)E \cos\theta}{(1 + (f/f_c)^2)^{1/2}} \tag{27}$$

and is shown schematically in figure 16.

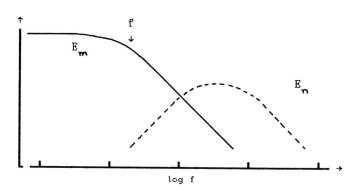

Fig.16. Frequency dependence of the transmembrane field acting across the plasma membrane (E_m) and the nuclear membrane (E_n).

As the frequency approaches f_c, then the external field begins to penetrate into the cell interior. If the cell contains a nucleus, then the nuclear membrane begins to be electrically stressed at this point, and the way in which the field across the membrane varies with the frequency of the applied field is also shown schematically in figure 16. The plots of figure 16 suggest that instead of a D.C. pulse, a pulse of A.C. voltage may give finer control over both plasma membrane and nuclear membrane poration and fusion. Also, since electrotation will interfere with the initial dielectrophoretic cell alignment, the frequency ranges (see figure 13) where electrorotation occurs should be avoided.

CONCLUDING REMARKS

There is now a good understanding of most of the underlying physico-chemical factors that control the ways in which cells (and colloidal particles in general) respond to A.C. electrical fields. The two principal phenomena open to biotechnological exploitation are dielectrophoresis (with the associated pearl chaining effect) and electrorotation. These two ponderomotive effects are linked through the complex (Clausius-Mossotti) factor

$$
\left(\frac{\epsilon_p^* - \epsilon_s^*}{\epsilon_p^* + 2\epsilon_s^*} \right) \text{ or } \left(\frac{\sigma_p^* - \sigma_s^*}{\sigma_p^* + 2\sigma_s^*} \right) \tag{28}
$$

with the dielectrophoretic force being governed by the real component of this factor and electrorotation by the imaginary component.

The principal characteristics of cell electrorotation, namely the sense and maximum angular frequency of rotation, are expressed over a relatively narrow frequency range of the applied electric field (see figures 13 & 14) and are particularly sensitive to changes in the electrochemical, physico-chemical and structural integrity of the cell membrane. Furthermore, measurements can be (and preferably are) made on single cells. These factors make the technique particularly exploitable for the real-time detection and monitoring of pharmacological agents and toxins.

The differences in dielectrophoretic behaviour of different cell types, or of chemically treated and untreated cells, can be enhanced through judicious choice of:
   a) The conductivity and/or permittivity of the cell suspending solution.
   b) The geometrical design of the electrodes so as to enhance either positive or negative dielectrophoresis.
   c) The frequency of the non-uniform field.

By controlling these various parameters a variety of effects can be achieved, such as the controlled translocation of a cell, the bringing together or separation of cells, and the monitoring of the heterogeneity or viability of cell cultures, for example. In figure 17 is shown the basic geometry of the

182

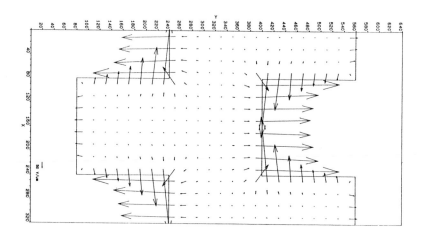

Fig. 17. The interdigitated, castellated, microelectrode structure used in some of our dielectrophoresis studies (refs. 27, 30, 32) with vectors showing direction and magnitude of the factor $\overline{\nabla E}^2$ for positive dielectrophoresis.

photolithographically produced microelectrodes used in some of our studies (refs. 27, 30, 32), with calculated vectors showing the direction and magnitude of the dielectrophoretic force for positive value of the Clausius-Mossotti factor given in expression (28) (ref. 50). For conditions where negative dielectrophoresis is in operation, then the vector arrows in figure 17 are reversed and cells (or particles) are directed away from the electrodes to regions where the factor $\overline{\nabla E}^2$ is a minimum. Our present efforts are concerned with the design and practical application of microelectrode geometries to maximise the effects of both positive and negative dielectrophoresis. Also, the overall dielectrophoretic response of a suspension of cells is related to the cell concentration, and this can be employed as an alternative approach to the dielectric measurement of biomass as developed by Kell et al (refs. 51 & 52).

Acknowledgements

I wish to acknowledge the valuable and appreciated collaborations with my colleagues Talal Al-Ameen, Julian Burt, Ying Huang, Jonathon Price, John Tame and Xiao-Bo Wang at Bangor, and Drs Frederick Becker and Peter Gascoyne at the M.D. Anderson Cancer Center, Houston. Our work has been supported by the UK Science and Engineering Research Council, Warren Spring Laboratory at Stevenage and the National Foundation for Cancer Research (USA).

REFERENCES

1   R. Höber, Messungen der inneren Leitfähigkeit von Zellen, Pfl.Physiol. Archiv. Mensch. Tiere, 150 (1913) 15 - 45.
2   H. Fricke, The electric capacity of suspensions of red corpuscles of a dog, Phys. Rev. 26 (1925) 682 - 687.
3   J.F. Danielli, The thickness of the wall of the red blood corpuscle, J. Gen. Physiol. 19 (1935) 19 - 22.
4   K.S. Cole and R.F. Baker, Longitudinal impedance of the squid giant axon, J. Gen. Physiol. 24 (1941) 771 - 788.
5   A.A. Teixeira - Pinto, L.L. Nejelski, J.L. Cutler and J.H. Heller, The behavior of unicellular organisms in an electromagnetic field, Exp. Cell Res. 20 (1960) 548 - 564.
6   J.L. Griffin and R.E. Stowell, Orientation of Euglena by Radio - Frequency Fields, Exp. Cell Res. 44 (1966) 684 - 688.
7   H.A. Pohl, Dielectrophoresis, Cambridge University Press, Cambridge, 1978.
8   M. Saito, H.P. Schwan and G. Schwarz, Response of nonspherical biological paricles to alternating fields, Biophys. J. 6 (1966) 313 - 327.
9   R.D. Miller and T.B. Jones, Frequency dependent orientation of ellipsoidal particles in A.C. electric fields, Ninth Annual IEEE - EMBS Conf., Boston, Nov. 1987.
10 J.C. Maxwell, A Treatise on Electricity and Magnetism, 3rd ed., Vol.1, Ch.ix, Clarendon Press, Oxford, 1891.
11 K.W. Wagner, Erklarung der dielektrischen Nachwirkungsvorgange auf grund Maxwellsher Forstellungen, Archiv. Elektrotechnik 2, (1914) S. 371 - 389.
12 H. Fricke, A mathematical treatment of the electric conductivity and capacity of disperse systems, Phys. Rev. 24 (1924) 575 - 587.
13 K.S. Cole, Electric impedance of suspensions of spheres, J. Gen Physiol. 12 (1928) 29 - 36.
14 H. Dänzer, Über das Verhalten biologischer Korper im Hochfrequenzfeld, Ann. Physik 20 (1934) 463 - 480.
15 H.P. Schwan, Electrical properties of tissue and cell suspensions, Adv. Biol. Med. Phys. 5 (1957) 147 - 209.
16 R. Pethig, Dielectric and Electronic Properties of Biological Materials, J. Wiley, Chichester, 1979.
17 H.P. Schwan, Dielectric properties of cells and tissues, in "Interactions between Electromagnetic Fields and Cells", A. Chiabrera, C. Nicolini and H.P. Schwan (Eds), New York, 1985 pp. 75 - 97.
18 R. Pethig and D.B. Kell, The passive electrical properties of biological systems: their significance in physiology, biophysics and biotechnology, Phys. Med. Biol. 32, (1987) 933 - 970.
19 A. Irimajiri, T. Hanai and A. Inouye, Evaluation of a conductometric method to determine the volume fraction of the suspensions of biomembrane - bounded particles, Experientia 31 (1975) 1373 - 1375.
20 A. Irimajiri, T. Hanai and A. Inouye, A dielectric theory of "multi-stratified shell" model with its application to a lymphoma cell, J. Theor. Biol. 78 (1979) 251 - 269.
21 K. Asami and A. Irimajiri, Dielectric analysis of mitochondria isolated from rat liver II. Intact mitochondria as simulated by a double - shell model, Biochim. Biophys. Acta 778 (1984) 570 - 578.
22 T. Hanai, Theory of the dielectric dispersion due to the interfacial polarization and its application to emulsion, Kolloid Z 171 (1960) 23 - 31.
23 H.P. Schwan, G. Schwarz, J. Maczuk and H. Pauly, On the low-frequency dielectric dispersion of colloidal particles in electrolyte solution, J. Phys. Chem 66 (1962) 2626 - 2635.
24 G. Schwarz, A theory of the low-frequency dielectric dispersion of colloidal particles in electrolyte solution, J. Phys Chem. 66 (1962) 2636 - 2642.

184

25 A. Garcia, R. Barchini and C. Grosse, The influence of diffusion on the permittivity of a suspension of spherical particles with insulating shells in an electrolyte, J. Phys. D: Appl. Phys. 18(1985) 1891 - 1896.

26 L. Benguigui and I. J. Lin, More about the dielectrophoretic force, J. Appl. Phys. 53 (1982) 1141 - 1143.

27 J.P.H. Burt, T.A.K. Al-Ameen and R. Pethig, An optical dielectrophoresis spectrometer for low-frequency measurements on colloidal suspensions, J. Phys. E: Sci. Instrum. 22 (1989) 952 - 957.

28 C.W. Einolf and E.L. Carstensen, Passive electrical properties of microorganisms. IV Studies of the protoplasts of Micrococcus lysodeikticus, Biophys. J. 9 (1969) 634 - 643.

29 H.A. Pohl and R. Pethig, Dielectric measurements using non-uniform electric field (dielectrophoretic) effects, J. Phys. E. Sci. Instrum. 10 (1977) 190 - 193.

30 J.A.R. Price, J.P.H. Burt and R. Pethig, Applications of a new optical technique for measuring the dielectrophoretc behaviour of micro-organisms, Biochim. Biophys. Acta 964 (1988) 221 - 230

31 K.V.I.S. Kaler, O.G. Fritz and R.J. Adamson, Dielectrophoretic velocity measurements using quasi-elastic light scattering, J. Electrostatics 21 (1988) 193 - 204.

32 J.P.H. Burt, R.Pethig, P.R.C.Gascoyne and F.F. Becker, Dielectrophoretic characterisation of Friend murine erythroleukaemic cells as a measure of induced differentation. Biochim, Biophys. Acta 1034 (1990) 93 - 101.

33 R. Pethig, P.R.C. Gascoyne and F.F. Becker, unpublished data.

34 W.M. Arnold and U. Zimmermann, Electro-rotation: Developments of a technique for dielectric measurements on individual cells and particles, J. Electrostatics, 21 (1988) 151 - 191.

35 A.A. Füredi and I Ohad, Effects of high-frequency electric fields on the living cell, Biochim. Biophys. Acta 79 (1964) 1 - 8.

36 H.A. Pohl and J.S. Crane, Dielectrophoresis of cells, Biophys J. 11 (1971) 711 - 727.

37 W.M. Arnold and U. Zimmermann, Rotating - field - induced rotation and measurement of the membrane capacitance of single mesophyll cells of Avena sativa Z. Naturforsch, 37 c, (1982) 908 - 915.

38 M.Mischel, A. Voss and H.A. Pohl, Cellular spin resonance in rotating electric fields, J. Biol. Phys. 10 (1982) 223 - 226.

39 R. Hölzel, Sine quadrature oscillator for cellular spin resonance up to 120 MHz, Med. Biol. Eng. Comput. 26 (1988) 102 - 105.

40 C. Holzapfel, J. Vienken and U. Zimmermann, Rotation of cells in an alternating electric field: Theory and experimental proof, J. Membrane Biol. 67 (1982) 13 - 62.

41 F.A. Sauer and R.W. Schlögl, Torques exerted on cylinders and spheres by external electromagnetic fields: A contribution to the thoery of field induced cell rotation. In "Interactions between Electromagnetic Fields and Cells", Eds. A. Chiabrera, C. Nicolini and H.P. Schwan, Plenum, New York (1985) pp 203 - 251.

42 G. Fuhr and P.I. Kusmin, Behavior of cells in rotating electric fields with account to surface charges and cell structures, Biophys. J. 50 (1986) 789 - 795.

43 R. Hölzel and I. Lamprecht, Cellular spin resonance of yeast in a frequency range up to 140 MHz, Z. Naturforsch. 420 (1987) 1367 - 1369.

44 X.Hu, W.M. Arnold and U. Zimmermann, Alterations in the electrical properties of T and B lymphocyte membranes induced by mitogenic stimulation. Activation monitored by electro-rotation of single cells, Biochim. Biophys. Acta 1021 (1990) 191-200.

45 U. Zimmermann, Electrical breakdown, electropermeabilization and electrofusion, Rev. Physiol. Biochem. Pharmacol. 105 (1986) 175 - 345.

46 U. Zimmermann and W.M. Arnold, Biophysics of electroinjection and electrofusion, J. Electrostatics, 21 (1988) 309 - 345.

47 S. Masuda, M. Washizu and T. Nanba, Novel method of cell fusion in field constriction area in fluid integrated circuit. IEEE Trans. Ind. Appl. 25 (1989) 732-737

48 M. Washizu, Electrostatic manipulation of biological objects in micro-fabricated structures, 3rd Toyota conf. Oct. 1989.

49 D. L. Farkas, R. Korenstein and S. Malkin, Electrophotoluminescence and the electrical properties of the photosynthetic membrane, Biophys. J. 45 (1984) 363 - 373.

50 Y. Huang, X-B. Wang and R. Pethig, unpublished work.

51 D. B. Kell, The principles and potential of electrical admittance spectroscopy: an introduction. In "Biosensors" (Eds. A. P. F. Turner, I. Karube & G. S. Wilson) Oxford Univ. Press (1987) pp. 427-468.

52 C. M. Harris, R. W. Todd, S. J. Bungard, R. W. Lovitt, J. G. Morris and D. B. Kell, Dielectric permittivity of microbial suspensions at radio frequencies: a novel method for the real-time estimation of microbial biomass. Enzyme Microb. Technol. 9 (1987) 181-186.

I. Karube (Ed.) *Automation in Biotechnology*
Proceedings of the 4th Toyota Conference, 21–24 October 1990
© 1991 Elsevier Science Publishers B.V. All rights reserved   187

MICROMECHANICAL SILICON DEVICES APPLIED FOR BIOLOGICAL CELL FUSION OPERATION

Kazuo Sato, Yoshio Kawamura and Shinji Tanaka

Central Research Laboratory, Hitachi, Ltd.
1-280, Higashi-Koigakubo, Kokubunji, Tokyo 185 (Japan)

SUMMARY
    Cell fusion has been conventionally carried out collectively in a large chamber containing millions of cell particles of different species. Consequently, a great deal of effort has been required to select a desirable hybrid from the total cell particles. A new concept of a cell fusion apparatus, named one-to-one, is presented. Two micromechanical components made of single crystal silicon have been developed for the apparatus. The first is a microchamber plate which has a matrix array of 44x36 fusion chambers. Each chamber contains the same type of cell pair composed of two different species. Cell fusion is performed on only desirable cell pairs. The second component is a carrier plate which supplies a single cell to every chamber. The apparatus is supposed to be applied to cell fusion operation for breeding plants. A variety of other applications are expected in the field of bio-technology, replacing conventional collective operations or individual operations under a microscope.

INTRODUCTION

    Cell fusion is expected to expand the breeding of useful plants for foods, pharmacy and many other purposes. For breeding applications, thousands of properly fused cell pairs from different plants are needed at the same time. Normally, when a number of processed cells are needed, scientists do not have the means to process them individually. Inevitably, cells are collectively operated and desirable cells are selected from the batch[1]. The selection means are not always available. Physical selection often damages fragile cells.

    The aim of our investigation was to devise a technique for operating thousands of biological cell particles individually at the same time. The scheme of our investigation is drawn in Fig.1 compared with the conventional collective operations or individual operations under a microscope. We propose a use of micromechanical silicon devices for the purpose.

    The authors have developed micromechanical silicon devices for

single cell : individual operation
under microscope

a large number : collective operation
of cells

individual operation of : operation with
a large number of cells   micromechanical
devices

Fig.1 Improvement in cell operation using micromechanical
devices.

cell manipulation and applied them to cell fusion processes.
Through the development of a "one to one fusion system" it has
become possible to assign the composition of the pair to be fused
in a mass operation.

CONCEPT OF ONE TO ONE FUSION
    The concept of the developed one to one cell fusion process is
shown in Fig.2 and compared with the conventional collective
method. Conventionally two groups of cell A and B to be fused are
mixed up in a large chamber dispersed in an isotonic solution.
Then a fusion medium is applied to the chamber. Polyethylene-
glycol(PEG) used to be a fusing medium. An electric pulse has
recently been used instead, because it is less toxic and is
easier to regulate fusing conditions with[2,3]. After the fusion
operation, the chamber contains a number of unfused cells and a
variety of fused pairs such as AA, BB, AAB and so on in addition
to the correctly fused AB pairs. The percentage of the fused AB

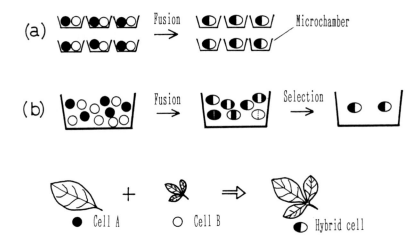

Fig.2 Concept of one-to-one cell fusion system(a), compared with conventional operation(b).

is less than 1% of the total particles[4]. A great deal of effort goes into searching for fused AB particles. For example, picking up particles under a microscope is very tedious work, and a selection culture where only the fused AB can survive needs years to develop.

The one to one fusion process proposed here is characterized by the fact that a number of microchambers, each containing just a single pair of cells from A and B, are prepared, and that electro-fusion is carried out in each chamber. It is expected that a number of desirable AB pairs will be obtained by a single operation and that selection efforts can be evaded altogether.

MICROMECHANICAL DEVICES

Two micromechanical components have been developed on single crystal silicon wafers. The first is a chamber plate having 1584 microchambers. The second is a carrier plate which acts as a tool for supplying a single cell to every chamber. The structures, the functions, and fabrication procedures are described below.

Microchamber plate

An enlarged view of the surface of a microchamber plate is

shown in Fig.3. Microchambers containing a single pair of cells are arrayed on a 36x44 matrix. The pitch distances of the chambers in the x and y directions are both 770µm. The structure of the microchamber plate is shown in Fig.4. It is composed of two stacked silicon wafers. In the figure, two pieces of wafer are illustrated apart for explanations. The upper plate has tapered pyramidal throughholes composing chamber cavities. Cell particles suspended in isotonic solution are driven downward to the center of each chamber guided by the tapered throughholes. The lower plate has apertures corresponding to each chamber. The aperture size is 10x80µm. A pair of cells whose diameters are 20-100µm is trapped on the aperture and keeps contact with each other by the effect of a weak suction force applied from the rear side of the lower plate. The aperture is located between the fusion electrodes patterned on the same plate. When a pulsed electric flux between the electrodes penetrates a pair of cells resting on the aperture, microscopic breakdown of cell membrane initiates the fusion at the contact area of the pair.

The microchamber plate is fabricated from single crystal silicon wafers 76mm in diameter and 390µm thick. The surface orientation is (100). Cavities and apertures of the chambers are micromachined by anisotropic etching using KOH aqua solution[5]. The accuracy of the aperture width is controlled within 2µm in total area. Electrodes of gold are deposited and patterned by a lift-off technique.

The upper and lower plates are mechanically stacked and clamped on a holder made of stainless steel. The holder has a recess to apply a suction force to the microchambers. These are demountable for the purpose of cleaning and sterilization in autoclave.

Carrier plate

In order to supply a single pair of cells to each microchamber, a carrier plate has been developed. It is fabricated on a silicon wafer. It has an array of 1584 cell absorption ports on the surface. Figure 5 shows enlarged views of the carrier plate. Each port has a throughhole connected to the rear side of the plate. A single cell is absorbed to and released from the port by controlling the pressure at the rear side of the wafer. The absorption ports are located in a 36x44 matrix array in the same manner as the microchambers. Each port has a 40µm

| 770 μm |

Fig.3 SEM photograph of the surface of a microchamber plate.

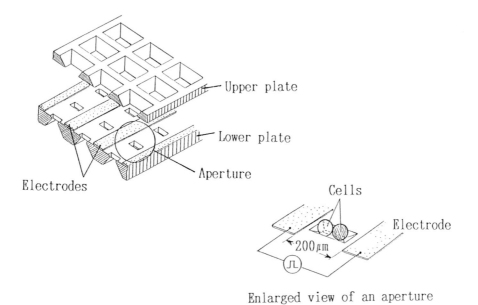

Enlarged view of an aperture

Fig.4 Structure of a microchamber plate.

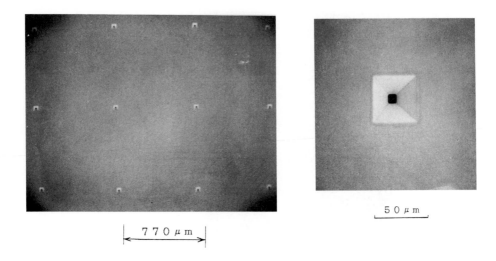

770 μm

50 μm

Fig.5 Arrayed absorption ports on a carrier plate and an enlarged view of a port.

square pyramidal cavity fabricated by KOH anisotropic etching. At the bottom of the cavity is a throughhole with an opening diameter of 10μm. Once a plant cell particle several tens of micrometers in diameter is absorbed to the port and chokes the hole, no other particles can come to the same port.

The cell delivery sequence to the microchamber using carrier plate is illustrated in a crosssectional view in Fig.6. The carrier plate is facing downward to a flat plate with a spacing of 300μm between them. Cells of group A are fed to the spacing suspended in isotonic solution. When a small amount of suction force is continuously applied from the rear side of the carrier plate, a single cell particle is absorbed to the carrier in matrix array. Then the carrier plate is transferred over the chamber plate and positioned so that the absorption ports meet the microchambers face to face. When the rear side of the carrier is pressurized, the absorbed cells are released simultaneously and fall into the chambers.

Repeating the same procedure for the cell group B, a single pair of cells A and B is supplied to each chamber.

The uniformity of the throughhole diameters is important for the cell operation. The overall accuracy of the throughholes

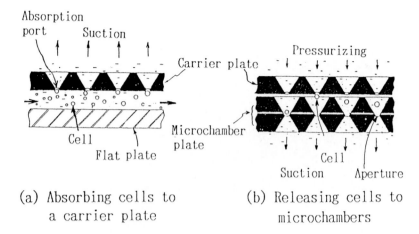

Fig.6 Crosssectional view of the cell delivery mechanism using carrier plate.

dimensions is controlled within 2μm in one wafer.

The carrier plate is mounted on a holder made of stainless steel which has a recess to apply pressure from the rear side of the wafer.

ONE TO ONE CELL FUSION SYSTEM

The developed prototype of the one to one cell fusion system is shown in Fig.7. It has dimensions of 700mm(W) x 250mm(H) x 500mm(D). The microchamber plate is located at the center of the vessel filled with isotonic solution. The system has two pieces of carrier plate for the cell groups A and B respectively in order to avoid contamination between the different cell groups. Each carrier plate is attached to a carriage arm which traverses over the vessel. Two sets of carrier mechanisms are separately installed to the right and left side of the microchamber plate. Cells are supplied from feeding stations located at the right and the left ends of the vessel respectively. The carriage arm transfers the carrier plate immersed in the solution from one end to the center of the vessel.

Figure 8 shows the schematic crosssection of a feeding station and a microchamber station in the vessel. Soft and fragile cells are handled with a small amount of flow of isotonic solution. Static pressure differences below several hundred

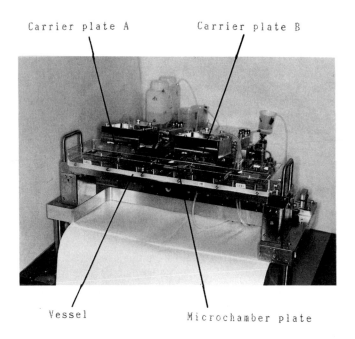

Carrier plate A   Carrier plate B

Vessel   Microchamber plate

Fig.7 Developed one-to-one cell fusion processor.

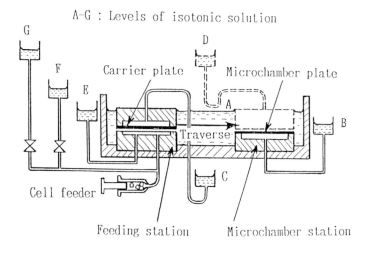

A-G : Levels of isotonic solution

Fig.8 Schematic crosssection of the cell fusion processor

pascals are supplied by heads of the solution. They are given by the difference in levels between vessel and cups connected by flexible siphons. Levels in Fig.8 are set at different heads from the level A of the vessel. Level B is applied to the rear side of the microchamber plate. Levels C and D are applied to the rear side of the carrier plate; C is for absorption and D is for release. Cells are fed to the carrier plate by the head F and the residual cells are washed away from the carrier plate by the head G.

The tubing in the system is made of silicon rubber and is easily demounted and autoclaved.

RESULTS AND DISCUSSION

Lettuce cells are used as specimens. They are processed by an enzyme to become individual particles of protoplast which is ready for fusion. A frequency chart of diameter sizes of protoplasts sampled from a single leaf is shown in Fig.9. Although the mean value is 35μm, sizes are widely distributed. Experiments are carried out without classifying the protoplasts by size or other characteristics.

Electric fusion conditions

Electric pulses are applied between electrodes whose spacing is 200μm immersed in an isotonic 0.5M sorbitol solution.

Electric pulse conditions are optimized according to voltage and duration. Figure 10 shows the relationship between pulse duration and fusion rate by varying the voltage between the electrodes. The fusion rate decreases when the electric excitation is too weak to initiate fusion, or when the excitation is too strong, causing cell membrane to burst. Maximum fusion rate of 52% is obtained under conditions of 20V and pulse width of 150μs. Nominal field strength is 1.0kV/cm in this case.

Figure 11 is a top view of a microchamber showing the fusing process of cells at a number of time increments. The fused pair becomes a single sphere in a couple of minutes.

Fusion in a microchamber has the additional merits that the applied voltage is about one tenth of that needed in conventional chambers because of the narrow spacing between the electrodes, and that dielectrophoresis preceding fusion is not necessary for keeping cell particles in contact.

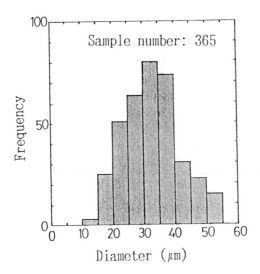

Fig.9 Distribution of diameters of the lettuce protoplasts operated in experiments.

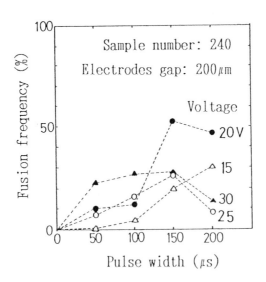

Fig.10 Frequency of fusion in a microchamber related to electric stimulation.

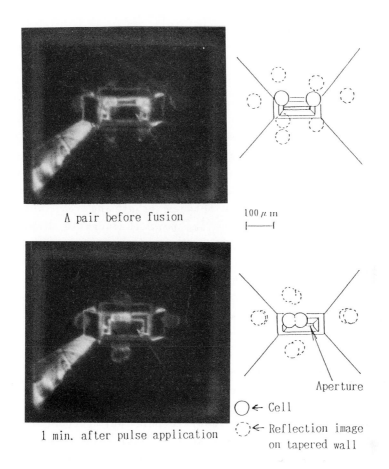

A pair before fusion

100 μ m

1 min. after pulse application

Aperture

○ ← Cell

◌ ← Reflection image on tapered wall

Fig.11 A pair of lettuce protoplasts fusing in a microchamber. Photographs taken with time increments.

Cell manipulation using liquid

A sufficient amount of cell particles must be absorbed to a carrier plate. The suction force applied from the rear side of the carrier plate must be strong enough to hold the cell particles at the absorption ports. On the other hand, cells in the form of protoplasts are easy to deform and burst. When the suction force is too excessive, cell particles are evacuated out from the throughhole or the cell membranes are fatally damaged.

Pressure conditions for holding the protoplasts safely at the absorption ports were also investigated. Figure 12 shows the optimum pressure difference between front and rear side of the carrier plate in relation to the port throughhole diameter. The number of safely absorbed cells was counted varying the pressure. The holding rate in Fig.12 is the ratio of safely absorbed cells to the total number(200) of inspected. When the pressure increases, the holding rate decreases because cells are evacuated out or burst. The reason why the holding rate does not decrease drastically during the increase in pressure, is that there is a variety of cell in sizes and characteristics. When throughhole diameter is large, cells are easily evacuated out of the port. In such a case, cells must be held at a low pressure difference. On the other hand, cells can easily drop out from absorption ports if pressure difference is insufficient. For lettuce cells, the authors have designed a throughhole diameter of 10μm and arrived at a pressure difference of 260Pa. These parameters must be appropriately chosen for cells having different diameters.

Cells are fed suspended in isotonic solution. The dispersion density of protoplasts is normally about $10^5/cm^3$. This value is too small to obtain a sufficient amount of absorption on the carrier. Cell particle must exist within an area of 50μm in radius from the absorption port in order to be caught by a flux evacuated from the port by a weak suction force. It is calculated that the dispersion must be ten times condensed to satisfy this condition. The authors have developed a technique to condense the dispersion locally in the system as follows.

Two kinds of isotonic solution, having different specific gravities have been used in the system. One is 0.5M sucrose solution whose specific gravity is 1.073 and the other is 0.5M sorbitol solution with specific gravity of 1.034. The specific gravity of the cells falls between that of the two solutions. At first, cells are supplied suspended in sucrose. Cells tend to

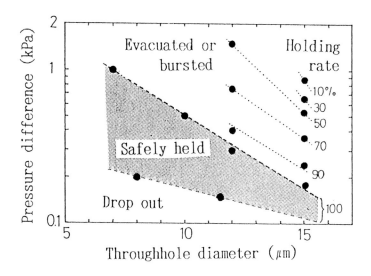

Fig.12 Safe absorbing conditions of cells to the carrier plate
related to the throughhole diameter and pressure difference
between front and rear side of the plate.

float up to the surface layer of the carrier plate. The
dispersion density is locally raised near the absorption port.
After a sufficient amount of absorption is obtained, residual
cells are washed away from the channel by feeding in sorbitol
solution.

Using the developed prototype system, the cell delivery
mechanisms with carrier plate were evaluated. The probability
that a sound single cell is successfully delivered to a
microchamber is 50-60% at present. The other possibilities
are:(1) the microchamber is vacant or two particles share it, (2)
delivered particle is a fragmented or burst cell improper for
fusion. These defects can be minimized by the following
techniques:(1) avoid creating turbulence in the solution during
operation, (2) minimize defects of micromechanical structures,
especially of apertures, (3) refine cell dispersed solution which
currently contains 20% of improper particles, the rest being
sound cells, (4) classify the cells which have widely distributed
sizes.

The life of protoplasts maintained in a buffer solution has

been recently measured using microchambers. The number of sound
protoplasts decreases according to the time increment as shown in
Fig.13. The decay depends on the cell preparation parameters such
as enzyme processing time and storing conditions. Optimization of
the process parameters will increase the number of sound cell
particles prepared for the fusion system.

Though the proposed cell operation system is not perfected,
the concept of the individual and mass operation of biological
cells is expected to be effectively applied in many other
processes besides fusion, such as gene injection and incubation.
A number of individual cells can be fixed in an array and easily
accessed by operators with the proposed system.

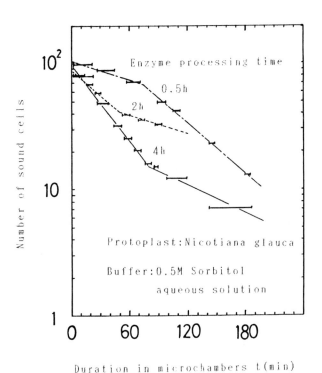

Fig.13 Decrease in the number of sound cell particles in
microchambers according to time increments.

CONCLUSION

Microchambers for one-to-one cell fusion have been developed on silicon wafers. A single pair of cells is individually fused by an electric stimulation in each chamber. In order to supply a cell particle to each chamber, a carrier plate has also been developed on a silicon wafer.

A variety of other applications besides cell fusion processes are expected to use the developed devices in bio-technology, where a number of cells need to be individually processed.

REFERENCES

[1] C.L. Afonso, K.R. Harkins, M.A. Thomas-Compton, A.E. Krejci and D.W. Galbraith, Selection of somatic hybrid plants in nicotiana through fluorescence-activated sorting of protoplasts, Bio/technology, 3 (1985), 811-816.
[2] M. Senda, J. Takeda, S.Abe and T. Nakamura, Induction of cell fusion of plant protoplasts by electrical stimulation, Plant Cell Physiol., 20(1979) 1441-1443.
[3] U. Zimmermann and P. Scheurich, High frequency fusion of plant protoplasts by electric field, Planta, 151(1981) 26-32.
[4] H. Morikawa, K. Sugino, Y. Hayashi. J. Takeda, M. Senda, A. Hirai and Y. Yamada, Interspecific plant hybridization by electrofusion in nicotiana, Bio/technology, 4 (1986), 57-60.
[5] D.L. Kendall, Vertical etching of silicon at very high aspect ratios, Ann. Rev. Mater. Sci., 9(1979) 373-403.

I. Karube (Ed.) *Automation in Biotechnology*
Proceedings of the 4th Toyota Conference, 21–24 October 1990
203

TRANSGENESIS OF ANIMALS

Hisato Kondoh[1], Kiyokazu Agata[2] and Kenjiro Ozato[3]
1. Department of Molecular Biology, School of Science, Nagoya
University, Nagoya 464-01 (Japan)
2. National Institute for Basic Biology, Okazaki 444 (Japan)
3. Department of Biology, College of Liberal Arts and Sciences,
Kyoto University, Kyoto 606 (Japan)

SUMMARY
    We discuss the principles of transgenesis, describe a model
gene suitable for analysis of the process of gene transfer, and
compare methodologies employed in the mouse, the chicken and the
medaka fish. We also discuss embryonic stem cells which will be
the major target of gene manipulation in transgenic technology in
the very near future.

INTRODUCTION
    Transgenesis of animals aims to analyze and modify the
genetic activities which support their life utilizing gene
transfer techniques. There are two basic aspects of
transgenesis: modification of gene activities of the somatic
cells in an animal, and permanent modification of a genetic trait
within a pedigree so as to create a transgenic animal line.

    In the majority of vertebrate species, three stages are
found in an animal's life in which manipulations for transgenesis
are most effective. The most popular is the stage of the egg,
shortly before or after fertilization. This stage is chosen in
anticipation of distribution of the transferred gene (transgene)
in a variety of cell types derived from the egg, including the
germ cells. Another stage in which gene manipulations are done
with the greatest ease is that of embryonic stem cells. The stem
cells are manipulated in culture and returned to their normal
embryonic environment allowing them to participate in
embryogenesis. The other, still theoretical, is the stage of
primordial germ cells (PGC). Stable integration of an exogenous
gene into a chromosome of PGC will ensure establishment of a
transgenic line.

    In this paper, we first describe a model gene with which the
analysis of the process of transgenesis has been facilitated. We

then compare techniques used for collecting and injecting eggs in
the mouse, the chicken and the medaka fish. We finally discuss
embryonic stem cell systems which are an imortant future prospect
of transgenic technology.

A MODEL TRANSGENE: MiwZ

Analysis of the transmission of the transgenes to descendant
cells after introduction into an egg or an embryonic cell is
facilitated if a gene is available which is easily detectable.
If a transgene is expressed without specificity in time and
space, it can be an excellent tag of the gene. Thus, we
developed a series of reporter genes coding for bacterial β-
galactosidase (refs. 1,2), exemplified by miwZ (Fig. 1). The
gene miwZ is expressed in most of the cell types and at a level
sufficient to be detected by a simple histochemical method.

There are two phases of expression of transgenes: transient
expression which occurs at a high level shortly after
introduction of a transgene, perhaps before integration into the
chromosomes of the recipient cells, and stable expression which
occurs at a moderate level in the population of cells which
incorporated the transgene as a part of their genomes. The gene
miwZ is so designed as to support efficient β-galactosidase
expression in both phases.

Transcription of miwZ gene depends on two elements: chicken
β-actin promoter and Rous sarcoma virus (RSV) enhancer/promoter.
The β-actin promoter is efficient, non-specific, and, more
importantly, exceptionally tolerant of the suppressing effect of
the chromosomes in which the gene is integrated (ref. 3). RSV
enhancer shows an activating effect in a greater variety of cell
types than other commonly used enhancers (ref. 1). Most
importantly, RSV enhancer is very active in the embryonic stem
cells to be described below in which other enhancers are poorly
active or even act as negative regulators.

Combination of these two elements resulted in a cumulative
effect. Transfection of various cell types in primary cultures
of mouse and chicken with miwZ resulted in high and non-specific
β-galactosidase expression. When miwZ was introduced into mouse
embryonal carcinoma cells, an analogue of embryonic stem cells,
we found that about half of the stable transfectants expressed a
siginificant level of β-galactosidase, in contrast to the case of

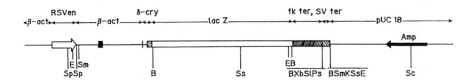

Fig. 1. Structure of the transgene plasmid pmiwZ. The segments
derived from chicken β-actin gene, chicken δ-crystallin gene,
Rous sarcoma virus enhancer, bacterial β-galactopsidase gene,
Herpes thymidine kinase gene termination region and SV40
termination region are indicated by β-act, δ-cry, RSVen, lacZ,
tkter and SVter, respectively. Restriction sites: B, BamHI; E,
EcoRI; K, KpnI; Ps, PstI; Sc, SacI; Sl, SalI; Sm, SmaI; Sp, SphI;
Ss, SstI; Xb, XbaI.

other gene constructs with which less than 1% of the stable
transfectants expressed β-galactosidase (I. Araki and H. Kondoh,
unpublished result).

TECHNIQUES OF COLLECTING AND INJECTING EGGS
Mouse
        Microinjection of molecular-cloned genes into fertilized
eggs has been widely used in transgenic mouse production. The
same technique should also be useful for the analysis of gene
regulation in early development.   We injected β-galactosidase-
coding recombinant genes, prototypes of miwZ, into nuclei of
fertilized eggs and cleavage stage embryos, and assessed
expression of the injected genes in the embryos by the enzyme
activity (ref. 1).
        Fertilized eggs were collected around noon of the plug day.
They were incubated in a culture medium containing hyaluronidase
to remove the cumulus cells, and further cultured to appropriate
developmental stages.
        Microinjection of embryo was done with the aid of holding
and injection pipettes held by micromanipulators under an
inverted microscope with Nomarski optics (Fig. 2).   After

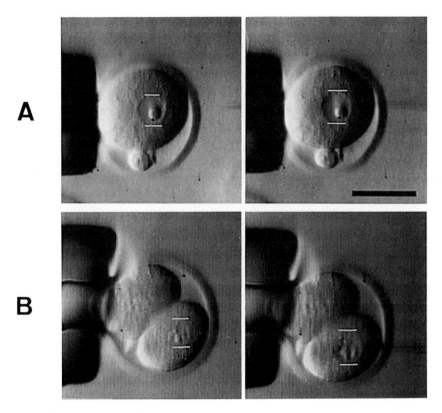

Fig. 2. Process of microinjection: Microinjection of male
pronucleus (A), microinjection of a nucleus of a blastomere of a
2-cell stage embryo (B). Left panels indicate embryos prior to
injection and right panels indicate the embryos just after
injection. The pair of lines touch the edge of the nucleus.
Note increase of the nuclear diameter after the injection. The
scale bar indicates 50 μm.

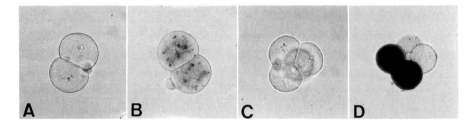

Fig. 3. Injected embryos were fixed after 24 hr and stained for
β-galactosidase activity with X-gal:(A) uninjected 2-cell embryo;
(B) 2-cell embryo injected at the early 1-cell stage; (C)
uninjected 4-cell embryo; (D) 4-cell embryo injected at the 2-
cell stage.

injection the embryos were returned to normal culture condition.

We found that expression of injected genes, which occurred transiently after injection, required the promoter sequences but without much distinction between the sources of enhancer/prompter complexes. This result was in a sharp contrast to transfection of mouse cell lines where the recombinant genes were variably expressed reflecting differential enhancer effects. By injection at the early 1-cell stage, expression of the injected transgenes was very low while the expression by injection at the 2-cell or later stages was several fold higher (Fig. 3), which correlates with the fact that most zygotic gene expression begins after the 2-cell stage.

Thus, any gene furnished with a promoter sequence may be expressed at least transiently upon injection into nuclei of blastomeres without much distinction of the kind of promoter. To produce transgenic mice with a gene whose product may interfere with early embryogenesis, it is essential that the injection be done at the early 1-cell stage to minimize transgene expression.

Chicken

Fertilized eggs were obtained from the magnum of the oviduct 2 h and 45 min after oviposition of the preceeding egg (Fig. 4). At this time the fertilized egg is still in the precleavage stage and the pronuclei are swelling. An egg was placed in a glass jar which contained the culture medium described by Perry (ref. 4), and transgene DNA was injected (Fig. 5). The micropipette was inserted vertically through the vitelline membrane into the central area of the germinal disc so that the DNA was injected into the cytoplasm close to the pronuclei. In the technique we employed, miwZ DNA solution was allowed to flow from the tip of the micropipette without interruption, so the injection volume may not always have been constant.

The injected eggs in the glass jar were incubated for 24 h at 41.5° C, and, after removal of the thick albumen and culture medium, transferred to recipient egg shells filled with thin albumen. The reconstituted eggs were sealed with cling film and incubated at 38° C with rocking over an angle of 90°. After three days, development of the embryos was terminated and expression of the injected transgene was analyzed.

Expression of miwZ was detected in about 60% of the embryos

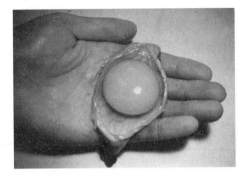

Fig. 4.   Isolated fertilized egg of the chicken.

Fig. 5.   Setup of the injection of a fertilized egg.

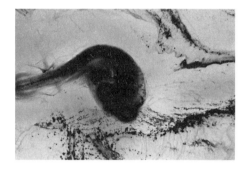

Fig. 6.   Expression of β-galactosidase activity in a 3-day
chicken embryo injected with pmiwZ at 1-cell stage.

which survived injection and incubation for 4 days. Most of the
embryonic expression was mosaic (Fig. 6). The extent of
mosaicism seems to be related to the efficiency of incorporation
of the transgene into the nucleus during the cell division. If
transgene DNA was injected to a site far from the first cleavage
plane, it was distributed in only a very small fraction of the
cells.

Medaka fish
     The problems in handling fertilized eggs of many fish
species lie in hardness of the chorion and difficulty in
observing the nucleus in these eggs. The problems are overcome
by using oocytes instead of fertilized eggs. At certain stages of
oocyte maturation the chorion of medaka oocytes is soft and the
large nucleus (germinal vasicle) is easily recognized (ref. 5).

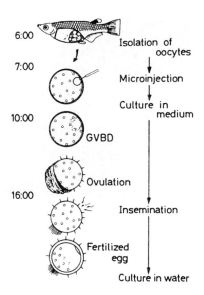

Fig. 7. Time schedule of microinjection of the medaka oocyte
nucleus.

A cultivated strain of medaka was bred under controled
lighting arrangements. Lighting started at 16:00 and continued
for 14 hours. The medaka lay their eggs at the beginning of
lighting (Fig. 7). At 6:00 o'clock, which was 10 hr before the
anticipated time of ovulation, intrafollocular oocytes were taken
out of females and put into culture medium. At 7:00 when oocytes
were in the prophase of the first meiotic division, DNA solution
was injected through a microneedle with sharpened tip until the
nucleus began to swell (Fig. 8). The injected oocytes were
cultured in the same medium at 26 °C for 8 hr until the
anticipated time of ovulation (16:00). Follicles were removed
spontaneously or mechanically with fine forceps. The oocytes
were then inseminated with a sperm suspension. The frequency of
fertilization was about 70%. Fertilized eggs were incubated
individually in distilled water in small wells at 26° C.

Most of the data on expression of transgene in medaka fish
have been obtained using tissue-specific δ-crystallin gene (refs.
6,7). Shown in Fig. 9 is an example of miwZ expression in a
transgenic medaka embryo at blastoderm stage.

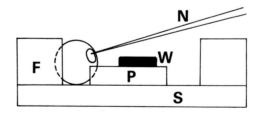

Fig. 8. Microinjection of the medaka oocyte nucleus with the aid
of the holding device schematically shown below. F, glass frame;
N, microneedle; P, glass plate; S, slide glass; W, metal weight.

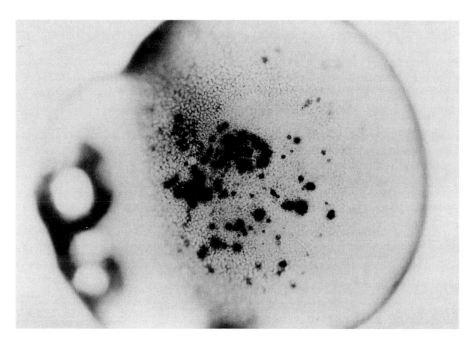

Fig. 9. Expression of β-galactosidase activity in a medaka
blastoderm embryo developed from an oocyte injected with pmiwZ.

EMBRYONIC STEM (ES) CELLS

In the past, efforts to obtain transgenic animals have been
primarily focused on handling fertilized eggs; however, there are
serious obstacles associated with this. Fate and expression of
the transgenes are in most cases unpredictable. Only after the
manipulated egg has developed into an animal are we able to know
the chromosomal integration and expression of the transgenes. In
addition, the frequency of obtaining transgenic animals is
generally low, and a low yield necessiates handling them in huge
numbers: to obtain eggs, to let embryo grow in foster mother
uteri in case of the mouse, and to obtain a sufficient number of
transgenic candidates to screen.

Use of embryonic stem cells as be described below is a way
to circumvent these problems and to allow more systematic
analysis of living systems utilizing transgenic technologies.

Mouse ES lines

In most vertebrate species the fate of the embryonic cells is not determined even after many cell divisions from a single zygotic cell, the fertilized egg. A group of cells without somatic lineage restriction is usually found in blastoderms and is called embryonic stem (ES) cells.

In the case of mouse, a number of experimental results indicate no restriction occurs during cleavage stage in the capacity to give rise to differentiated tissue types. The first restriction occurs at the blastocyst stage when the cells segregate into lineages of inner cell mass (ICM) and trophoectoderm. Among the cells of ICM, from which embryonic structures are derived, there appears no restriction of differentiation into particular cell types. If a single genetically marked ICM cell is injected into a blastocyst, the descendants of the injected cell marked genetically contribute to virtually any kind of tissue in the animal. Animals of this kind posessing dual genotypes are called chimeras. Most of the cells in ICM are multipotent and differentiate into a large variety of cells including germ cells. Thus, the cells of ICM have the properties of the ES cells.

It was the beginning of a breakthrough when ES cell lines were established from the mouse ICM, for these multiply almost infinitely under culture conditions and retain the capacity to form mouse chimeras in a way analogous to ICM cells (ref. 8). This opened up a way to introduce a transgene while ES cells are in culture, and to select for cell lines which have incorporated the transgene in an appropriate fashion (e.g., in a chromosomal locus favorable for expression) before introducing the transgene into mouse embryos.

ES cells contribute to the development of a variety of tissues in the same way as ICM cells. When an ES cell line and its lineage were marked with miwZ expression and chimeric embryos were produced, the embryos were heavily stained for β-galactosidase activity, often patchwise, throughout the body (ref. 2). If we analyze the relative contribution of ES-derived and recipient ICM-derived cells in such chimeras, the former account for an impressively large fraction of cells in various tissues, sometimes exceeding the portion of ICM-derived recipient cells. ES cells also contribute to germ cells to some degree, although the frequency depends on the particular ES lines used for the experiment.

Handling of ES line cells in culture enables us to select products of a rare recombination event between a transgene and the mouse genome, for instance, homologous recombinants. A rough estimate would be that integration of a transgene into mouse chromosome occurs once in 100-1000 instances of gene transfer, and integration by homologous recombination occurs once in 1000 integrations. Therefore, with appropriate selection strategies, such as those proposed by Capecchi (ref. 9) and Joyner and Rossant (ref. 10), one can obtain a reasonable number of homologous recombinants starting from 10^7 cells, which does not demand too much labor. If one were to work with fertilized eggs, at least 10^4 would be required to obtain a homologous recombinant.

Homologous recombination in ES lines has been regarded as a promising strategy to disrupt the target gene. However, we anticipate a greater variety of applications not thought of before homologous recombinants has come to the reach of mouse transgenesis.

Chicken ES cells

In the chicken embryos, blastodermal cells have the properties of ES cells. Immediately after oviposition, a blastoderm contains a number of embryonic stem cells, since transplantation of dissociated blastoderm cells to allogenic blastoderm resulted in extensive chimerism in the embryo (ref. 11). Transmission of the donor genotype to offspring through sperm has also been demonstrated for a chimeric chicken.

There has not yet been any cell line of chicken ES cells established. However, chicken blastoderm has a large number (more than 10^4) compared to less than 100 cells in mouse ICM. In addition, isolation of the chicken blastoderm is easily done with practice. Therefore, an adequate number of chicken ES cells can be obtained from a small number of newly layed eggs. In fact, successful transgenesis of chicken has been reported in which a transgene was introduced into blastodermal ES cells by means of lipofection followed by injection of the ES cells to the host egg blastoderm (ref. 12). Thus, the ease of collecting ES cells from a freshly layed egg without sacrificing a hen is a great advantage of the system as a means of obtaining transgenic chicken.

Fish ES cells

In teleost fish, the ES cell nature of early blastodermal cells was demonstrated by Kimmel's group (ref. 13). Using zebra fish embryos, they labeled individual cells of early blastoderm by means of fluorochromes and traced their movement and fate. Although highly ordered, a stereotypic pattern of cleavage occurs before blastodermal stage, and the cells later on undergo extensive intermixing simultaneously with multiplication. The fate of a cell is determined by its position in the late blastodermal stage.

Early blastodermal cells of fish embryos will prove to be excellent route of transgenesis in a way analogous to mouse and chicken embryos, although no report of this has yet been made.

TRANSGENESIS OF SOMATIC CELLS V.S. GERM CELLS

The use of ES cells as a means of transgenesis has two features. First, chimerism in the first generation facilitates analysis of cellular interactions in somatic cells. Second, incorporation of the transgene into the germ line through germ cell chimerism results in production of transgene heterozygotes in the second generation and homozygotes in the third generation if mating is appropriately controled.

If one were to obtain germ line transmission in a more straightforward way, direct introduction of a transgene into primordial germ cells or into pre-meiotic germ cells might be attempted. Although there are a number of problems to overcome before this kind of technique takes shape, germ cell transgenesis appears quite feasible and is bound to become practical in the very near future.

ACKNOWLEDGEMENTS

We thank all our colleagues, especially Dr. M. Naito and the members of the Aichi-ken Agricultural Research Center, for collaboration. Original works presented here were partly supported by grants from the Ministry of Education, Science and Culture, the Science and Technology Agency, and the Fisheries Agency of Japan.

REFERENCES

1 K. Ueno, Y. Hiramoto, S. Hayashi and H. Kondoh, Introduction
   and expression of recombinant β-galactosidase genes in
   cleavage stage mouse embryos, Dev. Growth Differ., 30 (1988)
   61-73.

2 H. Suemori, Y. Kadokawa, K. Goto, I. Araki, H. Kondoh and N.
   Nakatsuji, A mouse embryonic stem cell line showing
   pluripotency of differentiation in early embryos and
   ubiquitous β-galactosidase expression. Cell Differ. Develop.,
   29 (1990)181-186.

3 N. Fregien and N. Davidson, Activating elements in the
   promoter region of the chicken β-actin gene, Gene 48, (1986)
   1-11.

4 M.M. Perry, A complete culture system for the chick embryo,
   Nature 331 (1988) 70-72.

5 K. Ozato, K. Inoue and Y. Wakamatsu, Transgenic fish:
   Biological and technical problems, Zool. Sci., 6 (1989) 445-
   457.

6 K. Ozato, H. Kondoh, H. Inohara, T. Iwamatsu, Y.Wakamatsu and
   T.S. Okada, Production of transgenic fish: introduction and
   expression of chicken δ-crystallin gene in medaka embryos,
   Cell Differ., 19 (1986) 237-244.

7 K. Inoue, K. Ozato, H. Kondoh, Y. Wakamatsu, T. Fujita and
   T.S. Okada, Stage-dependent expression of the chicken δ-
   crystallin gene in transgenic fish embryos, Cell Differ.
   Develop., 27 (1989) 57-68.

8 M.J. Evans and M.H. Kaufman, Establishment in culture of
   pluripotential cells from mouse embryos, Nature 292 (1981)
   154-156.

9 M.R. Capecchi, The new mouse genetics: altering the genome by
   gene targeting, Trends Genet. 5 (1989) 70-76.

10 A.L. Joyner, W.C. Skarnes and J. Rossant, Production of a
   mutation in mouse En-2 gene by homologous recombination in
   embryonic stem cells, Nature 338 (1989) 153-156.

11 J.N. Petitte, M.E. Clerke, G. Liu, A.M.V. Gibbins and R.J.
   Etches, Production of somatic and germline chimeras in the
   chicken by transfer of early blastodermal cells, Development,
   108 (1990) 185-189.

12 C.L. Brazolot, J.N. Petitte, R.J. Etches and A.M.V. Gibbins,

The establishment of efficient gene transfer and expression in chicken embryonic stem cells, Cell Differ Develop. 27 (1989) S90.

13 C.B. Kimmel and R.M. Warga, Cell lineage and developmental potential of cells in the zebrafish embryos, Trends Genet., 4 (1988) 68-74.

I. Karube (Ed.) *Automation in Biotechnology*
Proceedings of the 4th Toyota Conference, 21–24 October 1990
217

AUTOMATION OF PLANT TISSUE CULTURE PROCESS

Y. MIWA, Ph.D.

Professor, Department of Mechanical Engineering,
School of Science and Engineering,
Waseda University, 3-4-1, Okubo, Shinjukuku, Tokyo 169 (Japan)

SUMMARY
    The present authors have developed a position detector and a
growth state discriminator for a seedling during the previously
conducted fundamental research on an automatizing plant tissue
culture process. Furthermore, we have developed a micro robot for
transplanting young seedling into culture medium. In this study,
we will discuss a fully automated lily's bulb tissue culturing
system that was developed as a trial of automation of biotechnology performed in a flower production. This system was build with
integrating subsystems which are developed to perform automatically in each process of supplying a bulb, cutting its root, separating the bulbscales, transplanting bulbscales one by one, recognizing the shape of each bulbscale, and planting the bulbscale into
culture medium. Individual subsystem was designed to cope with
irregularity in size and shape of a bulb or bulbscale. At present,
the neither a virus contamination nor a detrimental effect to a
genetic trait by the mechanical stressing due to the system was
recognized in a trial test. It is also found that a completion of
the present system was within one minute. It is,therefore, thought
that the system can be used in a practical stage. Meanwhile, the
culture robot of a miniature capsule enclosed structure for the
home and/or personal purposes was made for a trial performance,
extending our techniques towards a developing the aforementioned
system. Furthermore, a protoplast positioning system using a
dielectrophoresis effect in medium chamber having electrodes will
be described. This system was designed to be applicable to a cell
fusion and gene injection.

INTRODUCTION
    Recently the demands for an improved food-productivity by
employing the plant biotechnology are recognized. Moreover, an
amenitization and resort development promotes an ever-increasing
requirements on flowers and greens. However, according to the
current trends of the plant biotechnology, the majority of the
biotechnology is relied upon the manual performances. For example,
approximately seventy percent of the labor cost on the orchid
culture for commercial purposes comes from the wage. Furthermore,

since such task requires extensively developed skill and intensive works, too much works are demanded to labors. Accordingly, interests in FA and LA on the plant biotechnology become more recognized.

Based on these current trends in the biotechnology field, the present author has initiated studies on the automation of the plant tissue culture as a typical example of the biotechnology, and conducted the feasibility studies of development of robots which are designed for especially transplanting of seedlings between the culture systems. Furthermore, the present author has developed the full-automatization of the culture process in the bulb culture tissue of lily which is evaluated as one of the most valuable flower for ornament. At the same time, the capsule-type culture robot, to which the aforementioned system is installed with a small-scaled portable structure, was also developed. The latter robot system was designed and manufactured particularly to meet the future demands for the individuals' leisure purposes.

In this paper, the design details and the performance capability of the above systems as well as the position control system in the solution chamber for protoplast - which can be applied especially to the cell fusion or gene injection - are described. Furthermore, the unique communication system between the human beings and the plants will also be discussed with respect to the biological-information networks in the plants.

FULLY-AUROMATIZED SEEDLING TRANSPLANTING SYSTEM BY USING THE "MERICLONE ROBOT"

It is suggested to refer the proceeding of the MOET-HENNESSY International Conference for detailed information on this system which the present author prefers to call a "mericlone robot" (refs.1,2).

This system is constructed of a seedling transplanting robot, a position control device for planted seedling, and a monitoring equipment for evaluation of the degree of the growth of the seedling. The significance of the seedling planting robot is that an actuator is made of the shape memory alloy. Therefore, the seedling having 20 mm long and 0.5 mm diameter can be softly caught without causing any undesired scratches within the relatively narrow space of the culture vessel. Moreover, such seedling can also be pulled from the culture medium and replanted on the other

Fig.1 Mericlone robot driven by a shape memory alloy actuator

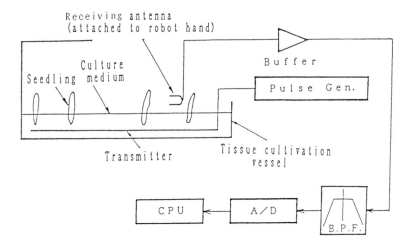

Fig.2 The plant-antenna sensory system for
detecting the seedling position (Ref.2)

sites. Fig.1 demonstrates a planting performance. For detecting
the position of the seedling, the very weak pulse current was
applied to the culture medium, and intensity of the electric
signal transmitted from the seedling was sensed (see Figs.2 and
3). This system posses a quite different feature from the conven-

220

tional method with respect to that the seedling itself can act as
a type of the antenna, therefore as a sensor. The position detecting method of the conventional technology utilizes an image processing. The monitoring the degree of the growth of the seedling

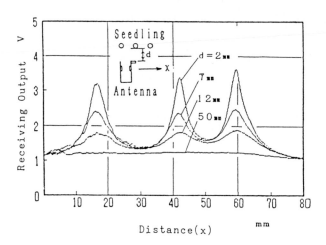

Fig.3 The distribution of the receiving output
by the plant-antenna sensory system (Ref.2)

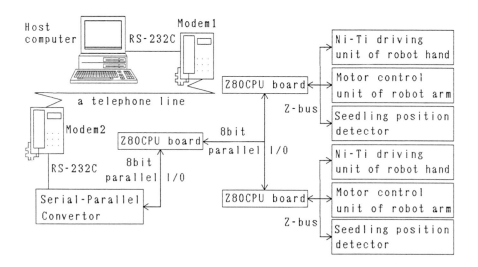

Fig.4 Remote control system for the mericlone robot

was carried out to examine the length and color of the seedling by using the corrimeter method comprising the RGB output type color sensing element, slit and the light source unit.

The above individual components are assembled to form a complete unit. By multiplying such units with a simultaneous controlling, the working time for seedling planting can be remarkably shortened. This system was also designed for the man-less operation in the culturing room. This can be done by performing a remote-control through the public circuit (see Fig.4 for general outlines of the system).

FULLY-AUTOMATIZED CULTURE SYSTEM OF LILY'S BULB

Fig.5 shows the cell culture process of the lily's bulb. Firstly, a root is removed from the bulb, followed by removing the bulbscale. After each removed bulbscale is planted in the culture medium, the bulb will grow afterwards. The system developed in this study was simulated and corporated with the actual field to achieve its compatibility. The bulb of the lily has about 5 to 15 mm in diameter and is consisted of 7 to 8 sheets of bulbscale. Size of each bulbscale as well as the bulb itself differs widely. Therefore, it is required that the constitutive components of the system should be respond to the wide scattering in size and configurations of the bulbs. Furthermore, it is also necessary that the performing speed should be equivalent to manual works, and that the whole operation should not affect any unexpected effects on the inherited character.

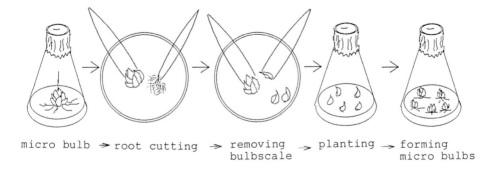

micro bulb → root cutting → removing bulbscale → planting → forming micro bulbs

Fig.5   Tissue culture process of lily's bulb

Bulb supply unit

When bulbs are stocked layer by layer, it is extremely difficult to separate individual bulb. Hence, in this system, the stocked bulbs are loosened by rotating rubber-shaped sheets with 20 rpm. Such rubber-shaped sheets with 10 mm width are attached to the shaft side by side in parallel, as seen in Fig.6. Thus separated bulbs drop gradually into a V-shaped tray. After individual bulb is separated through the V-shaped tray, it is fed into a root removing unit.

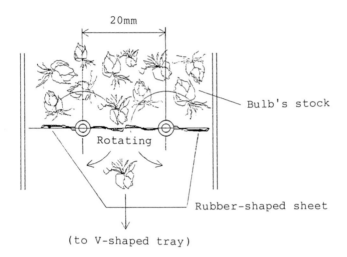

Fig.6 Bulb supply unit

Root removing unit

Approximately 10 to 30 roots usually grow on the bulb, each having 10 to 20 mm length. Generally, bulb and root are held with two pairs of forcepses and they are manually separated. In order to let the robot to perform the similar operation, it is necessitated to recognize the three-dimensional geometry of the bulb. However, the geometry of the bulb is extremely nonuniform, therefore it is quite difficult to separate individual root from the bulb even if the image processing is successfully performed.

While the bulb is rotating, being independent of the size and growing condition of the individual bulb, it was found that the bulb can rotate horizontally along its center axis. Hence, the roller-method was investigated to examine its performance of a winding root of bulbs. Namely, when the bulb is put onto two rollers which are rotating along a winding direction of the root, bulbs start to rotate parallel to the roller's axial direction. By keeping this condition for a while, only root portions are wound in the rollers. Then, the root portions are removed from the bulbs through the vertical plate which moves along the axial direction on the rollers. In this study, a pair of rollers (one having 11 mm diameter and the other having 4 mm diameter) used. It was found that the root portions are removed successfully, despite of variations in the size of bulbs. Fig.7 shows the system.

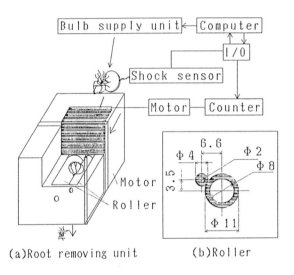

(a)Root removing unit   (b)Roller

Fig.7 Root removing unit

Bulbscale removing unit

    Although the bulbscale is usually removed while the bulb being fixed, in this system the bulb is not fixed for removing the bulbscale to develop an automatized system. As seen in Fig.8, the bulb is developed onto a rotating disks (approximately 100 mm in diameter) having obstacles. While said disk rotates at maximum 5000rpm, the bulbs are in contact several times with the inner surfaces of a cylinder located around the rotating disk. By using this mechanical shock, bulbscales are easily removed.   Being independent of the size of bulbs, bulbscale (except those situat- ed at a core portion) can be removed within about 10 seconds.

The time required for the removing operation depends on the size and geometry of obstacles as well as the revolution speed. Since there might be undesired effects of mechanical shock stressing on the inherited character, additional culture tests on removed bulbscales were carried on. Up to now, no detrimental effects have been observed. Besides, this system is simply constructed and offers an excellent efficiency.

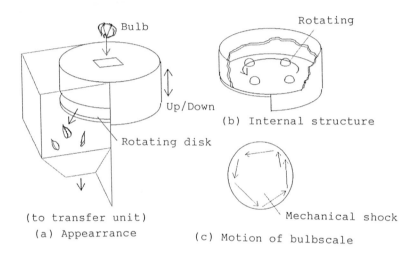

Fig.8 Bulbscale removing unit

Transfer/Planting equipment

In the next step, it is necessary to transfer removed bulbscale individually in order for the robot to hold each bulbscale. For this purpose, removed bulbscale is rotated inside the declined (with 6 degree) rotating cylinder, which has 25 mm of an inner diameter. If such rotating cylinder is separated into 2-steps, the separation can be completely achieved.

Since the bulbscale possesses a polarity, the root side should be planted into the culture medium. Therefore, by employing an image processed unit, the direction of planting is controlled by using the characteristic asymmetry of the bulbscale as seen in Fig.9. Namely, the centroid position of the bulbscale from its image, and a boundary line between the bulbscale's image and others are firstly detected. The position, while the centroid is located farthest away from its boundary line, is defined as a germination direction to determine the position of the root-side.

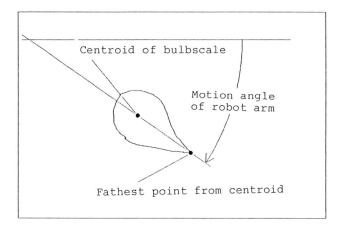

Fig.9 The method of image processing
to the direction of planting

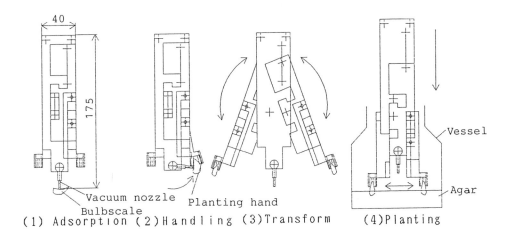

(1) Adsorption (2)Handling (3)Transform (4)Planting

Fig.10 Structure of robot hand developed for grasping and
planting the bulbscale

A plant-robot is formed by a hand installed into a commercially
available micro-robot-arm (Mitsubishi Electronics, trade name:
Move-Master). At an early stage of the system development, a
shape memory alloy was used for hands-actuator. However the operational speed by using the shape memory alloy was not satisfied.
Therefore, in the final version of the system, an outer/inner
wire   method   was   employed   as   a   driving   mechanism.   Fig.10

226

illustrates the structure and the operational mechanism. Detailed
design of this system is answered to meet requirements from the
field such that plants are needed to be planted in the culture
vessel (inlet diameter: 60 mm, depth: 80 mm) which is currently
used. The bulbscale is adsorbed by a transferring action of the
arm along the longitudinal direction to planting which is determined by an image processing. Then the bulbscale is replaced to

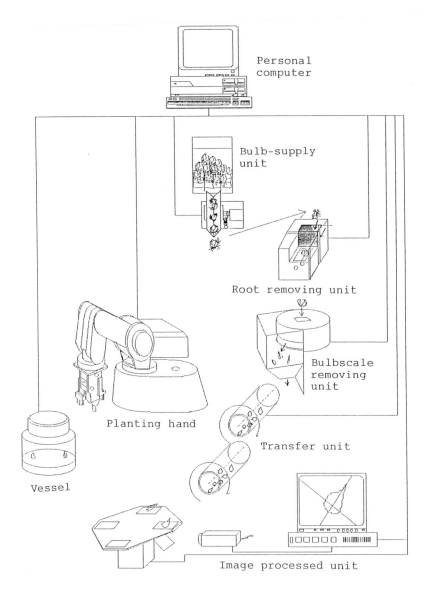

Fig.11 Outline of total system with integrated individual units

the planting arm and passed through the culture vessel, in which a planting-operation is completed by opening the hand position with a predetermined distance. It will be possible to plant plurality of bulbs with using a multi-hand system. The depth of planting is determined by detecting the changes in resistance when the bulb-scale contacts with the culture medium.

The above equipments are finally sub-systemized. Figs.11 and 12 show a construction and an outer view of a hardware portion of the fully-automatized culture system for lily's bulb. As a result of performance tests, a bulbscale can be planted into the culture medium within about 50 seconds. It is, therefore, that the presently developed system can be evaluated to be applicable and operative. "Mitsubishi-Mericlone-Robot System" which was demonstrated at the International Garden and Greenery Exposition (Osaka, Japan, 1990) was manufactured, based on this system.

Fig.12 Appearance of the fully-automatized culture system for lily's bulb

A Miniature culture-robot for personal uses

It is anticipated that an automation of biotechnology for industrial and research uses is ever-increasing to great extent. In the future, the automatized biotechnology could be applied to house appliances. It will include culture vessel which can be used by individuals although the guide-lines for user's manual should be well documented. Moreover, a bio-amenity-systemization in which an individual will be able to design and enjoy the living environment surrounded by the flower and greens will become popular. It is, therefore, needed to develop a culture system in an individual-level or a home-level.

The present author has initiated to develop a capsule-structured culture robot to which the aforementioned system is scaled-down. Fig.13 shows an image sketch of a miniature culture-robot. A fundamental idea is based on that when the bulb is thrown into the culture vessel, a bulbscale which is planted into a culture medium will appear as if an artificial seed.

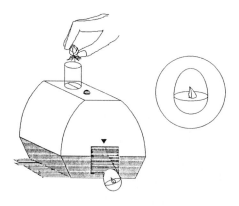

Fig.13   Image sketch of a miniature culture robot

As seen in Fig.14, each equipment is packaged inside a capsule. The bulb is fed into the feeding device by an one-touch operation. The root portion is wound by a holding device and is oriented vertically. It is then conveyed to a center position of the roller. In the next step, root portion of the bulb is replaced to a replacing hand, causing that one sheet of bulbscale is removed through a removing hand. The removed bulbscale is planted into the culture medium of a miniature culture vessel by rotating the hand. The bulb is conveyed forwardly out of the system, after the cap of the culture vessel is closed. The bulb, from which one bulbscale was previously removed, is immediately returned back to a holding equipment. By rotating the holding equipment at a certain degree, the second sheet of bulbscale can be removed and planted in a

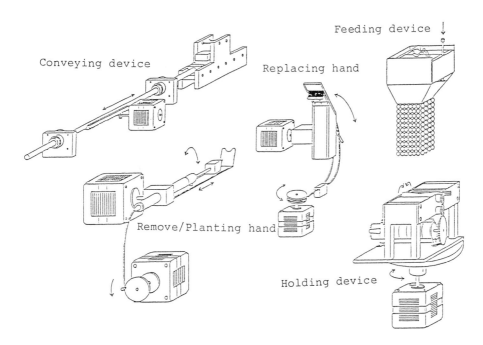

Fig.14 Several units packaged in a miniature culture robot

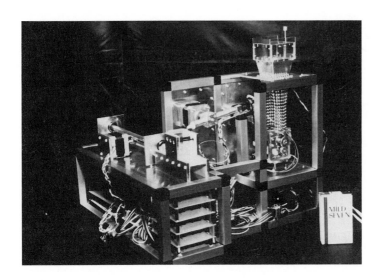

Fig.15 Appearance of a miniature culture robot

similar way. The above procedures can be controlled by an onboard micro-computer through the Z80-CPU system. The whole dimen-

sion of the system is 400 mm long, 350 mm wide, and 250 mm high. The total weight is 2.5 Kg. General view of inner devices is presented in Fig.15. Although the reliability of this system is not high as expected at this moment, there are possibilities of applying this unit for practical uses.

MOVEMENT-CONTROL SYSTEM OF PROTOPLAST

The technology including a transferring the protoplast to a certain location of the solution chamber or holding the above can offer versatile and useful techniques supporting the biotechnology; for example, an application to a micro-manipulation. The present author and his research group have developed a system to control the movement of the protoplast by using an externally applied electric manipulation. Fig.16 shows two pairs of electrode; each of them is facing to the other. By using this type of electrode, one can convey the protoplast to either X-axis or Y-axis direction by controlling an applied the electric field.

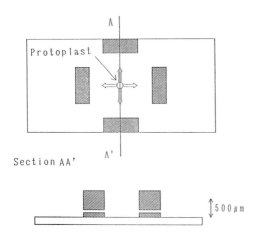

Fig.16 Electrodes for conveying the protoplast
to either X-axis or Y-axis direction

Furthermore, one can also transfer the protoplast to a relatively long distance without contacting with the electrode, as seen in Fig.17. The positioning of the protoplast can be monitored and controlled automatically by an image processing software, which was previously designed for this purpose. It is known that

the protoplast flows to a highest side of the current density due
to the dielectrophoresis phenomenon under an A.C. electric field.

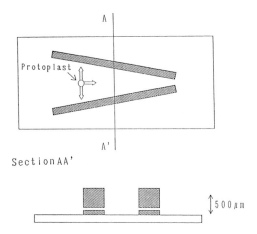

Section AA'

Fig.17   Electrodes to transfer the protoplast to
a relatively long distance without contact

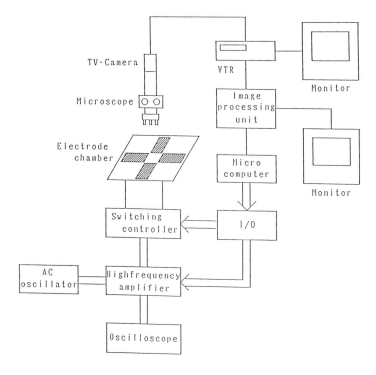

Fig.18   Movement control system of protoplast

Hence the protoplast can be transferred to a desired position by using properly designed electrode. In this study, the effects of electric field strength and frequency on the moving speed of the protoplast were investigated, by using various electrodes with different configurations. A membership function was employed to recognize the protoplast as a target. The system structure controlling the movement of protoplast is demonstrated in Fig.18.

Results of research and development on an automation of plant tissue culture were discussed in this paper. In general, any type of living substances behave in different ways because they possess nonuniform structure and complicated nature. Hence, there are many cases where the existing machine technology, which are dominantly employed to industrial products, can not be applied directly to living substances. In the mean future, a development of a control software as well as hardware is urgently needed. It is also necessary to establish a design concept for machines and robots which will be operative to a handling living substances.

ACKNOWLEDGEMENTS

Many thanks are due to graduate students including T.Yamamoto, K.Kodama, T.Murase, S.Yoshimura and K.Baba, and undergraduate senior students working their graduating thesis at my laboratory.

PROLOGUE

In our laboratory, beside the automation of the plant tissue culture, there are another on-going research activities including a designing an environmental psychological space by using plant biological information system or plant itself. Developments of visual and auditory expression of changes in biological potentials at plant surfaces, and flower-wall and botanical performance robot by using the mechatronics are in progress. These activities are related to applicability studies of flower and/or greens which are produced by the tissue culture (Refs.3,4).

REFERENCES
1 Y.MIWA, Plant tissue culture robot operated by shape memory alloy actuator and a new type of sensing, In: Moet-

Hennessy(Ed) Electronics and management of living plants, Abstract pp.7 (1987)

2   Y.MIWA, T.YAMAMOTO, Y.KUSHIHASHI and H.KODAMA, Study on automation of plant tissue culture process (1st Report), JSPE 54 (6) (1988) 1107-1112

3   Y.MIWA, Biological information system in plant, JSME 90-818 (1987) 76-82

4   Y.MIWA, T.YAMAURA, N.SHIMADA, Y.KUSHIHASHI and Y.KOZU, The development of the audio-visual system for bio-electropotential variation of plant under field condition, The 2nd Bioengineering Division Conference, JSME 900-51 (1990) 29-31

I. Karube (Ed.) *Automation in Biotechnology*
Proceedings of the 4th Toyota Conference, 21–24 October 1990
© 1991 Elsevier Science Publishers B.V. All rights reserved

235

TRENDS IN AUTOMATION FOR CLONAL PROPAGATION BY TISSUE CULTURE

JENNY AITKEN-CHRISTIE

Forest Research Institute, Private Bag 3020, Rotorua, New Zealand

SUMMARY
    Clonal propagation by tissue culture is frequently more expensive than other forms of propagation using cuttings or seed because it is labour intensive and more specialised. The aim of automation is to reduce the cost per plantlet by reducing labour input. Bulk handling of tissues and plantlets is essential. The main areas of clonal propagation by tissue culture that have been automated include nutrient media preparation, handling of containers in the laboratory and greenhouse, misting and watering of plantlets *ex vitro*, and management in the laboratory and greenhouse by computer. These aspects are more straightforward and have been easier to automate than the *in vitro* stages. Where *in vitro* automation has been attempted, the methods chosen were dependent on the growth and multiplication habits of species and where and if tissues were cut during subculture. Various aspects of *in vitro* automation, including liquid feeding, support systems, hedging, homogenisation, nodule culture, encapsulation and sugar-free micropropagation are discussed. Organogenic cultures are being grown, multiplied, and processed in some cases, on a pre-commercial scale in bioreactors. For shoot cultures with an upright growth habit, robotic and mechanised systems have been developed for cutting and planting nodal segments with leaves and/or meristems. Automated and robotic systems have been developed for handling (grading, trimming and transplanting) small seedlings and cuttings in the greenhouse. These systems could also be applied to plants propagated by tissue culture at the greenhouse stage. Important considerations when evaluating and developing automated systems include the cost, yields and quality of plants, contamination, damage to the tissues and vitrification. Some of the automated systems developed to date will be reviewed with respect to these points.

1.   INTRODUCTION

    Tissue culture, or micropropagation, is being used for the vegetative multiplication of

millions of superior plants annually for horticulture and forestry. Tissue culture is

frequently more expensive than other forms of propagation using cuttings or seed (ref. 1)

because it is labour intensive and more specialised. This can be justified when conventional

methods do not work or the end product is valuable. For ornamentals and herbaceous

species the price per plantlet varied between US$0.32 and $1.15 (1989 price list from

commercial laboratory for 50 clones from 17 species) while for forest trees the price per

plantlet varied between US$0.38 and $0.48 (ref.2, adjusted for inflation and converted from

NZ$ to 1990 US$). For broccoli (*Brassica oleracea*), where rooting was done *in vitro*,

approximately 75% of the cost was for *in vitro* stages and 25% for *ex vitro* stages (ref. 3). For

radiata pine (*Pinus radiata*), where rooting was done *ex vitro*, approximately 40% of the cost

was for *in vitro* stages, 40% for *ex vitro* stages, and 20% for growing on in the nursery (ref. 2,

J. Gleed & B. Nairn, personal communication). Many ornamentals and herbaceous crops

may fit into the *in vitro* rooting category while woody plants and forest trees may fit into the

ex vitro category. There are, of course, exceptions to both scenarios. The aim of automation

is to reduce the cost per plantlet by reducing the labour input. Labour usually accounts for

70-80% of the *in vitro* and *ex vitro* costs. Steps to automate tissue culture procedures should be targeted at areas where the costs are greatest and these would vary depending on species.

Recent years have witnessed an increased interest in and awareness of automation possibilities in tissue culture by biologists, plant physiologists, and agricultural and horticultural engineers. The majority of references in this article are less than 5 years old and new publications are appearing rapidly. The need to devise suitable equipment and to exploit the biology of growth of tissues so that cultures are more amenable to large-scale processing has been suggested (refs. 4-8). Further details of where automation could be used in the *in vitro* and *ex vitro* stages have been given (refs. 4, 9). A good descriptive review of this area has recently been done by Young et al. (ref. 10). Some of the more straightforward aspects of tissue culture, such as media preparation, handling of containers in the laboratory, and handling plantlets in the greenhouse, are easier to automate than the *in vitro* stages.

This article will explore recent progress on automation and steps towards automation relating to plantlet production via the organogenesis method. Embryogenesis is discussed by Preil (ref. 11). With the exception of automated media preparation and computerised management systems, very few of the developments in automation technology outlined in this article are being used in commercial practice. When the term 'automation' is used some man-power is always required.

2. AUTOMATION IN LABORATORY AND GREENHOUSE MANAGEMENT

Computer systems have been developed to assist tissue culture laboratories and greenhouse managers with the smooth handling and flow of large numbers of plants, clones, and containers at both the *in vitro* and *ex vitro* stages (refs. 12-15). Once programmed and if kept up-to-date, computer systems can provide immediate information on clones and subculturing records, and allow for easy retrieval of information on the status of orders and numbers of plants lost due to contamination, and can also save time.

The minimum amount of information needed for each container at the *in vitro* stage was found by Wolf & Hartney (ref. 12) to be:

(a)   what medium is in the container;

(b)   when the medium was prepared;

(c)   what plant is in the container;

(d)   when the plant was placed in the container;

(e)   when the plant is due for subculturing to fresh medium;

(f)   where the plant containers are located, e.g., position on shelves in the propagation room, plant containers in cool storage.

This system was developed for tissue culture when many clones were being propagated, e.g., for forestry species. Wolf & Hartney also found that by having the computer system, errors, such as placing plants on the wrong medium and misidentification of clones, were more easily detected where a laboratory was producing certified lines or cultivars of a particular plant.

Hemple's computer model had added features over <u>Wolf & Hartney's</u> and could account for both *in vitro* and *ex vitro* stages for any micropropagation method (ref. 15). The program enabled:

(a)   control of production, e.g., daily labour and material requirement;

(b)   control of production parameters, e.g., efficiency of workers;

(c)   production prognostics — answers to questions such as:

   (i)   how many plants can be produced in a particular time period from stock cultures present?

   (ii)   how many stock cultures must the laboratory have to produce the required number of plants in a particular time period?

   (iii)   when can the required number of plants be produced from the current stock cultures?

   (iv)   what changes in technology can be introduced, and at what stage to enable (a), (b), or (c) above.

Another model (ref. 13) has further optimised the control of flow of plant material to better meet fluctuating market demands while adhering to constraints set by contamination and cost. The computer system must be compatible with the organisation and handling systems used in a laboratory and greenhouse. Some commercial laboratories have their own personal computer systems for management and for providing production information.

3.   AUTOMATED ASSESSMENT OF TISSUE CULTURES AND PLANTLETS

Assessment of numbers, health, and growth during shoot initiation, elongation and multiplication stages *in vitro* and during rooting or acclimatisation *ex vitro* are often time consuming activities whether in a research or a commercial facility. Automated image analysis by machine vision has been developed for the objective analyses of plant experiments *in vitro* and *ex vitro* (ref. 16), for the grading of tissue-cultured plantlets (ref. 17) and for use in conjunction with robotics and the automated system developed by Commonwealth Industrial Gases (C.I.G.), Australia (<u>see</u> Section 4.11). A machine vision system for grading pine tree seedlings was developed and tested at a commercial nursery by <u>Rigney & Kranzler</u> (ref. 18). Automated measurements of stem diameter, shoot height, projected root area, projected shoot area, shoot-root ratio, and sturdiness ratio were done. Image analysis was also used to measure the area increment growth by meristematic nodules in an experiment on nutrition by <u>Aitken-Christie et al</u>. (ref. 19).

Other features that could be analysed by machine vision include disease recognition by detecting colour variations (ref. 17). Open vision systems used in conjunction with personal computers could be used for monitoring growth and recording experiments as well as vision processing. They also have versatility in their application, for example,

(a)   at the output end of a bioreactor for statistical quantification,

(b)   in a greenhouse for grading,

(c)   on a microscope for computer-aided inspection.

An area for research and development would be a system for inspection of cultures for contamination, browning, and death. Also, better systems for analysing complex 3-dimensional, variable growth of cultures are needed.

4.   AUTOMATION: LABORATORY AND *IN VITRO*

4.1   Sterilisation of explants

Sterilisation of explants is the critical first step for the successful tissue culture of any species. In comparison with the rest of the *in vitro* and *ex vitro* tissue culture operations, the labour requirement for sterilisation is generally small. Sterilisation of limited supplies of plant material or seeds may normally be carried out several times a year when material is available or in the right stage of development. Successful sterilisation is more important than automation of this step. Moreover the cost of sterilisation relative to the total cost of micropropagation is not high.

In some situations where a large amount of material needs to be sterilised, e.g., large numbers of clones or large amounts of starting material, then automation may save time and costs providing it can be done with at least the same success rate as manual sterilisation.

An automated system for the surface sterilisation of date palm (*Phoenix dactylifera*) and pumpkin (*Cucurbita pepo*) seeds and for dill (*Anethum graveolens*) petioles was developed by Maurice et al. (ref. 20). Agitation of explants, changing of sterilising solution, sterile water rinses, and timing of sterilisation were automatically controlled by a microcomputer. This system consisted of a sterilising chamber, large reservoirs of sterilising solution and sterile water, pumps, tubing, and a computer controller. It was found to be as successful as manual sterilisation for pumpkin and dill explants, and more effective (less contamination and more germination) for date palm seeds than the manual method when Alcide and NaOCl were used as the sterilising agents. Though no cost for the system was given, the authors claim that it was inexpensive. Handling of health-hazardous sterilising agents such as sodium hypochlorite and mercuric chloride was eliminated. There do not appear to be any other reports of automated sterilisation systems. This may be an indication of the low level of interest in automating surface sterilisation of explants.

4.2   Media preparation

Media preparation is ideally suited for automation, especially where large volumes of media are prepared and thousands of small containers are used in commercial production laboratories. An automated media preparation and dispensing system (ClonMatic) and new Vegbox container, which were initially developed by Fari (refs. 21, 22), have been further developed by Plant Production Systems B.V., The Netherlands (Figure 1). With this system a productivity increase of 50% over the traditional manual method was possible (A. Broekman, personal communication). Up to 120 different ingredients and 999 recipes could be handled by the nutrient and chemical dispensing and mixing system. From 50 to 500 litres of media could be automatically prepared from concentrated media and sterile water (1:10). The ClonMatic dispensed media in three stages at a rate of 700 Vegbox containers per hour at a sterile workbench:

(a) sterilised media was dispensed into the base of the tissue culture container which came from another dispenser;

(b) the filled vessel was passed along a moving belt;

(c) a third dispenser placed the lid on.

The volume dispensed into the container could be adjusted from 1 to 125 ml and 50 litres of media per hour could be processed. A specially designed non-drip tap was used to avoid splashing medium on the sides of the container. The Vegbox container (6 cm high x 7 cm wide x 11 cm long) was made of clear Durovinyl, recyclable by gas sterilisation. A 13-cm-high container of the same width and length was currently being produced as a result of demands from the commercial market.

Fig. 1. Automated kitchen and dispensing system initially developed by Dr M. Fari and staff and further developed by P.P.S. BV, The Netherlands (courtesy of A. Broekman).

Lab Associates, The Netherlands have also developed an automated system for the preparation, sterilisation, and dispensing of media (J. Bambelt, personal communication). Medium is prepared and autoclaved in a large 50 or 100 litre stainless steel vessel on wheels. The lid has a stirrer attached to it and openings for the introduction of ingredients. Medium is sterilised by steam and a valve at the base of the vessel allows sterile filling of tissue culture containers after connection to flexible autoclavable tubing. Medium is dispensed into containers with the aid of a "Touch-o-Matic"-type of switch. This system has the advantage that only one vessel is used for media preparation and sterilising and that it is transportable around the laboratory.

Another automated system for media preparation, sterilising, and dispensing has been developed by Floratiss, a commercial tissue culture laboratory in The Netherlands (G. Zuidgeest, personal communication). Additional features of the Floratiss system over the ClonMatic system were:

(a)   chemical weights for nutrient stock solutions were checked automatically by a computer attached to the balance;

(b)   autoclaving of media was pre-programmed to start in the early hours of morning so that when staff arrived the media was ready to pour.

For the annual production of over 5 million plants only one person was required in the media kitchen.

Although their methods are unpublished, other laboratories may have developed their own automated methods for media preparation suited to their individual needs. Currently there are automatic dispensing units available for pouring media into petri dishes that rely on peristaltic pumping action to deliver media. These systems require manual container handling and are slow when volumes greater than 100 ml per container are dispensed.

Media prepared for the above automated systems were all sterilised by autoclaving in large vessels. Media could also be filter sterilised and still used in some of the automated dispensing systems. Recently, sterilisation by an electrical method of inserting an electrode into the culture vessel and culture medium, as an alternative to autoclaving, has been developed (ref. 23). A current flowing through the liquid killed any contaminants. Another electrical method which heats and sterilises media in a continuous flow system has been developed by Tasman Forestry Laboratory, New Zealand, and was recently commissioned (B. Nairn, personal communication). These other methods for sterilising media may be more amenable to automated media preparation and dispensing because there is no need for large autoclave units. However, there is the disadvantage that a separate system to sterilise glassware and various containers will normally be required as well.

If bioreactors or other automated systems for tissue culture in larger containers are developed, automated media preparation may diminish in importance because there would no longer be the need to dispense media into thousands of small containers. Media would still need to be prepared, sterilised, and dispensed to feed the bioreactors.

4.3   Liquid media and supports

Shoots, organs, and plantlets grown in liquid nutrient media, with or without supports, are suitable for automation because spent media can be changed to fresh media automatically, handling costs can be decreased, and large-scale liquid cultures in bioreactors can be developed. Thus, research on new supports and machines for shoot culture in liquid media which facilitate good nutrient uptake and aeration of tissues and which avoid vitrification and contamination, is important for automation. There have been several advances in this area recently.

Grape (*Vitis vinifera*), *Fuchsia*, *Amelanchier alnifolia*, and tobacco (*Nicotiana*) shoots were grown and aerated in liquid media in 50- and 125-ml Erlenmeyer flasks on two prototype machines developed by Harris & Mason (ref. 24). Agitation of cultures was obtained by tilting for one machine and by rocking for the other. Shoot tips were first grown on agar, then transferred to liquid when they were larger. Small shoot tips cultured directly in liquid medium appeared to vitrify. Growth was seven times better in liquid than on agar after 90 days. Explants grew rapidly in liquid media on the tilter machine and soon became too large to remove through the neck of the flasks and nutrients became depleted. Obviously a larger container was necessary. In addition, the costs were lower with liquid since the volume of media was reduced by 50% and the agar by 90% or more. Non-agitated liquid in jars has been used to culture poplar (*Populus*) shoots without support (B. Christie, personal communication) and also horticultural crops in commercial laboratories (W.C. Anderson & P. Myers, personal communications). These methods of growing shoots in liquid are suitable for species that are not sensitive to vitrification and can be cultured continuously in liquid.

Liquid shake cultures have been developed for numerous species including *Lilium* and *Begonia* (refs. 25, 26) and *Gladiolus* (ref. 27). In the latter, growth inhibitors and a polymer mixture "K" were used to stimulate cormlet development to avoid vitrification and the need for acclimatisation.

A double-layer technique of placing a liquid nutrient layer on top of the spent agar base was used to replenish nutrients for growing *Cordyline*, *Philodendron*, *Magnolia*, and *Spathiphyllum* shoots (ref. 28), for pear (*Pyrus*) shoots (ref. 29), for shoots of 28 other species (ref. 30), and for *Pinus radiata* (ref. 31). In all experiments shoot growth was better than controls on agar and rooting was improved with Maene & Debergh's cultures. With radiata pine, in order to avoid vitrification liquid was not left on the agar (see 4.4). Vanderschaeghe & Debergh (ref. 32) emphasised the need to use bottom cooling with their cultures to avoid vitrification. It has been suggested that increased availability of nutrients and washing away of toxic substances may have been responsible for the improved growth response with liquid feeding.

A "sandwich" system of growing shoots in microporous polypropylene membrane supports (Celgard by Hoechst Celanese, USA) floating on liquid medium and training the growth through polypropylene netting was devised by Young et al. (ref. 10). This system was designed as a step towards automation of liquid feeding and mechanised harvesting of shoots that grow in an upright fashion. During the early stages of the study, 30 support materials ranging from paper to plastics to wire screens were tested. The five materials subsequently selected for further testing were polyurethane foam, non-woven polypropylene fibre, glass beads, microporous polypropylene membrane, and polypropylene netting. A combination of the last two proved to be the most promising. During these experiments with tomato (*Lycopersicon esculentum* cv. Rutgers) and sweet potato (*Ipomoea batatus*) shoots, particular attention was paid to sinking of supports, sterilisation techniques, the absorption capacity of supports, and avoiding contamination and desiccation. The authors envisaged that the sandwich device might become sufficiently cost-effective to be disposable after two to five mechanical harvests of shoots and after the shoot bases had aged.

The height of shoots produced by the sandwich system after mechanical harvesting may be variable resulting in wastage of shoots. Tissue-culturists and nurserymen need a relatively uniform product to produce a "quality" plant and for streamlining operations. Small shoots and the tips of shoots and leaves harvested by a mechanical process could be discarded if not suitable for rooting or recycled for multiplication. This problem was also encountered by Aitken-Christie & Jones (ref. 31) when they simulated mechanised harvesting by a mowing device for radiata pine shoot hedges. Selective harvesting of shoots with scissors was done to avoid wastage of shoots. However, elongation may become more uniform after regular cropping. The long-term growth of tissue cultures for at least 6 months on the membrane boats with liquid media needs to be assessed for vitrification and for plant growth and behaviour in such a system.

A variety of other tissue culture supports for use with liquid media have been reported for other species, e.g.,

(a)   non-woven material for supporting banana (*Musa*) protocorm-like bodies (ref. 33);

(b)   microporous polypropylene membrane (Celgard 3500) for supporting asparagus (*Asparagus officinalis*) cells (ref. 34) and *Atropa belladonna* shoot cultures (ref. 35);

(c)   capillary support of water-absorbing material with small wells in the bottom of the culture vessel for tissue cultures (ref. 36);

(d)   viscous slurry or granule support covered with cellulose filter for supporting callus, shoots, and plantlets. A hole was made in the support with a slit towards the outside for shoots, and plantlets (ref. 37);

(e)   floating water-absorbing supports for *Liliaceae* cultures (ref. 38).

This is not a comprehensive list but shows a few examples of different types of supports that could be used for automation of liquid media exchange.

The use of membranes to increase gas exchange of the container, similar to that described by Kozai (refs. 39, 40) may enable greater flexibility with the use of liquid media. Increasing the number of air exchanges using membranes in the lid, may avoid shoots becoming vitrified with long-term continuous liquid culture. Species that may have previously been sensitive to vitrification may grow better in liquid medium with the increased gas exchange between shoots or plantlets and the air.

4.4   Liquid feeding

A number of automated systems based on the addition and removal of liquid media have been developed. One of the first systems, described by Tisserat & Vandercook (ref. 41), consisted of a culture chamber for plant growth with inlets and outlets for nutrients. Nutrients were pumped in and out by peristaltic pumps and the system was controlled by a microcomputer. The same principle has been used in other automated systems under development since:

(a)   the Mega-yield system (ref. 42);

(b)   the Phytocultor system (ref. 43);

(c)   Moët-Hennessy's automated system (J.P. Barbe & F. Brenkmann, personal communication);

(d)   the automated system for use with the *in vitro* hedging system (ref. 44);

(e)   the automated system by Bio-tron Company, Japan, which was demonstrated at the Hortimation Exhibition, Tokyo, 9– 11 May 1988;

(f)   the automated system by Komatsu Ltd, Japan, which was demonstrated at the Hortimation Exhibition, Tokyo, 9– 11 May 1988.

These systems enabled shoot growth and multiplication in some trials to be continued without the labour, disturbance, and trauma associated with transfer and cutting of shoots.

Another system for the automatic introduction of liquid media to spent media was developed by Vanderschaeghe & Debergh (ref. 45) for *Prunus avium* shoot cultures. A transportable injection apparatus was developed to allow the injection of liquid medium directly through the lid of the container. There was no need to transport containers to the laminar flow cabinet and injection of media could be done in a culture room without contamination. This method could be used for the addition of nutrients to spent media for shoot cultures that would not need cutting, even if it could avoid one or two transfers. It could also be used to add liquid media containing auxin to single shoots at the pre-rooting stage.

An automated misting system, the Mistifier, was developed to grow shoots and plantlets *in vitro* (ref. 46). A fine nutrient mist, which was produced by a submerged ultrasonic transducer, sprayed tissues growing on a mesh support, with excess media draining away continuously. Gas mixtures were used to alter humidity and distribute mist throughout the growth chamber. Higher shoot numbers and lengths were recorded for banana (*Musa* spp.), *Cordyline*, and Boston fern (*Nephrolepsis* spp.) in the Mistifier after 18 days than for traditionally grown cultures on agar (ref. 47). The effects of nutrient misting and increased gas exchange on *Pinus radiata*, a vitrification-sensitive species, has recently been tested (C. Gould & P.J. Weathers, personal communication). With both wet (partially vitrified) and waxy shoots, the levels of vitrification could be reduced with increased gas exchange. Hairy root cultures have now been grown in the Mistifier (ref. 48). Growth of hairy roots was better in the Mistifier than in flasks or in an airlift bioreactor. This was thought to be because the Mistifier provided a more gentle culture environment for the growth of fragile root tissue and root hairs and for the production of secondary products. Automated misting and aeration systems require further trials.

Results on long-term effects of growing shoots or plantlets in automated liquid or mist systems are generally lacking. Several points and questions need addressing before commercial applications are more common:

(a)   what are the effects of multiplying shoots for several cycles under these conditions?

(b)   do they vitrify with time?

(c)   dynamics of crowding and shoot form

(d)   contamination experience and harvesting

(e)   can these plantlets survive transfer to soil?

The short-term effects of liquid medium on growth of tissue cultures are generally beneficial and are well documented for many species. However, although shoot height, leaf growth, multiplication, and rooting may increase with liquid feeding *in vitro*, there can be subtle changes in leaf structure and function.   These changes can be worse in sensitive species and later will result in vitrification, poor acclimatization, and sometimes death. For sweet gum (*Liquidambar styraciflua* L.) shoots, the percentage rooting *in vitro* and number of roots per plantlet were better in liquid medium than on agar medium, but the plantlets in liquid had a higher water content and succulent leaves (ref. 49). During the development of the liquid feeding system for *Pinus radiata* shoot hedging, particular attention was paid to avoiding vitrification and growing top quality shoots for long periods (ref. 31). This was achieved by not leaving the shoot bases permanently in liquid medium. Liquid nutrients were left on top of the agar base for about 6 hours at a time, either once or three times a week. Containers for liquid feeding with gas exchange membranes, air inlets, or special filters may avoid vitrification in species where it is a problem and may improve leaf wax development, stomatal structure, and stomatal function.

There are many differences in the automated liquid feeding systems: methods of sealing and type of lid; location of entrance to container; relationship of lid size to container size; type of support system for shoots or plantlets; form of nutrients supplied; positions of entrance and withdrawal of nutrients; frequency of nutrient application; whether nutrients were recycled or not; type of pump; type and size of nutrient reservoirs; and controls for the system. Features that all containers have in common are that they are clear, autoclavable, and larger than conventional tissue culture containers. It is too early in the development of automated liquid feeding systems to know which features are important. None of the systems has undergone extensive testing yet; therefore, valid comparisons cannot be made. However, a good descriptive review has been done by Young et al. (ref. 10).

Research to assess shoot growth and quality (vitrified or not), numbers of shoots or plantlets produced, rootability of shoots, survival in the field or greenhouse and the cost of plantlet production still needs to be carried out for automated liquid feeding systems. The production rate from an automated system should be equal to or higher than conventional tissue culture to offset the development cost. The quality of shoots and plantlets must be equal or superior to those produced by conventional tissue culture methods.

All of the automated liquid feeding systems described here rely on nutrient media containing sugar. Automated liquid feeding under sugar-free conditions may be worthwhile investigating to reduce problems with contamination and to avoid the need for clean-rooms (refs. 39, 40).

Automated liquid cultures in bioreactors or fermentors have also been used to culture lily (*Lilium*) bulblets (Mitsui Petrochemical Company, personal communication), potato

(*Solanum tuberosum*) microtubers (ref. 50) and *Gladiolus* bulbs (ref. 51). Aeration was used
in both systems. Bioreactors are covered in more detail by Dr Preil (ref. 11) and by
Takayama (ref. 52).

4.5   Hedging

The concept of hedging and stoolbeds for the maintenance of mother clones and for the
production of cuttings is well known in plant propagation. Using *in vitro* cultures as small
stoolbeds was first suggested by Zimmerman (ref. 53), and was done with rhododendron
(*Rhododendron* sp.) and apple (*Malus domestica*) cultures. Hedging with the aid of liquid
feeding for continuous culture has also been done with *Pinus radiata* shoots (ref. 31).
Monthly crops of rootable shoots were produced continuously in the same container for up
to 18 months. A similar system was developed for oak (*Quercus robur*) (ref. 54). Growing
shoots in an upright mode from a base appears to be amenable to automation because the
base can be fed nutrients automatically and rootable shoots can be selected and cut
horizontally using either a mechanical harvesting device or a robot. Species that are
multiplied by topping or have nodes with axillary buds are suitable for hedging.

When scaling up the *in vitro* hedging system, from a 600 ml jar to an automated liquid
feeding system in a 25 x 39 x 12 cm container, the growth habit and health of shoot hedges
differed for both radiata pine and apple shoots in the larger container (J.Aitken-Christie &
H.E. Davies, unpublished results). There did not appear to be any nutritional or health
problems with radiata pine hedges because preliminary work using small jars had been
done (ref. 31). However, there were nutritional problems with the apple cultures. Growth
habit differed in the larger container for both species. Therefore, for adapting the system to
larger containers suitable for automation, further research on the dynamics of shoot growth
and nutrition of hedges is necessary. The composition of the gas phase may also be
important (P. Debergh, personal communication).

4.6   Homogenisation

As an aid to automation of the *in vitro* process of multiplication, homogenisation has
been successful with ferns, in particular. The method is now being more widely applied.
Homogenisation is successful when tissues are cut or broken at random and subsequent
plantlet regeneration occurs at an acceptable rate. This technique was first done with
staghorn fern (*Platycerium*) and rabbitsfoot fern (*Davallia*) (ref. 55). Fern shoots and
runners grown on agar medium were placed in a blender with lukewarm (40°C) ungelled agar
medium and these were homogenised into small pieces and then poured into containers to
develop into another crop of fern shoots and plantlets. Fern fragments were scattered
throughout the agar. Some damage and tissue death occurred, but this was not detrimental
to the growth of another cycle of shoots. The same results were obtained by blending with
sterile water and dispensing an aliquot over solidified medium. The pieces containing
meristems and runner tips proliferated to form more shoots and the older more
differentiated fragments of fronds, which were not organogenic, died. Survival of fern

plantlets of both species in the greenhouse was 80%. Cutting of fern tissues in the laminar flow cabinet was avoided by homogenisation, although media preparation, the use of small containers, and transfer of tissue from the blender to new containers was still required.

This simple form of automation has been successfully applied to Boston fern and fishtail fern (ref. 56). Liquid Murashige & Skoog multiplication medium (MSMM) was used to homogenise the shoot and runner cultures. Small aliquots of the mixture were transferred to fresh solid MSMM medium. Fern cultures could be transferred to the greenhouse after 4 to 6 weeks. Janssens & Sepelie (ref. 57) have also recently applied the homogenisation technique to two additional fern species, *Pelea rotundifolia* and *Blechnum brasiliense*. Two to three grams of cultured prothalli were homogenised at 3000 rpm for 5-10 seconds in 20 ml of liquid Knudson's medium. Homogenisation of fern cultures has also been demonstrated in Australia (H. Van der Staay, personal communication).

Fast-growing healthy shoot cultures of *Begonia x hiemalis* were roughly cut to give smaller clumps of material and then homogenised in a blender (ref. 58). The best time and speed were 5 seconds at 1000 rpm. Homogenised tissue was filtered off and then inoculated on to solid or liquid medium. A homogenisation liquid of Murashige & Skoog (MS) medium gave 70% survival of tissue. After homogenisation, growth on liquid or solid MS medium gave 80% survival and agitation of tissue was essential to prevent vitrification. An output of 600 transferrable shoots was obtained from 2–3 g of tissue in approximately 8 weeks. This was twice that of conventional *in vitro* methods. Planting out in the greenhouse gave 90% survival and true-to-type offspring.

Homogenisation of callus was performed at a high speed of 10,000-30,000 rpm by Oji Paper (ref. 59) and plantlets were subsequently regenerated. This was done with a variety of forestry, fruit-tree, flower-tree, flower, and vegetable species.

Wider application of the simple homogenisation technique to woody species has not been demonstrated. One of the possible reasons for this may be that woody species and higher plants may produce more phytotoxic or phenolic compounds than other species as a result of tissue damage and trauma caused by homogenisation. Modifications to the blender or method, such as (a) placing silicone tubing over the blades for blunt blending to avoid cutting but still mixing and separating clumps of tissue, (b) using a more viscous homogenising fluid like the hydrolysed agar liquid nutrient solution developed by Boxus & Paques (ref. 43), and (c) choice of tissue in the right developmental state, may be worthwhile. Preliminary trials with radiata pine meristematic nodule cultures have shown that use of blunt blades for blending for 2 minutes in Lepoivre (LP) nutrient medium containing BAP and hydrolysed agar was best (B. Bergmann & J. Aitken-Christie, unpublished results). However, further research is necessary to improve survival of initial homogenised pieces of nodule tissue plated out, and their subsequent proliferation, to more than 25% after 10 weeks. Survival of material treated without tubing on the blades and without hydrolysed agar was 0% after a similar period. The addition of charcoal or other compounds to absorb phytotoxic material and the washing of tissue after homogenisation may be helpful. The technique can also be applied to *Forsythia intermedia*, another woody plant (A. Stradiot, unpublished results).

After homogenisation the tissues are covered in a liquid film. This may prevent good outgrowth and induce vitrification, even on agar medium. Steps to dry out the liquid film before placing tissue on the growth medium, or to culture tissues in a vessel with increased gas exchange, may be worthwhile.

4.7 Nodule culture

Meristematic nodules are ideal tissue for automation for a number of reasons:

(a) They can be proliferated in liquid medium.

(b) The organisation of tissues lends itself to grouping, mechanical division, separation and screening for size uniformity.

(c) The mode of proliferation and separation is by the formation of new nodules, therefore the tissue is relatively homogeneous from one subculture to the next.

(d) Nodules often do not need or like cutting, therefore one less step is involved.

The potential of nodules for automated tissue culture was reported for poplar (Populus) (ref. 60) and for *Pinus radiata* (ref. 19). Meristematic nodules or similar structures have been successfully grown in liquid culture for garlic (*Allium sativum*) (ref. 61), for poplar (ref. 60) and for daylily (ref. 62). Meristematic nodules or similar structures of most other species, *Prunus* (ref. 63), lily, *Freesia*, tomato, pea (*Pisum sativum*), *Nicotiana rustica* (ref. 4), banana (Musa) (ref. 64), white spruce (*Picea glauca*) (ref. 60), hybrid poplar, and *Eucalyptus deglupta* (K. Ito, personal communication), *Eucalyptus grandis* (ref. 65), and *Pinus radiata* (ref. 19), were best grown on agar-gelled media. Radiata pine nodules grown in liquid media became vitrified and died within 4 weeks (ref. 19). Improvements were made with modified media and the addition of hydrolysed agar (J. Aitken-Christie & B. Bergmann, unpublished results). For species where nodules were prone to vitrification in liquid, automated systems or bioreactors with increased aeration and semi-continuous liquid may be worthwhile, for example, the Mistifier developed by Weathers & Giles (ref. 46) or the culture vessel with a spraying system and a funnel-shaped support developed by Mitsui Petrochemical Industries in Japan (ref. 66). However, to date, nodules or clumps of meristematic tissue have been grown in only one automated system (ref. 67), section 4.10.

4.8 Encapsulation

Encapsulation of somatic embryos to produce artificial seeds is a method being widely pursued because the seeds could be sown by a mechanical seed sower in a glasshouse or nursery. Shoot buds have also been encapsulated and plantlets regenerated (ref. 68). Mulberry (*Morus indica*) axillary buds with a portion of stem present, were encapsulated in 4% sodium alginate prepared with MS medium and rooted on agar medium, filter paper, or vermiculite. Addition of a fungicide, Carbendenzim, prevented contamination and increased survival of encapsulated buds when transferred to soil (ref. 69). The authors claimed that encapsulated buds could be handled in a similar manner to seeds and that they could be efficiently transported from one place to another in small bottles. Large parcels of mulberry cuttings were more bulky and difficult to transport, and only 30–40% survived the pruning, transportation, and final transplanting.

This encapsulating method for shoot tips or buds has also been applied to *Valeriana wallichii* and growth from encapsulated buds was achieved *ex vitro* (ref. 70). It also used less media and less space, and saved time at the hardening-greenhouse stage compared with conventional tissue culture methods. This approach is new for shoot buds and in future could be applied to a wider range of species.

4.9   Sugar-free (photoautotrophic) micropropagation

Contamination and vitrification are two of the main problems facing commercial tissue culture laboratories and automation and liquid feeding research. These problems have been addressed in more detail elsewhere (refs. 71, 9, 32 and 72). Sugar-free tissue culture in containers with a special gas-permeable membrane may help overcome these problems (Kozai, this volume) and allow automated liquid feeding in larger containers with a variety of tissue cultures.

Automated sugar-free systems were developed in larger containers by Fujiwara et al. (refs. 73, 74). They could be used to culture shoots or plantlets for multiplication, rooting, and acclimatization under aseptic or semi-aseptic conditions using a microhydroponic propagation system with environmental control. These systems consisted of a large culture box (28 cm wide by 53 cm long by 12 cm high), a gas flow assembly, and a nutrient feeding assembly. Strawberry (*Fragaria*) shoots cultured in the system under high light and sugar-free conditions rooted, grew faster than controls on agar medium containing sugar, and had a higher net photosynthetic rate than controls after 28 days. Further work was necessary to modify the humidity control system, nutrient solution composition, and structural design because shoot growth was not optimal. Robotics are being developed for use in conjunction with this system (ref. 40).

To counteract the extra costs of lighting and cooling incurred when using high light, a novel lighting system that used optical fibres for providing light to cultures was developed (ref. 75). Cooling costs could be reduced and the number of shoots or plantlets per culture room could be increased. Natural light could also be used as a light source to reduce lighting costs. Shoots without roots grew to be rooted plantlets faster in a controlled growth unit under natural light than under artificial light in a growth room (ref. 76).

So far, sugar-free tissue culture with or without automation has been restricted to herbaceous species; application of the system to fruit trees and forest trees has not been demonstrated. The potential of this type of automated system may be restricted to shoot cultures with leaves and chlorophyll present because they are already in the right physiological state for photoautotrophic growth. Small axillary buds, meristematic tissue, and nodules do not respond well to increased light and CO_2 on sugar-free media (R. Levin, personal communication).

4.10 and 4.11 are examples of different types of automated systems that have been developed for the entire tissue culture process from tissue multiplication *in vitro* through to plantlet growth *ex vitro*. There will be many more examples in future but, because they are often commercially sensitive or patents are pending, information is not readily available.

4.10 Automation at Plant Biotech Industries and at Albright and Wilson Ltd

Dr Levin and colleagues of Plant Biotech Industries (PBI), Israel, have developed an automated system based on scaling-up the homogenisation procedure (ref.67). The handling of plantlets in the greenhouse is integrated with the laboratory multiplication process. When scaling-up the homogenisation, it was necessary to dilute the homogenate for easier dispensing, increase uniformity of the inoculum by sieving out debris, and dilute toxic substances out by washing. Details of this process have been described by Levin (ref. 77). A 10- to 50-litre liquid culture bioreactor with aeration was used for growing tissue, a bioprocessor sorted and dispensed tissue and media at a rate of 100,000 propagules per 8 hours, and a transplanting machine placed plantlets in soil at a rate of 8000 plantlets per hour. This system has worked with a wide variety of species including ferns, asparagus, potato tubers, African violets (*Saintpaulia ionantha*), *Spathophyllum*, *Syngonium*, *Philodendron*, lilies, cherry tomatoes, and radiata pine. Field testing of many plants has been carried out. It was estimated that this system saved 60% of the total cost per plantlet. The average cost to produce the most difficult plant was US$0.12 per plant in 1989 (R. Levin, personal communication).

Avoiding vitrification and contamination would be two main concerns of the PBI process. When thousands of propagules were handled in the bioreactor and bioprocessor the risk of losing them to contamination became high. Contamination may not be apparent at the time of transfer and may not be evident for several weeks. Therefore, precautions and careful screening were necessary. PBI has developed its *in vitro* system for operation in a clean-room where all the air has been Hepa-filtered and staff wear sterile gowns and masks. Cultures were screened for contamination several weeks before homogenisation by adding 0.5 ml aliquot of liquid or solid medium that surrounded tissue from the multiplication stage to 5 ml of three separate indexing media. The three separate indexing media were: yeast extract plus dextrose (10 g per litre) broth; Sabouraud liquid and activated charcoal broth; and trypticase soy broth (ref. 77). These extra costs should be taken into account.

Some species were more sensitive to liquid culture than others and became vitrified easily. Therefore multiplication in liquid medium in a bioreactor even with aeration and mixing might not always be suitable. Steps to avoid vitrification by using agar beads in combination with aeration and liquid gave encouraging results (R. Levin, personal communication).

Clumps of meristematic nodules were selected as being the right state of development for automation by the homogenisation, sieving, and dispensing method in the Israeli process. Meristematic nodule cultures were also selected for their uniformity and homogeneous nature, and for having good potential for automation by Aitken-Christie et al. (ref. 19) and McCown et al. (ref. 60).

Albright and Wilson Ltd (ref. 78) have also developed an automated process for growth of cells in liquid medium, sorting and placement of cells capable of plantlet regeneration, and separating plantlets. Interesting features of the process were that homogenisation was used to separate the cells ready for sorting, cellulosic pulp support was used as the plant propagation medium, and rotating or reciprocating blades could be used as an option to recycle shoot tips. It was not mentioned which species worked with this process.

4.11   Automation at Commonwealth Industrial Gases and at Twyfords

Commonwealth Industrial Gases (CIG), Australia, has developed an *in vitro* automated system that has machine vision image analysis, with automatic cutting of shoots and transfer to new containers or to a transplant plug (B. Johnson, personal communication). The shoot must have an identifiable top and bottom and should have visually identifiable junctions (meristems or nodes) in order for the computer to make appropriate choices and cuts at viable distances towards the base from nodes. Cut pieces are transferred to a sealed, sterile, growing area by a pneumatic system and a grasping tool (ref. 79). The CIG system has been developed primarily for *Pyrethrum* and woody plants for forestry, e.g., *Eucalyptus*. To be cost effective, such an automated system would need to operate at rates in excess of 20 million plants per annum. Individual clones handled would need to be on a scale of at least 1 million per annum to warrant the development and quality assurance work involved. CIG are developing the system and are planning to construct their first factory, mainly for *Pyrethrum*, by the end of 1989.

Critical issues requiring further research are:

(a)   contamination control;

(b)   reproducibility – plants should be grown in a manner as repeatable as possible; and

(c)   understanding the physiology of the cultivar or clone to achieve commercial optimisation.

A manual tissue culture facility for plant initiation, process and method development, quality assurance and testing, and mother stock maintenance was found to be still necessary.

Twyford Plant Laboratories Ltd, UK, now Twyford International Inc, USA, also developed new apparatus and methods for cutting and moving plant tissue (ref. 85). Their system was comprised of a tool with unique cutting and planting devices which could, after cutting, orientate an explant in an upright position in the medium. This device could be used as a manual tool or in conjunction with a semi-automated robotic system.

Of particular interest with both of these systems would be the ease of changing, cleaning and sterilising the cutting systems to avoid cross contamination, by bacteria in particular.

4.12   Robotization *in vitro*

4.12.1   *In vitro*. One of the first tissue culture robot prototypes was developed in France by Deleplanque et al. (ref. 80). It was used to pick up and transfer *Eucalyptus* plantlets *in vitro*. It was based on an image analysis vision system for the detection of suitable plantlets and a manipulator transfer arm to pick and plant the plantlets. Special spongy material was used on the robotic "hand" to avoid tissue damage. The prototype could theoretically transfer 1000 plantlets per hour (or 3.6 seconds per plantlet) but there is no record of the actual performance. This speed was similar to the speed for an *ex vitro* or non-sterile robot described by Miles (ref. 81) and differs from the *in vitro* tissue culture robots developed in Japan that transferred, or cut and transferred, at a speed of 20-60 seconds per plantlet (refs. 82, 83).

At the Moët-Hennessy conference in 1987, Dr Y. Miwa of Waseda University, Japan, received first prize in the competition for the development of a tissue culture robot that had a new type of recognition system, a shape memory alloy actuator for detecting the plant position in the agar medium (refs. 82, 84). Shoots produced an electric charge that was detected by the alloy plate in the base of the container. The robot could remove a 15-mm or larger shoot from one container of agar medium and transfer it to another container of fresh medium in an upright position. Colour image analysis was used to separate healthy shoots from unhealthy ones. A special "grasper" with three fingers like tweezers was used to hold the shoot. Using this robotic transfer system, it took 1 minute to pick up and replant a shoot. With improvements it could possibly be reduced to 30 seconds. This robotic system may be too slow at present for commercial use and is limited to operations where direct transfer of shoots is done.

Considerable progress towards detecting, cutting, and planting shoots using a robot has been made by Toshiba Corporation, Japan (refs. 39, 75, 83). The six-jointed robot located in a clean room 1) detected the position of nodal sections of the shoot, 2) cut off one nodal section at a time, and 3) planted each section into one fibre support (Figure 2). The average length of time to detect, cut, and plant each nodal segment was 20 seconds in 1989 (ref. 83) and is now 10 seconds with the new improved robot shown in Figure 2. Recognition of growth nodes was achieved by moving the sensor from the bottom to the top of the plant. The robot sterilised its grasping tool and cutter automatically. The Toshiba robot has been developed and tested for species that have upright shoot growth and clearly distinguishable nodes and leaves which need cutting horizontally, e.g. potato, carnation (*Dianthus caryophyllus*), *Eucalyptus*, and redwood (*Sequoia sempervirens*). Further research is expected to speed up the robot and check the quality of the nodal segments.

The application of robotics and vision systems for an automated tissue transfer system was also studied by Rowe et al. (ref. 86). Specifications for a robotic work station and process line were developed. Results with potato shoots grown *in vitro* indicated that once a shoot was identified, the manipulator should locate, cut, and transfer it without damage in less than 3 seconds. Difficulties were encountered with the proper sequence of shoot

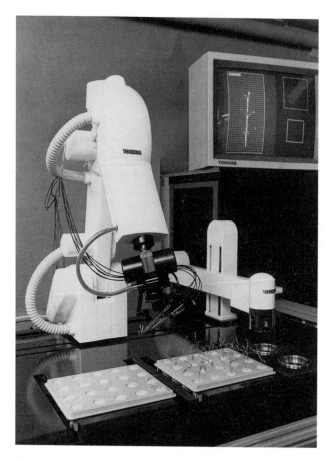

Fig. 2. Tissue culture robot developed by Toshiba Corporation, Japan (Courtesy of H. Watake).

excision from apical shoot downwards to the base of the shoot or plantlet and with the variable time needed to identify nodal cuttings.

An *in vitro* robotic transplanting system (Zymark-robot system) which counts, identifies, and transports tissue-cultured plant parts, which have been manually cut, to a robotic arm has been developed by Plant Production Systems, BV, The Netherlands (A. Broekman, personal communication). The robotic arm plants the plant parts (shoots or parts of shoots) into gelled media inside the Clonmatic (see 4.2). This system has been successfully tested for planting shoots of various species and up to 15 laminar flow cabinets can be connected to the transport system.

Other robotic systems have also recently been developed for the separation and transfer of tissue cultured lily bulb scales (Miwa, this volume), by Komatsu Ltd for bulk automated micropropagation using nodal sections (Kozai, this volume) and by Tillet (ref. 87) for the location, cutting and planting of tissue cultured *Chrysanthemum* nodes. Tillet addressed some of the problems associated with the handling of the many variable shapes of nodal explants. Of the 62% of nodes located correctly, 84% were planted correctly. The tool for planting nodes was similar to that used by Brown (ref. 85) of Twyford's.

4.12.2 General considerations. Robotics are currently one of the more elaborate and expensive ways of automating steps in the tissue culture process of plant production. Robotic systems involve the use of a vision or recognition system and an automatic transfer system for *in vitro* stages in the laboratory or for *ex vitro* stages in the greenhouse. Two plant tissue culture robots developed in Europe and described by De Bry (ref. 88) were quoted as costing at least US$100,000 in 1985. Sluis & Kahn (ref. 89) suggested that robots will cost at least US$2.5 million to develop and implement, while Kozai (ref. 40) mentions a robot for use with an autotrophic micropropagation system for several hundreds of thousands of dollars. Robots for packaging items in food production lines cost US$170,000 each (ref. 90). Despite their cost, tissue culture robots offer the elimination of unpleasant, boring, or unsafe work, reduced labour costs, increased productivity, reduced contamination, and continual service. A team of robots, however, would still need to be programmed and supervised by skilled persons.

During multiplication of shoots by conventional tissue culture, cutting of tissues with a scalpel or scissors is necessary and its use differs depending on species and on growth habit, e.g.,

(a)   to make nodal segments from elongated shoots;

(b)   to top shoots to stimulate further shoot development from the base and to use the tops for rooting;

(c)   to remove dead shoots or leaves that may affect healthy tissue;

(d)   to trim the base of shoots in contact with the agar to remove excess callus and unhealthy tissue;

(e)   to divide clumps of shoots into smaller pieces to facilitate good uptake of nutrients and phytohormones.

It would be difficult to develop a robot to do all these steps, particularly when decisions on whether to cut or not and where to cut can vary from one explant to another even in the same container. The perfect situation would be uniform shoots that never died.

The robots described so far have been developed to handle shoots that were grown using conventional tissue culture methods in small containers. The many steps that are necessary may cause problems for the robot. Recognition of the growth habits and new systems, some of which are described in this paper, that may be more amenable to a robotic system, speeding up the robot, and avoiding or reducing dead or contaminated tissue are of importance for future progress. The effect of cutting and transferring by a robot on shoot growth, the number of shoots that can be handled per hour, and the quality of plantlets produced by the robot warrants further research. The expense of a robot may not be justified if the production rate of shoots or plantlets is inferior. Some of the expense may be justified if a more uniform plant is produced.

5. AUTOMATION: *EX VITRO*

5.1 Rooting and growing-on

Very little work has been done on automation for handling unrooted tissue-cultured shoots, shoots with small root initials, or plantlets at the greenhouse or field stage, even though 25–40% of the cost involves these stages. This could be because current technology for cutting propagation and seedling production can be applied to tissue-cultured plantlets with minor modification. For example, automated misting or fogging systems for watering, fertilising, and maintaining high humidity, automated preparation and dispensing of rooting or soil mix into trays or plugs, mechanised trolley or pulley systems for moving large batches of plantlets in and out of the greenhouse, and automatic field transplanting systems are already available. Lannen Plant Systems, Finland, is one company that has developed many of these systems (A. Rogers, personal communication). There is scope for further efforts to mechanise the handling and planting of small, fragile tissue-cultured shoots or plantlets and make it cost-effective. Future developments in automation for cuttings and seedlings could also be applied to tissue-cultured shoots and plantlets.

The automated planting systems which have been developed for use with shoots and plantlets *ex vitro* by Levin et al. (ref. 67) and Johnson (see Section 4.11) may be suitable for transplanting other tissue-cultured shoots or plantlets produced conventionally.

5.2 Robotization *ex vitro*

The greenhouse or *ex vitro* stage of handling tissue-cultured plantlets by robotics should be simpler than the *in vitro* stages because there is no need for sterility. Savings should also be able to be made because by this stage plantlets are also more uniform than at the various stages of development during *in vitro* elongation and multiplication. Often during a sterile transfer, single rootable shoots have to be isolated from clumps of smaller shoots of varying size that require further elongation or multiplication, whereas by the time plantlets have formed only one stage of development has to be handled.

Robotic transplanting systems for small seedlings in the greenhouse are being developed to reduce the cost of transferring from one container to another (refs. 91, 81, Ting, this volume). These systems could also be applied to tissue-cultured plantlets in the greenhouse (ref. 92). Transplanting seedlings from a seedling flat to a growing flat using a Puma 560 robot was technically feasible (ref. 81). The average cycle time to transplant a 36-cell growing flat was 3.3 minutes or 5.5 seconds per seedling. The robot successfully transplanted 96% of the tomato and marigold seedlings. Machine vision and image analysis were necessary to detect plant features that were related to quality. Sorting and handling of tissue-cultured plantlets could be done using similar systems for determining leaf area, stem diameter, stem length, root area, number of roots, health (colour), etc. An advantage of this robot was that it could be quickly reprogrammed for other species.

At Williames Hitech International Pty Ltd, Australia, a robotic system has been developed for grading and transplanting millions of bedding plants, e.g., flower and vegetable seedlings (G. Williames, personal communication). A drum seed sower, with 30 different drums with different sized holes, was used to automatically sow the seeds into thimble-sized containers for germination. Eight rows could be planted simultaneously. After germination, seedlings were graded for diameter, size, and germination misses using machine vision analysis based on fibre optics and subsequently were selected and transplanted to larger containers using a robotic transfer system. This system has been developed in conjunction with Woodlyn Nurseries Pty Ltd, Australia, and some units have already been sold to lettuce (*Lactuca sativa*) growers in the USA. The robotic transfer system can handle a large range of seedling sizes and could possibly be adapted for tissue-cultured plantlets.

A robotic system is also being developed for processing geranium cuttings used in vegetative propagation (ref. 93). Operations include retrieving cuttings from a conveyor, trimming to size, stripping select petioles, and setting the finished product into a plug tray cell.

One robotic system developed specifically for transplanting tissue cultured plantlets in plugs has been developed by the Japan Tobacco Company (Kozai, this volume).

Most of the recent developments in greenhouse robotics have been for conventional propagation using seedlings or cuttings. These systems may be adaptable for handling tissue-cultured plantlets in the future.

6.   CONCLUSIONS AND FUTURE POSSIBILITIES

There appears to have been more emphasis on automation *in vitro* than on automation *ex vitro*. This does not truly reflect the relative percentages of cost for each stage. Further efforts to automate steps *ex vitro* should be easier to develop than *in vitro* automation because there is no need to carry out operations under sterile conditions.

The main areas of plant production by tissue culture that have been automated include automated media preparation and dispensing, and management in the laboratory and greenhouse by computer. Improved laboratory designs and equipment leading to streamlining of operations have also played a significant role.

Where *in vitro* automation has been attempted, the methods chosen were dependent on the growth and multiplication habits of species. There appeared to be several different types of growth in culture that were amenable to automation: meristematic nodules, small meristems, or similar structures; protocorms or microtubers; outward shoot growth (particularly bulbous-type plants); and upright shoot growth with clearly visible nodes or axillary buds. Depending on the species and mode of growth, cutting or division of tissue for multiplication could be random, vertical, horizontal, or not necessary. The choice of automated method(s) will vary from one species to another and more extensive studies to interrelate equipment and machine components with plant growth options are necessary. One common factor in the development of automated methods was the evaluation of liquid cultures for both meristem and shoot growth. A better understanding of the physiology and biochemistry of growing cultures in liquid media should avoid problems with vitrification and shoot and plantlet quality.

So far there is insufficient information on the yields and quality of shoots and plantlets from automated systems, or on the efficiency and costs of such systems, to be able to say whether automation will reduce the cost per plantlet or not. Very few automated systems for tissue culture are being used commercially. A heightened understanding of the integration between the biology of shoot and root growth and engineering possibilities is necessary to successfully automate steps in the tissue culture process. Levels of contamination and vitrification need to be reduced and more uniform development obtained. The genetic stability and formation of off-types produced by any new method also needs evaluation.

Future prospects for robots in tissue culture will depend on a thorough understanding involving engineers and plant physiologists or biologists in the matching of robotics to plant growth and development (ref. 90). Robotization of the tedious steps to produce new improved plant varieties by genetic engineering or mutation is also a future possibility (ref. 90).

Ultimately the choice of whether to use an automated or semi-automated system for the *in vitro* or *ex vitro* stages of tissue culture propagation depends on cost-effectiveness of the system. Market demands and what the market will pay for tissue-cultured plants are important considerations. Alternatives to automation to reduce the cost of tissue culture must also be considered; these include other methods of propagation, cheaper rent, labour and electricity, streamlining laboratory and greenhouse operations, reducing contamination losses, improving multiplication rates and shoot quality *in vitro*, and reducing the labour content by reducing the number of steps, e.g., liquid feeding to avoid one or two transfers, carrying out rooting and acclimatization together, and improving rooting and reducing losses at this stage.

7. ACKNOWLEDGEMENTS

The author would like to thank H. Watake, B. Johnson, G. Williames, T. Kozai, and A. Broekman for providing up-to-date information for this article and K. Horgan and H. Davies for helpful comments and discussion. Kluwer Academic Publishers are thanked for their permission to reprint this article in the Toyota proceedings.

8. REFERENCES

1 C.L. Brown and H.E. Sommer, Vegetative propagation of dicotyledonous trees, in: J.M. Bonga and D.J. Durzan (Eds), Tissue Culture in Forestry, Martinus Nijhoff/ Dr W. Junk, The Hague, 1982, pp. 109– 149.

2 D.R. Smith, Radiata pine, in: Y.P.S. Bajaj (Ed.), Biotechnology in Agriculture and Forestry, Vol. 1: Trees I, Springer-Verlag Berlin Heidelberg, 1986, pp. 274– 291.

3 W.C. Anderson and G.W. Meagher, Cost of propagating broccoli plants through tissue culture, HortScience 12(6), 1977, pp. 543– 544.

4 G. Hussey, Problems and prospects in the *in vitro* propagation of herbaceous plants, in: L.A. Withers and P.G. Alderson (Eds), Plant Tissue Culture and its Agricultural Applications: Butterworths, London, 1986, pp. 69-84.

5 D.R. Constantine, Micropropagation in the commercial environment, in: L.A. Withers and P.G. Alderson (Eds), Plant Tissue Culture and its Agricultural Applications: Butterworths, London, 1986, pp. 175– 186.

6 C.J. Sluis and K.A. Walker, Commercialization of plant tissue culture propagation, IAPTC Nwsl 47, 1985, pp. 2– 12.

7   R. Harrell and W. Simonton, Automation opportunities in plant tissue culture operations, ASAE Paper No. 86– 1596, ASAE, St. Joseph, MI49085– 9659, USA, 1986.

8   K.A. Walker, Automation of *in vitro* plant propagation: has its time finally arrived?, Genet Eng News July/August, 1986, p. 52.

9   P.A. Debergh, Recent trends in the application of tissue culture to ornamentals, in: C.E. Green, D.A. Somers, W.P. Hackett and D.D. Biesboer (Eds), Plant Tissue and Cell Culture, Alan R Liss Inc, New York, 1987, pp. 383-393.

10   R.E. Young, A. Hale, N.D. Camper, R.J. Keese and J.W. Adelberg, An alternative, mechanized plant micropropagation approach, ASAE/CSAE Paper No. 896092, ASAE, St. Joseph, MI49085– 9659, USA, 1989.

11   W. Preil, Application of bioreactors in plant propagation, in: P.C. Debergh and R.H. Zimmerman (Eds), Micropropagation, Technology and Application, Kluwer Academic Publishers, The Netherlands, 1991, pp. 425-445.

12   L.J. Wolf and V.J. Hartney, Computer system to assist with management of a tissue culture laboratory, N Z J For Sci 16, 1986, pp. 392– 402.

13   S. Humphries, W. Simonton and C.N. Thai, Computer aided management of plant tissue culture production, ASAE/CSAE Paper No. 89– 4050, ASAE, St. Joseph, MI49085– 9659, USA, 1989.14   M. Hemple, Microcomputer program for the management of micropropagation, in: Moët Hennessy (Ed), Electronics and Management of Plants, Paris, 1987, p. 35.

15   M. Hemple and K. Meszka, Modelling of micropropagation, Acta Hort 230, 1988, pp. 137– 144.

16   M.A.L. Smith and L.A. Spomer, Direct quantification of *in vitro* cell growth through image analysis, In Vitro Cell Dev Biol 23, 1987, pp. 67– 74.

17   A. Grand d'Esnon, N. Sujian, S. Faure and F. Sevila, On line evaluation by vision systems in biotechnologies, ASAE/CSAE Paper No. 89 7057, ASAE, St. Joseph, MI49085– 9659, USA, 1989.

18   M.P. Rigney and G.A. Kranzler, Performance of machine vision based tree seedling grader, ASAE/CSAE Paper No. 89– 3007 ASAE, St. Joseph MI49085– 9659, USA, 1989.

19   J. Aitken-Christie, A.P. Singh and H.E. Davies, Multiplication of meristematic tissue: a new tissue culture system for radiata pine, in: J.W. Hanover and D.E. Keathley (Eds), Genetic Manipulation of Woody Plants, Plenum Press, New York, 1988, pp. 413– 432.

20   V. Maurice, C.E. Vandercook and B. Tisserat, Automated plant surface sterilization system, Physiol Veg 23, 1985, pp. 127– 133.

21   M. Fari, Berendezés növenyek mickroszaporitására, Hungarian patent No. 15752/84, 1984.

22   M. Fari, Development of technical and biological basis of micropropagation toward automation and industrialization, in: Moët Hennessy (Ed.), Electronics and Management of Plants, Paris, 1987, pp. 55– 56.

23   Hitachi, Sterilization of culture vessel and culture medium by an electrical method – as an alternative to autoclaving, Japanese Patent No. J62122– 579, 1987.

24   R.E. Harris and E.B.B. Mason, Two machines for *in vitro* propagation of plants in liquid media, Can J Plant Sci 63, 1983, pp. 311– 316.

25   S. Takayama and M. Misawa, Mass propagation of *Begonia hiemalis* by shake cultures, Plant Cell Physiol 22, 1981, pp. 461-467.

26   S. Takayama and M. Misawa, The mass propagation of *Lilium in vitro* by stimulation of multiple adventitious bulb scale formation and by shake culture, Can J Bot 61, 1983, pp. 224-228.

27   M. Ziv, Enhanced shoot and cormlet proliferation in liquid cultured gladiolus buds by growth retardants, Plant Cell, Tissue Organ Culture 17, 1989, pp. 101-110.

28   L. Maene and P. Debergh, Liquid medium additions to established tissue cultures to improve elongation and rooting *in vitro*, Plant Cell Tissue Organ Culture 5, 1985, pp. 23– 33.

29   J. Viseur, Micropropagation of pear, *Pyrus communis* L., in a double-phase culture medium, in: ISHS (Ed.), *In vitro* problems related to mass propagation of horticultural plants, Gembloux, Belgium, 1985, p. 20.

30   G. Molnar, A new method for mass propagation of shoot cultures, Hungarian Patent No. 183.987, 1985.

31   J. Aitken-Christie and C. Jones, Towards automation: radiata pine shoot hedges *in vitro*, Plant Cell Tissue Organ Culture 8, 1987, pp. 185– 196.

32   A. Vanderschaeghe and P.C. Debergh, Technical aspects of the control of the relative humidity in tissue culture containers, Med. Fac. Landbouww. Rijksuniv. Gent, 52, 1988, pp. 1429-1437.

33   K. Matsumoto and H. Yamaguchi, Nonwoven materials as a supporting agent for *in vitro* culture of banana protocorm-like bodies, Trop Agr 66, 1989, pp. 8– 10.

258

34   C.K. Chin Y. Kong and H. Pedersen, Culture of droplets containing asparagus cells and protoplasts on polypropylene membrane, Plant Cell Tissue Organ Culture 7, 1988, pp. 59–65.

35   R. Hamilton, H. Pederson and C.K. Chin, Plant tissue culture on membrane rafts, Biotechniques 3, 1985, p. 96.

36   Komatsu, Plant tissue culture device – includes capillary member of water-absorbing material with small voids in bottom of culture vessel, Japanese Patent No. J63044–816, 1988.

37   Humboldt University, Nutrient carrier for use in *in vitro* plant and callus culture – comprises viscose preferably as slurry or granules covered by a cellulose filter, German Patent No. DD255–439, 1985.

38   Mitsubishi Petrochemical Company, Tissue culture of plants – Liliaceae culture on solid culture medium floating on liquid culture medium, Japanese Patent No. J63074–421, 1988.

39   T. Kozai, High technology in protected cultivation – from environmental control engineering point of view, in: Horticulture in High Technology Era, Tokyo, Japan, 1988, pp. 1–43.

40   T. Kozai, Micropropagation under photoautotrophic conditions, in: P.C. Debergh and R.H. Zimmerman (Eds), Micropropagation, Technology and Application, Kluwer Academic Publishers, The Netherlands, 1991, pp. 447-469.41   B. Tisserat and C.E. Vandercook, Development of an automated plant culture system, Plant Cell Tissue Organ Culture 5, 1985, pp. 107–117.

42   M.A. Farrell, Liquid medium system for plant micropropagation, in: Moët Hennessy (Ed.), Electronics and Management of Plants, Paris, 1987, p. 32.

43   P. Boxus and M. Paques, Propagation of woody plants in a culture medium containing hydrolyzed agar to prevent vitrification, Belgian Patent No. 904661, 1987.

44   J. Aitken-Christie and H.E. Davies, Development of a semi-automated micropropagation system, Acta Hort 230, 1988, pp. 81–87.

45   A.M. Vanderschaeghe and P.C. Debergh, Automation of tissue culture manipulations in the final stages, Acta Hort 227, 1987, pp. 399–401.

46   P.J. Weathers and K. Giles, A novel method of *in vitro* culture adaptable to many varieties of plants, in: Moët Hennessy (Ed.), Electronics and Management of Plants, Paris, 1987, p. 45.

47   P.J. Weathers, R.D. Cheethan and K.L. Giles, Dramatic increases in shoot number and lengths for *Musa*, *Cordyline* and *Nephrolepsis* using nutrient mists, Acta Hort 230, 1988, pp. 39–44.

48   R. Cheetham, P. Weathers, A. DiIorio, M. Glubiak and D. Hofling, *In vitro* growth and development of roots using nutrient mist culture, In Vitro Cell Dev Biol 25, 1989, pp. 59A, 194.

49   N. Lee, H.Y. Wetzstein and H.E. Sommer, The effect of agar vs liquid medium on rooting in tissue-cultured sweetgum, HortScience 21, 1986, pp. 317–318.

50   M. Akita and S. Takayama, Mass propagation of potato tubers using jar fermentor techniques, Acta Hort 230, 1988, pp. 55–61.

51   M. Ziv, Morphogenesis of gladiolus buds in bioreactors – implication for scaled-up propagation of geophytes, in: H.J.J. Nijkamp, L.H.W. van der Plas and J. van Aartrijk (Eds), Progress in Plant Cellular and Molecular Biology, Kluwer Academic Publishers, The Netherlands, 1990, pp. 119-124.

52   S. Takayama, Bioreactors for plant cell tissue and organ cultures, in: Handbook of Fermentation Technology, Noyes Publishing Co, USA (in press), 1990.

53   R.H. Zimmerman, Application of tissue culture propagation to woody plants, in: R.R. Henke, K.W. Hughes, M.J. Constantin and A. Hollaender (Eds), Tissue Culture in Forestry and Agriculture, Plenum Press, New York, 1984, pp. 165–177.

54   E. Vermeer and P. Evers, Continuous production of *in vitro* propagated plants, Agricell Report 9(3), 1987, p. 20.

55   R.C. Cooke, Homogenization as an aid in tissue culture propagation of *Platycerium* and *Davallia*, HortScience 14, 1979, pp. 21–22.

56   J.D. Caponetti and T.E. Byrne, Rapid propagation of Boston ferns by tissue culture, R.R. Henke, K.W. Hughes, M.J. Constantin and A. Hollaender (Eds), Tissue Culture in Forestry and Agriculture, Plenum Press, New York, 1984, pp. 309–310.

57   J. Janssens and M. Sepelie, Rapid multiplication of ferns by homogenization, Scientia Hort 38, 1989, pp. 161–164.

58   J. Janssens, Homogenization *in vitro* of *Begonia* x *hiemalis* for propagation, Meded Fac Landbouwwet Rijksuni Gent 52, 1987, pp. 1501–1503.

59   Oji Paper, Mass propagation of plants from seedling primordium derived callus – meristem culture for crop improvement, Japanese Patent No. J01047-318, 1989.

60   B.H. McCown, E.L. Zeldin, H.A. Pinkalla and P.R. Dedolph, Nodule culture: A developmental pathway with high potential for regeneration automated

micropropagation and plant metabolite production of woody plants, in: J.W. Hanover and D.E. Keathley (Eds), Genetic Manipulation of Woody Plants, Plenum Press, New York, 1988, pp. 149– 166.

61 A. Nagasawa and J.J. Finer, Development of morphogenic suspension cultures of garlic (Allium sativum L), Plant Cell, Tissue Organ Culture 15, 1988, pp. 183– 187.

62 A.D. Krikorian and R.P. Kann, Plantlet production from morphogenetically competent cell suspensions of daylily, Ann Bot 47, 1981, pp. 679– 686.

63 P. Druart, Plantlet regeneration from root callus of different Prunus species, Scientia Hort 12, 1980, pp. 339– 342.

64 S. Cronauer-Mitra and A.D. Krikorian, Adventitious shoot production from calloid cultures of banana, Plant Cell Rep 6, 1987, pp. 443-445.

65 E. Warrag, M.S. Lesney and D.J. Rockwood, Nodule culture and regeneration of Eucalyptus grandis hybrids, Plant Cell Rep., 1991, pp. 586-589.

66 Mitsui Petrochemical Industries, Culture apparatus for plant cell culture – has receivers with gradually increasing sections, Japanese Patent No. J61– 139381, 1986.

67 R. Levin, V. Gaba, B. Tal, S. Hirsch, D. Nola and I.K. Vasil, Automated plant tissue culture for mass propagation, Bio/Technology 6, 1988, pp. 1035– 1040.

68 V.A. Bapat, M. Mhatre and P.S. Rae, Propagation of Morus indica L. (Mulberry) by encapsulated shoot buds, Plant Cell Rep 6, 1987, pp. 393– 395.69   V.A. Bapat and P.S. Rao, In vivo growth of encapsulated axillary buds of mulberry (Morus indica L.), Plant Cell, Tissue and Organ Culture 20, 1990, pp. 69-70.

70 J. Mathur, P.S. Ahuja, N. Lal and A.K. Mathur, Propagation of Valeriana wallichii DC using encapsulated apical and axial shoot buds, Plant Science 60, 1989, pp. 111– 116.

71 P. Debergh and L. Maene, Pathological and physiological problems related to the in vitro culture of plants, Parasitica 40, 1984, pp. 69– 75.

72 M. Ziv, Vitrification: morphological and physiological disorders of in vitro plants, in: P.C. Debergh and R.H. Zimmerman (Eds), Micropropagation, Technology and Application, Kluwer Academic Publishers, The Netherlands, 1991, pp. 45-69.

73 K. Fujiwara, T. Kozai and I. Watanabe, Fundamental studies on the development of a propagation system for in vitro plantlets, Proc Ann Meet Agric Met Japan, 1987, pp. 202– 203.

74 K. Fujiwara, T. Kozai and I. Watanabe, Development of a photoautotrophic tissue culture system for shoots and/or plantlets at rooting and acclimatization stages, Acta Hort 230, 1988, pp. 153– 158.

75 T. Kozai, Autotrophic micropropagation, in: Y.P.S. Bajaj (Ed.), Biotechnology in Agriculture and Forestry, Springer-Verlag Berlin, Heidelberg (in press), 1989.

76 M. Hayashi, M. Nakayama and T. Kozai, An application of the acclimatization unit for growth of carnation explants and for rooting and acclimatization of the plantlets, Acta Hort 230, 1988, pp. 189– 194.

77 R. Levin, Process for plant tissue culture propagation, European Patent No. 0132414A2, 1985.

78 Albright and Wilson Ltd, Plant micropropagation – by suspension cell culture or tissue culture, European Patent No. 0303472A1 or Great Britain Patent No. 023615, 1989.

79 D.N. Schonstein and B.J. Johnson Method and apparatus for dividing plant materials, Australian Patent PCT/AU86/00136, World Patent WO86/06576, 1986.

80 H. Deleplanque, P. Bonnet and J.P.G. Postaire, An intelligent robotic system for in vitro plantlet production, Rovisec 5, 1985, pp. 305– 314.

81 G.E. Miles, Robotic transplanting, in: Horticulture in High Technology Era, Tokyo, Japan, 1988, pp. 75– 86.

82 Y. Miwa, Plant tissue culture robot operated by a shape memory alloy actuator and a new type of sensing, in: Moët Hennessy (Ed.), Electronics and Management of Plants, Paris, 1987, p. 7.

83 N. Fujita, Application of robotics to mass propagation system, In Vitro Cell Dev Biol 2, 1989, pp. 22A– 47.

84 Y. Miwa, T. Yamamoto, Y. Kushihashi and H. Kodama, Study on automation of plant tissue culture process – Development of automatic seedling transplantation system, Journal Japan Soc. Precision Engineering 54(6), 1988, pp. 99– 104.

85 F.R. Brown, Apparatus for and methods of cutting and/or moving plant tissue, Great Britain Patent No. GB87/00913, World Patent WO88/04520, 1988.

86 W.J. Rowe, N.M. Rowe and H.P. Roepe, The application of robotics for an automated tissue transfer system, in: Moët Hennessy (Ed.), Electronics and Management of Plants, Paris, 1987, p. 41

87 R.D. Tillet, Vision-guided planting of dissected microplants, J. Agric. Engng. Res. 46, 1990, pp. 197-205.

88 De Bry, Robots in plant tissue culture: an insight, IAPTC Newsletter 49, 1986, pp. 2– 22.

260

89   C. Sluis and R. Kahn, Economic parameters governing application of robotics to micro-propagation, in: Moet-Hennessy (Ed.), Electronics and Management of Plants, Paris, 1987, p. 6.

90   De Bry, Prospects in the mating of robotics to plant culture and research, in: Horticulture in High Technology Era, Tokyo, Japan, 1988, pp. 61–72.

91   G.E. Miles, Artificial intelligence in greenhouse production: robotics, machine vision, and expert systems, in: Moët Hennessy (Ed.), Electronics and Management of Plants, Paris, 1987, p. 4.

92   G.E. Miles, Robotic transplanting for tissue culture, In Vitro Cell Dev Biol 25, 1989, pp. 22A–46.

93   W. Simonton, Geranium stock processing using robotic system, ASAE/CSAE Paper No. 89–7054 ASAE, St. Joseph, MI49085–9659, USA, 1989.

Reprinted by permission of Kluwer Academic Publishers and updated with new references.

Full bibliographic details:

J. Aitken-Christie, Automation, in: P.C. Debergh and R.H. Zimmerman (Eds), Micropropagation, Technology and Application, Kluwer Academic Publishers, The Netherlands, 1991, pp. 363-388.

I. Karube (Ed.) *Automation in Biotechnology*
Proceedings of the 4th Toyota Conference, 21–24 October 1990

ROBOTIC WORKCELL FOR FLEXIBLY AUTOMATED HANDLING OF YOUNG TRANSPLANTS

K.C. Ting

Biological and Agricultural Engineering Department, Rutgers University, P.O. Box 231, New Brunswick, New Jersey 08903 (U.S.A.)

ABSTRACT

Automated handling of young transplants in the form of plugs has become an important process in meeting their increasing market demand. The research team at Rutgers University has been studying the implementation of robots for plug transplanting. The objective is to develop a flexibly automated plug transplanting workcell which will handle a wide range of plug species and container sizes. In this workcell, the plugs are extracted from one container, and transported and planted into another container. The end-effector used to manipulate individual plugs is equipped with a capacitive proximity sensor. The function of the sensor is to detect the presence of a plug, after extraction and before planting by the end-effector, to insure that the finished container is filled with plugs. The source and destination plug containers are transported by two overpassing conveyer belts. The belts are capable of making indexed advancement so that the distance for plug transportation between the containers may be minimized. The characteristics and parameters associated with the workcell are systematically analyzed.

INTRODUCTION

Transplants have long been used in the plant production systems. The advantages of using transplants have been realized in terms of uniform plant quality, effective production scheduling, and efficient materials handling. Recently, one particular form of transplant, namely plugs, has gained substantial interest in the greenhouse industry. A plug is an actively growing young plant, which consists of an aerial portion (stems and leaves) and a root zone portion (roots and growth medium). Plugs may be produced utilizing seeds (i.e. seedlings) or micropropagation techniques (i.e. micropropagules); therefore, they have the potential to play an important role in bridging biotechnology development and production agriculture (ref. 1-2). At a certain stage, individual plugs are normally transferred from one container to another at wider spacing for further development. This transplanting operation is extremely labor intensive. Automated handling of this type of young transplant is a necessary process in meeting the increasing demand for plugs.

The research team at Rutgers University has been studying the implementation of robots for plug transplanting (refs. 3-5). The objective is to develop a flexibly automated plug transplanting workcell which will process a wide range of plug species and container sizes. In this workcell, the plugs are extracted from one container, and transported and planted into another container by an innovatively designed end-effector. The end-effector is composed of a sliding-needles gripper and a capacitive proximity sensor. The function of the sensor is to detect the presence of a plug, after extraction and before planting by the gripper, to insure that the finished container is filled with plugs. The source and destination containers are transported by two overpassing conveyer belts. The belts are capable of making indexed advancement so that the distance for plug transportation between the containers may be minimized. A computer model has been developed to simulate the performance of SCARA robot based workcells. The performance indicators emphasized in this model are the workability and productivity of workcells. Several deterministic and stochastic factors which influence the workcell performance were studied both experimentally and by using the simulation model.

It can be realized, based on the above description, that a robotic workcell is composed of a number of interrelated "components" for accomplishing certain production goals. The effectiveness of a workcell is inevitably affected by the compatibility of the components. A basic understanding of the characteristics of each component and the interactions among them will be helpful in the workcell design and operation. This article will provide a systematic description of robotic workcells for plug transplanting.

WORK OBJECT

The objects to be processed in the workcell are transplants in the form of plugs. At present, more than fifty percent of all bedding plants in North America are grown from plugs (ref. 6). In 1988, this amounted to more than 2.75 billion plugs produced in the greenhouses. In addition, plugs are also used in field vegetable production during mechanized transplanting. The technology of plug production has become a popular topic of research and development. Koranski and Laffe (ref. 7) gave an overview of cultural technologies which could be readily applied to plug production. The advancement in biotechnology creates the possibilities of generating high quality seed materials including pregerminated seeds, somatic embryos, etc. A good plug production system can further develop these seed materials into actively growing young plants with well-defined morphological parts such as the aerial (stems/leaves) and the root zone (root/growth medium) portions.

It is always an engineering challenge to design systems for manipulating individual transplants effectively. Hwang and Sistler (ref. 8) and Kutz, et al. (ref. 9) used two different approaches in handling transplants in their robotics work. Hwang and Sistler's gripper handled pepper seedlings by grasping the plant foliage and stems. The parallel fingers were equipped with circular plates having spongy materials covered with vinyl tape. The purpose was to avoid damaging the plant stems. Care was taken to account for individual variations among the plants during picking and placing operations. Kutz, et al. modified a Unimation model 510 pneumatic, parallel-jaw gripper into a seedling gripper. Two flat fingers made of galvanized sheet metal were mounted on the gripper to vertically penetrate, horizontally grasp and release, and vertically depart the root zone portion of a seedling. The seedling shapes, the gripper's approach/departure paths relative to a seedling, the tip speed, and the depth of penetration into root zone affected the success rate of transplanting. Furthermore, the depth of penetration created opposite effects on successful grasping and releasing of the seedlings.

In general, plugs are easier to handle than other types of transplant such as bare-root seedlings. Plugs are grown in an upright orientation and located orderly in multiple-cell trays. As a result, the root zone portion of each plug is confined in a cell, and assumes a molded shape and a fixed location. Therefore, the root zone portion of a plug becomes an ideal geometric reference for its location and orientation. Using this reference, the object presentation scheme within the robotic transplanting workcell is simplified. However, due to the numerous variations of plug shape and size, the success of the transplanting process will be heavily influenced by the robot's capability of manipulating plugs. One additional variability is added by the various materials used as the growth media which eventually become a major component of the root zone portion. Some commonly seen growth media include soil mix and rockwool cubes. When soil mix is used, due to its lose form, the formation of the root system significantly influences the rigidity of the root zone portion of a plug.

EQUIPMENT

The mechanical and electrical/electronic hardware and software to be included within a robotic plug transplanting workcell are discussed in this section. Fig. 1 is a schematic diagram of the workcell under development at Rutgers University. The items presented here are in most cases generic. The emphasis of this discussion is on the design concept of the workcell. Therefore, mention of a specific product is for illustration purpose and does not constitute an endorsement by the author.

Plug containers

There is no standardization of plug containers in the U.S (ref. 10). The basic function of a container is to facilitate plug development and handling. Most containers are plastic trays capable of holding an array of plugs in individualized cells. Many factors affect the choice of container shapes and sizes (ref. 11), including plant growth requirements, space utilization, mechanical handling compatibility, and marketplace preference. Growing plugs in containers eases the task of singularizing and orientating plugs during automated transplanting. But, in designing a transplanting system capable of processing different type containers, the concept of flexible automation, as opposed to fixed automation, appears to be more promising. In other words, the system must have some intelligence so that it can respond to variable situations. Normally, the transplanting process is to relocate plugs from a source container to a destination container. In this article, for the purpose of identification, the source and destination containers will be designated as plug tray and growing flat, respectively. Fig. 2 shows typical examples of a plug tray and a growing flat (ref. 12).

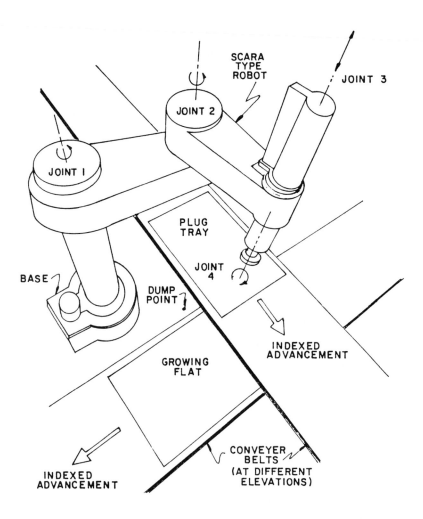

Fig. 1. Schematic diagram of the workcell under development at Rutgers University.

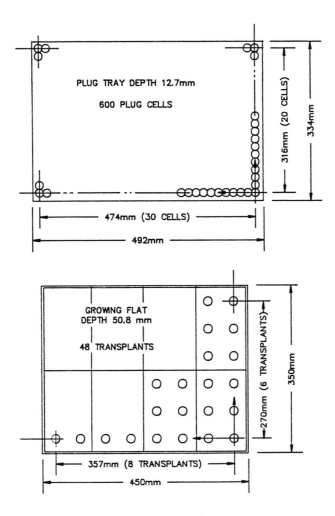

Fig. 2. Typical examples of a plug tray and a growing flat.

Conveyer belts

Plug trays and growing flats are transported on two separate conveyer belts within the workcell. The movement of the belts is monitored and controlled in coordination with the robot motion during plug transplanting. A bi-level overpassing arrangement and indexed (i.e. programmed stop-and-go) belt motions were found to be necessary in order to minimize the overall travel of the robot manipulator (Fig. 1).

Ending-effector

During transplanting, an end-effector is the interface between a robot and a plug. The function of an ideal end-effector is to grasp, hold, insert and release a plug regardless of plant species and container sizes. In our study, we established some design guidelines to aid in the development of the end-effector (ref. 13). The guidelines were that:

1. the end-effector be structurally simple, and powered by low cost actuators,
2. the end-effector accommodate a wide variety of plugs in different sizes of containers,
3. the end-effector avoid the aerial portion and operate on the root zone portion of the plug,
4. the "fingers" of the end-effector be able to effectively (without sacrificing the speed of the robot) penetrate, grasp, hold and release a wide variety of growth media with minimum damage to the roots,
5. there be a sensor to detect the presence of a plug on the end-effector ensuring that the destination container be completely filled with transplants.

A series of end-effectors were designed based on a fundamental concept that a plug was to be manipulated by its root zone portion by grasping with a pair of needles. The first design was a gripper consisting of two slanted needles. The needles were kept parallel when penetrating or releasing the root zone portion of a plug. A mechanism was included in the gripper so that the tips of the needles could be rotated toward each other for firmly grasping the root zone portion during plug extraction and transportation. In the second gripper design, the rotating actions used in the first design for opening and closing the needles were changed so the two needles were extended/retracted inside two collars (Fig. 3). Therefore, for grasping a plug, the two needles became fully extended to penetrate the root zone. To release a plug, the two needles were simply retracted into their individual collars. After an extensive testing of this second gripper design, it was found that there were two characteristic angles important to the effectiveness of the gripper. They were (1) the included angle between the needles when they were fully extended (alpha) and (2) the angle between the plane defined by the two needles and the horizontal (beta). On the third design version of the gripper, both alpha and beta angles were made fully adjustable. Fig. 4 shows the drawings of this final gripper design (ref. 14).

In order to have a high assurance that a finished growing flat would be filled with transplants, the gripper was equipped with a capacity proximity sensor (Model no. CJ10-30GM-E2, manufactured by Pepperl & Fuchs, Inc., Twinsburg, OH). It was found that the sensor was capable of detecting a plug held by the gripper (ref. 15). The physical location and sensitivity of the sensor could be adjusted to accommodate different plug shapes and sizes. The sensor output was a binary signal which was acceptable by most robot controllers.

With this sensing capability, a robot would complete a transplanting process only when a plug was correctly held by the gripper. This enabled the robot to respond properly to the inherent uncertainties associated with the transplanting operation. The uncertainties were mainly caused by the missing plugs of the source tray and the occasional unsuccessful plug extraction by the gripper.

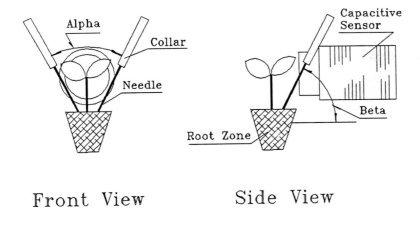

Fig. 3. Schematic diagram of a gripper holding a plug.

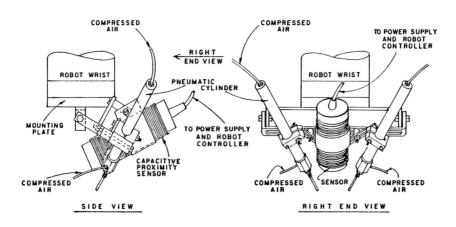

Fig. 4. Design drawing of the Sliding Needles with Sensor (SNS) gripper.

Robot

Traditionally, automation has been considered to be economical only for mass production of standard items due to investment requirements and technology limitations. This is the concept of fixed automation. Recently, the rapid technological advancement and the ever diversifying marketplace demand inspired the thought of flexible automation. In such a system, computers and generic manufacturing devices are integrated to perform batch production of customized goods while maintaining the economic advantage of fixed automation. The system change-over is mainly done through computer software revision, and requires minimum hardware modification. Robots are ideal generic devices to be included in a flexible manufacturing system.

A robotic system normally consists of a manipulator, a controller and a software package. The function of the manipulator is to position and orient an end-effector in 3-D space. The controller monitors and controls the status and motion of the manipulator and the peripheral equipment such as conveyer belts. The software, including a computer operating system and a programming language, provides an interface between the robotic system and its users. In most cases, the software can be operated through the use of a pendant and/or a keyboard/monitor.

The robot system used in this workcell development is an AdeptOne SCARA robot manufactured by Adept Technology, Inc., San Jose, CA (Fig. 1). The robot manipulator has four degrees of freedom. Joints 1, 2 and 4 are rotary joints and joint 3 is prismatic. All the joint axes are parallel to each other, and perpendicular to the base of the manipulator. Fig. 5 shows the top view of the robot work envelope. Also shown in the figure are the first and second links of the robot, a plug tray and a growing flat. Since the third joint is a prismatic joint, the work envelope has a depth in the direction of the joint 3 axis.

Machine vision system

To make the workcell intelligent, some sensing capability is very important. One example is the above mentioned proximity sensor installed as a part of the end-effector. The proximity sensor performs satisfactorily in detecting the presence of plugs. However, if more information about plugs, such as shape, size, maturity and vigor, is needed, a different type of sensing capability will be required. Machine vision systems have been known to be able to provide a wide possibilities of "intelligence" within a robotic workcell. Currently, we are developing machine vision algorithms for robot guidance during transplanting (ref. 16) and plant quality grading for uniform growth (ref. 17). The system used in this project is an ITI Series 151 manufactured by Imaging Technology Inc., Woburn, MA.

Host computer

Most components in a robotic workcell, such as the robot, the machine vision system, and the conveyer belts, can operate as individual units through their own controllers. However, in order to incorporate these components into an integrated system, it is functionally useful to have a centralized processor. This centralized processor is a host computer which has two-way communications to the controllers of component units. The software residing in this host computer facilitates its role as the system commander at the highest hierarchical level. We are using a 386 DOS machine (32-bit, 25 MHz) as the host computer for the plug transplanting workcell.

OPERATIONS

The process of robotic transplanting requires certain operations to be carried out within the workcell. These interrelated operations are performed either on the plug containers (plug trays and growing flats) or on the individual plugs. The effectiveness of a transplanting workcell is dependent upon the success of each operation and the coordination among the operations.

Transport of containers

Plugs enter the workcell on plug trays and leave it on growing flats. The pattern of container flow through the workcell needs to be well planned in order to keep the transplanting process uninterrupted and efficient. The robot manipulator's work envelope imposes a constraint on the available space that the containers can use. We studied a bi-level, overpassing arrangement of conveyer belts. The arrangement was found promising, especially when used in combination with the scheme of indexed movement.

Locating containers

When the plug containers are transported through the workcell, they are moved in a stop-and-go fashion. It is necessary to advance containers after a given number of transplantings to move the concerned cells to the most advantageous positions. It is considerably easier for the robot to extract and plant a plug while the containers are stationary. Once a container has been moved, the new location of the container needs to be determined so that the robot manipulator can access the plugs accurately. The required accuracy in locating a container is within 0.5 mm for the plug tray in Fig. 2 and to within several millimeters for the growing flat in Fig. 2. Due the inherent variability of trays and the accuracy requirement, an economical solution to this problem is not trivial. Several alternatives, including machine vision techniques (ref. 16), are being investigated at this time.

Extraction, sensing, holding, transporting, and planting

The manipulation of an individual plug during robotic transplanting includes the following steps:

1. approaching a plug on the plug tray,
2. grasping the plug,
3. lifting the plug out of the cell,
4. holding the plug in its upright position,
5. detecting the presence of the plug held by the gripper
6. transporting the plug along a straight-line path above a cell on the growing flat,
7. detecting the presence of the plug held by the gripper
8. lowering the plug into a pre-dibbled hole on the growing flat,
9. releasing the plug,
10. departing from the plug planted in the growing flat, and
11. returning along a straight-line path to a point above the next plug in the plug tray and repeat the process.

These actions may be categorized into manipulator actions, end-effector actions, and combined actions. For a given manipulator, the outcome of the categorization is influenced by the end-effector design. A good end-effector should be reliable and minimize the need to restrict the manipulator motions. The above procedure will be allowed to continue in cycles as long as the signals generated in steps 5 and 7 are positive. If either one gives a negative signal then the remaining steps will be replaced by the process of preparing the end-effector for the next attempt to approach a plug on the plug tray (step 1). This preparation process is described in the following section.

Discarding the rejects

Two possible situations will cause the sensor on the end-effector to return a negative signal: (1) the end-effector is empty and (2) the end-effector is holding something of non-detectable quality. In the interest of producing finished flats with minimum missing transplants, either of the above situations is considered as undesirable for the planting operation to proceed. In stead, another operation which empties the end-effector is carried out. In this operation, the end-effector travels to a convenient location (e.g. dump point in Fig. 1) and performs the action similar to releasing a plug. After this, the end-effector will move toward the plug tray and resume the transplanting process from step 1 as describe in the preceding section.

PERFORMANCE INDICTORS

The performance of a robotic transplanting workcell can be determined by evaluating several tangible and intangible aspects. We consider the following aspects as important indicators of the workcell performance.

<u>Workability and productivity</u>

The components in the workcell must be compatible with each other physically and functionally. In other words, they must work together to perform the assigned duties within the imposed constraints. Computer simulation techniques have been proven to be effective in analyzing a systems' workability. To use the techniques requires the identification of design parameters which describe component characteristics and equations which govern component behavior. While the governing equations are unique to individual components, most design parameters are shared by two or more components. A particular system design is represented by a set of parameter values. The system is considered workable if all the governing equations are satisfied by using the parameter values.

The immediate goal of the workcell is to transplant plugs from plug trays to growing flats. The most direct indication of the system performance would be the productivity of the workcell. The productivity is defined as the number of finished flats over a fixed length of time. The other side of this definition is how much time it requires to finish a specified type of flat, the throughput time of a flat. Since the transplanting processes of all flats are not necessarily identical, the productivity is expressed based on statistics. The measure of productivity may be done by actually operating the workcell. However, it will become a formidable task if a large number of design alternatives are to be studied. The computer simulation technique described above can be used to facilitate the estimation of workcell productivity.

Micro-computer software has been developed in this study to stochastically simulate the performance of SCARA robot-based plug transplanting workcell (ref.18). This QuickBasic based, animated, user-friendly software package "WORKCELL" is capable of evaluating various workcell designs for their workability and productivity. The model has been validated by actual operation of the robotic system.

<u>Plug quality preservation and quantity conservation</u>

The quality of the plugs must be maintained while they are processed through the workcell. Transplanting of plugs involves direct machine-plug contacts. Using our SNS gripper, the major contact point is the root-zone portion of a plug. Experiments were conducted to evaluate the vigor of the plugs transplanted by an SNS gripper. After robotic transplanting, the flats filled with transplanted plugs were watered and placed in a greenhouse. Later observations revealed that all the transplants were alive and growing with no observable damage due to transplanting.

Ideally, all usable plugs from every source container should be properly transplanted into destination containers. However, in a real situation, some usable plugs may be discarded due to unsuccessful handling by the gripper or the sensor. The ratio between the number of successfully transplanted plugs and the number of usable plugs supplied to the workcell is an indication of how effective the workcell is in terms of plug quantity conservation.

Reliability, complexity and safety

One of the advantages of process automation is to have a continuous operation with minimum interruption. The environment for plant production is different from most manufacturing factories which are more structured. The equipment used to process plant materials needs to be able to tolerate situations not conducive to normal equipment maintenance. In addition, the system must not require complicated operating procedures to be followed by the users. The measures for protecting personnel and property should also be considered when evaluating a system.

Recently, the industrial robots have become very advanced and their reliability has been substantially improved. In the plug transplanting workcell, the operations which need to be emphasized are the end-effector design and the techniques for positioning plug containers. The software for the workcell can be developed so that users with a minimum knowledge of computers and robotics will be able to operate with confidence. The devices for workcell safety are mostly commercially available and can be readily incorporated into the design.

Engineering economics

The proof of technological feasibility is only a part of an engineering project. The result of feasibility study will identify workable systems. However, for a system to be accepted by the users, the return on investment needs to be calculated. Engineering economic analysis is a technique for determining the potential return on investment (ROI) of a new project. A computer program "EEGA" was developed by Ting, et al. for economic analysis of greenhouse production systems (ref. 19). With the aid of the simulation model WORKCELL, the ROI's of workcell design alternatives can be analyzed by using EEGA. This work is currently in progress.

FACTORS AFFECTING WORKCELL WORKABILITY AND PRODUCTIVITY

The workability and productivity of a plug transplanting workcell are the two most important performance indicators which are fundamentally related to the technological and economical feasibility of the system. Work has been done to identify and quantify the factors affecting these two indicators (refs. 20-21). Both laboratory tests and computer modeling were performed in this study. The laboratory tests were conducted to investigate the effectiveness of the SNS gripper and validate the model WORKCELL. Figs. 5 and 6 show two examples of the WORKCELL monitor display. Fig. 5 displays the workcell layout as specified by the user. Fig. 6 is a snapshot of the animated simulation of plug transplanting. The factors which influence the performance of a workcell are grouped into two categories: (1) deterministic factors and (2) stochastic factors.

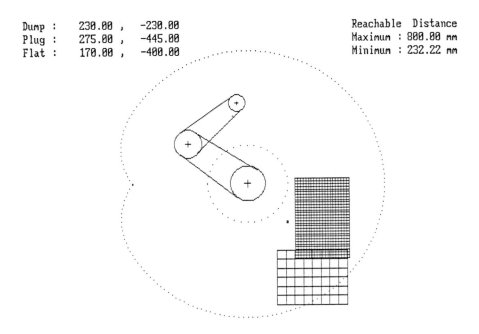

```
Dump :    230.00 ,   -230.00                    Reachable  Distance
Plug :    275.00 ,   -445.00                    Maximum  : 800.00 mm
Flat :    170.00 ,   -400.00                    Minimum  : 232.22 mm
```

Fig. 5. Computer monitor display of a workcell layout by the model WORKCELL.

Deterministic factors

These are the factors which normally do not involve probabilities. They are the parameters specified during the design and construction stage of the workcell.

(i) Layout. The robot work envelope and the arrangement of plug containers relative to the manipulator constitute the major portion of the layout. The required motions of the

274

manipulator are influenced by the layout. As a result, the time required for transplanting each individual plug is also affected.

(ii) <u>Robot kinematics</u>. The built-in motion control for a robot governs the path planning and kinematics of the robot. For most robot systems, the lengths of time for the motions under identical conditions are repeatable. However, the kinematics of different robot systems can be significantly different.

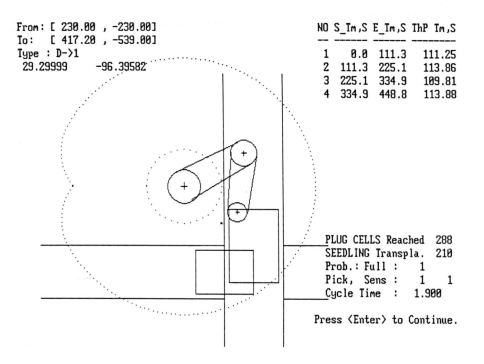

Fig. 6. A snapshot of the animated simulation of plug transplanting by model WORKCELL.

(iii) <u>End-effector design</u>. The design of the end-effector affects the strategy of plug manipulation. This will in turn affect the effectiveness of transplanting process.

(iv) <u>Materials flow</u>. The flows of the plug trays and the growing flats relative to the robot are an important aspect of a workcell design. The timing and distance of each advancement of a tray or a flat within the workcell during transplanting were found to significantly affect the transplanting time. In a study, twelve different combinations of layout and materials flow patterns were investigated, and the average cycle time per transplanting was found to vary from 2.62 s to 3.25 s. If the number of transplantings is on the order of millions, this result emphasizes the importance of the workcell layout and materials flow design.

(v) Operational algorithms. Flexible automation systems provide the advantage of allowing the users to select or change the operational algorithms with minimum efforts. On the other hand, this flexibility also imposes a challenge for the software designers to develop the most efficient algorithm possible. In many cases, this computer programming task exceeds the effort spent for hardware design.

Stochastic factors

During plug transplanting, several factors of probabilistic nature will affect the performance of the workcell. Some of them are related to the robot's capability of successfully manipulating plugs and the others are associated to the reliability of system components.

(i) Success rates of extraction, sensing and planting. Many mechanical and horticultural factors were found to affect the percentage of successful transplanting, during the laboratory testing of the SNS gripper. The mechanical factors were: (1) the gripper needles' angles, (2) plug extraction speed, and (3) the sensor sensitivity. The horticultural factors included (1) empty cells on the plug trays, (2) plant species, (3) root connections, (4) adhesion between roots and cell walls, (5) root zone moisture, and (6) the number of plugs in one cell. Fig. 7 is shown as an example for illustrating the effect of the successful plug extraction rate on the time required for completing one 48-cell growing flat. This result was obtained by using the simulation model WORKCELL. Nine cases were studied by varying only the extraction success rate and keeping all other conditions the same. The extraction success rate was the ratio of the number of plugs lifted from the trays by the gripper to the number of attempted extractions at the locations where plugs were present in the cell. Since the transplanting times are not necessarily identical for all growing flats, the statistical distribution (the mean and one standard deviation each way) of each of the nine cases is presented. The curve in the figure passes through the average transplanting times of all the cases.

(ii) Machine failure and downtime. It is obvious that the probability of machine failure and the time needed to bring the system back to service can significantly influence the workcell operation.

SUMMARY

The concept and design of robotic plug transplanting workcells are discussed in this article. Specific results obtained from this workcell development project are also included. The practical experiences gained by incorporating an industrial robot within a workcell for an agricultural application such as plug transplanting can be utilized for other plant production tasks. The underlying concept of flexible automation and robotics will play an important role in bringing the results of biotechnology research up to the full scale production. Table 1 gives an overview of a robotic plug transplanting workcell.

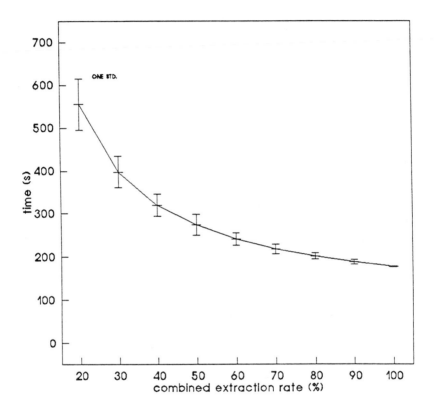

Fig. 7. The effect of the successful plug extraction rate on the time required for completing one 48-cell growing flat.

TABLE 1

Overview of a robotic plug transplanting workcell.

I. Work Object

1. plugs (transplants in the forms of seedlings or micropropagules)

II. Equipment

1. plug containers (trays, flats, pots, etc)
2. conveyer belts (including motion monitor and controller)
3. end-effector (gripper & sensor)
4. robot (manipulator, controller & software)
5. machine vision system
6. host computer

III. Operations

1. transport of containers (indexed advancement)
2. locating containers
3. extraction
4. sensing (object presence and quality)
5. holding
6. transport of individual plugs
7. planting
8. discarding the rejects

IV. Performance Indicators

1. workability
2. productivity (throughput time of finished containers)
3. plug quality preservation
4. plug conservation
5. reliability (robot, gripper, sensor, and belts)
6. complexity (operation, maintenance, management)
7. safety
8. engineering economics

V. Factors Affecting Performance

1. Deterministic
    a. layout
    b. robot kinematics
    c. end-effector design
    d. materials flow
    e. algorithms (robot motion/image processing programs & process control program)

2. Stochastic
    a. success rates of extraction, sensing, and planting
        i. mechanical: characteristic angles of the sliding-needles gripper, extraction speed, and sensor sensitivity
        ii. horticultural: empties in the source container, plant species, root connections, adhesion between roots and container walls, root zone moisture, and number of plants in one container
    b. machine failure and downtime

REFERENCES

1   T. Kozai, S. Sase, G.A. Giacomelli, K.C. Ting and W.J. Roberts, Future seedling production systems, Agriculture and Horticulture [in Japanese], 65(1) (1990) 97-103.

2   P. de Groot, The "Centipede" automatic transplanter, International Summer Meeting of ASAE, Columbus, Ohio, U.S.A., June 24-27, 1990, Paper No. 901035.

3   K.C. Ting, Automate flexibly with robots, Greenhouse Grower, 5(11) (1987) 24-28.

4   K.C. Ting and G.A. Giacomelli, A robot for transplanting plugs, Greenhouse Grower, 6(12) (1988) 58-60.

5   K.C. Ting and G.A. Giacomelli, Robotics research on transplanting of seedlings, in: Proc. the First Int. Workshop-Robotics in Agriculture and the Food Industry, Avignon, France, June 14-15, 1990, IIRIAM and Teknea, Marseilles, France, pp. 85-90.

6   D. Hamrick, 2.75 billion plugs a year and growing, GrowerTalks, 52(8) (1988) 25.

7   D.S. Koranski and S. Laffe, Checking out plugs up close...an update on what's new in cultural technology, GrowerTalks, 52(8) (1988) 28-44.

8   H. Hwang and F.E. Sistler, A robotic pepper transplanter, Applied Engineering in Agriculture, 2(1) (1986) 2-5.

9   L.J. Kutz, G.E. Miles, P.A. Hammer and G.W. Krutz, Robotic transplanting of bedding plants, Transactions of the ASAE, 30(3) (1987) 586-590.

10   K.Z. Peppler, Container standardization, Greenhouse Manager, 9(4) (1990) 46-50.

11   D.R. Roberts, How growers choose containers, Greenhouse Manager, 9(4) (1990) 52-54.

12   K.C. Ting, G.A. Giacomelli and S.J. Shen, Robot workcell for transplanting of seedlings part I-layout and materials flow, Transactions of the ASAE, 33(3) (1990) 1005-1010.

13   K.C. Ting, G.A. Giacomelli, S.J. Shen and W.P. Kabala, Robot workcell for seedling transplanting part II-end-effector development. Transactions of the ASAE, 33(3) (1990) 1013-1017.

14   K.C. Ting, G.A. Giacomelli, D.R. Mears, W.P. Kabala, S.J. Shen and S.E. Williamson, Piercing Element Gripping Apparatus, U.S. Patent Application, (1989) Serial No. 335,109.

15   K.C. Ting and G.A. Giacomelli, End-effector for seedling transplanting, in: Proc. International Automation Conference, Detroit, Michigan, U.S.A., June 5-7, 1990, pp. 7:27-7:39.

16   P.P. Ling, Y.W. Tai , K.C. Ting, Vision guided robotic seedling transplanting, International Winter Meeting of ASAE, Chicago, IL, U.S.A., December 18-21, 1990, Paper No. 907520.

17   P.P. Ling, G.A. Giacomelli and K.C. Ting, Feature measurement of germinated tomato seeds for uniform flowering, International Summer Meeting of ASAE, Columbus, Ohio, U.S.A., June 24-27, 1990, Paper No. 907056.

18   K.C. Ting and W. Fang, Computer model of SCARA robot-based workcell for plug transplanting, (1990) in preparation.

19   K.C. Ting, J. Dijkstra, W. Fang and M. Giniger, Engineering economy of controlled environment for greenhouse production, Transactions of the ASAE, 32(3) (1989) 1018-1022.

20   Y. Yang, Performance of sliding-needles gripper and SCARA robot-based workcell for seedling transplanting, M.S. Thesis, Biological and Agricultural Engineering Department, Rutgers University, New Brunswick, New Jersey, U.S.A. (1990).

21   K.C. Ting, Y. Yang and W. Fang, Stochastic modeling of robot workcell for seedling transplanting, International Winter Meeting of ASAE, Chicago, IL, U.S.A., December 18-21, 1990, Paper No. 901539.

ACKNOWLEDGMENT - New Jersey Agricultural Experiment Station Publication No. J-03232-19-90, supported by State funds.

I. Karube (Ed.) *Automation in Biotechnology*
Proceedings of the 4th Toyota Conference, 21–24 October 1990
© 1991 Elsevier Science Publishers B.V. All rights reserved

ENVIRONMENTAL CONTROL AND AUTOMATION IN MICROPROPAGATION

T. KOZAI

Department of Horticulture, Faculty of Horticulture, Chiba University, Matsudo, Chiba 271 (Japan)

SUMMARY

Reasons for the high production cost of micropropagated plantlets are discussed. CO_2 concentrations in the culture vessel and photosynthetic characteristics of plantlets in vitro (in tissue culture vessels) are described. The effect of CO_2 enrichment under high photosynthetic photon flux conditions on the growth of plantlets in vitro (in tissue culture vessels) is shown. Four prototype robotic/automated micropropagation systems recently developed in Japan are introduced. Both may contribute to a reduction in cost of micropropagated plants in future.

INTRODUCTION

The use of micropropagated plantlets has been expanding worldwide in horticulture, agriculture and forestry. However, because of the high production cost, its wide spread use is still restricted mainly to horticultural crops despite its advantages.

Reasons for the high costs include: 1) a high labor cost due to manual operation, and 2) the low growth and multiplication rates of plantlets in vitro.

In the author's opinion, the above reasons are mainly the result of the presence of sugar in the culture medium, because 1) sugar in the medium promotes biological contamination, 2) small air-tight vessels must be used to reduce the loss of plantlets in vitro due to contamination, 3) the relative humidity and ethylene concentration thus tend to be high, 4) the abnormal in-vitro environment induces physiological/morphological disorders such as vitrification and growth/development retardation, 5) plant growth regulating substances are therefore often necessary to overcome these problems, and 6) automated micropropagation systems with the use of sugar in small vessels are difficult to develop (ref. 1). Then, why do we have to supply sugar in the medium? Everyone knows that plants grow without sugar in the soil under favorable environmental conditions!

Explants/shoots/plantlets in vitro have been considered to have little photosynthetic ability and to require sugar in the medium as a carbon source for their heterotrophic (sugar-dependent) or photomixotrophic (partly sugar-dependent) growth.

Our recent research (refs. 1, 2 and 3) revealed, however, that chlorophyllous plantlets in vitro, in general, have relatively high photosynthetic ability and they may grow faster under photoautotrophic (sugar-free medium and photosynthesis-dependent) conditions than under hetero- or photomixotrophic conditions, provided that the physical and chemical environments are properly controlled for photosynthesis.

The carbon dioxide concentration, air movement and light intensity (or PPF, photosynthetic photon flux) in conventional (small and air-tight) vessels for heterotrophic micropropagation are too low for promoting photosynthesis. The relative humidity is too high for normal transpiration (and associated mineral nutrient uptake) and normal stomatal functioning. Furthermore, the presence of sugar in the medium itself seems to increase sugar concentration in the plantlet and thus inhibit the photosynthetic activity of plantlets in vitro.

In this article, possibilities of photoautotrophic micropropagation are discussed based upon our recent research on the interrelationship between the environment and the growth of plantlets in vitro. The automation of micropropagation processes is also discussed.

REASONS FOR THE HIGH PRODUCTION COSTS OF MICROPROPAGATED PLANTLETS

Labor costs in commercial micropropagation are said to account for 60% or more of the total production costs. A significant reduction in labor costs could be achieved only by automation or robotization based upon a novel concept. Other reasons for the high production costs include (ref. 1):

1) a long time period (usually several weeks or more) required for each culture stage and a low multiplication (proliferation) rate at the multiplication stage; 2) sometimes a relatively high death percentage of plantlets due to physiological disorders and/or biological (bacterial or fungal) contamination during the successive stages in vitro, and due to serious environmental stresses during the acclimatization stage; 3) large variations in size, quality and morphology of plantlets; 4) significant energy costs for lighting and cooling or air conditioning, sterilization,

washing, etc.; 5) significant costs for sugar, gelling agent, culture vessels, macro- and micronutrients, plant growth regulating substances, etc.; 6) an overproduction or shortage of plantlets at various times of the year, due to unpredictable circumstances; and 7) a significant space (and its related costs) required for culture room.

If the growth rate is doubled and the percentage of death is halved by proper environmental control, the reduction in overall production cost will be significant even if the energy cost for environmental control may be increased to a certain extent.

CO_2 CONCENTRATION IN THE VESSEL

It has been found recently that the CO_2 concentration in an airtight vessel containing chlorophyllous plantlets is as low as CO_2 compensation points (50-80 ppm) during most of the photoperiod (ref. 4, Fig.1), approximately 250-300 ppm lower than normal atmospheric CO_2 concentration (345 ppm). It increases with time during the dark period up to 3000-9000 ppm but decreases sharply with time close to the compensation point after the onset of photoperiod and remained the same until the dark period begins.

Based upon the results mentioned above, the following hypotheses could be made:

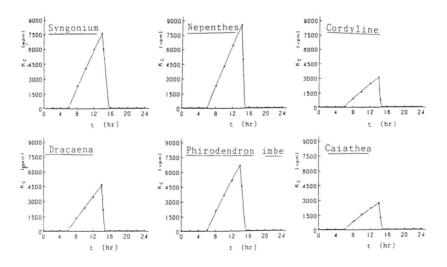

Fig. 1. Time variation of CO_2 concentrations (Kc) in the closed vessels containing various plantlets (ref. 4.).

1) The chlorophyllous explants/plantlets have photosynthetic ability, 2) Their photosynthesis is restricted by low CO_2 concentration during the photoperiod, 3) They are forced to develop heterotrophy or photomixotrophy and a higher PPF will not increase net photosynthetic rates under such low CO_2 concentration conditions, 4) The plantlets may develop photoautotrophy and grow faster under photoautotrophic, high CO_2 and high PPF conditions than under hetero- or photomixotrophic conditions, 5) The initial growth rate should be greater for an explant with a larger area of highly chlorophyllous tissues under photoautotrophic conditions, and 6) The presence of sugar in the medium itself modifies the photosynthetic characteristics of plantlets _in vitro_.

In order to confirm these hypotheses experimentally, we need, first of all, to know the photosynthetic characteristics of plantlets _in vitro_.

PHOTOSYNTHETIC CHARACTERISTICS OF MICROPROPAGATED PLANTLETS
AND GROWTH PROMOTION UNDER HIGH CO_2 AND HIGH PPF CONDITIONS

Photosynthetic response curves for intact _Cymbidium_ plantlets _in vitro_ cultured on _Hyponex-agar_ medium with 20 g/l sucrose are given in Fig. 2 as a function of CO_2, PPF and temperature (ref. 3).

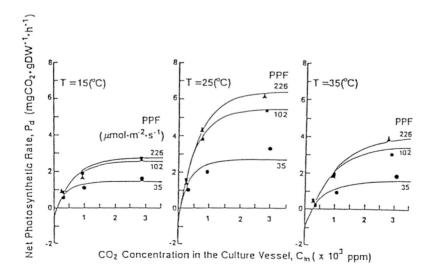

Fig. 2. Net photosynthetic curves _in situ_ for _Cymbidium_ plantlets _in vitro_ (ref. 5.).

The photosynthetic response curves do not differ greatly from those for the plantlets grown under shade in the greenhouse.

Furthermore, literature shows that the net photosynthetic rate of plantlets $\underline{in}$ $\underline{vitro}$ decreases with increasing sugar concentrations in the medium (refs. 6 and 7, Fig.3). This characteristic is considered to be related to the enzyme activities of RUBPCase (refs. 6 and 8) and PEPCase (ref. 8) in chloroplasts.

Significant growth promotion of plantlets cultured $\underline{in}$ $\underline{vitro}$ with or without sugar in the medium under high CO_2 and high PPF has been reported recently for many species such as $\underline{Cymbidium}$ (ref. 9), statice (ref. 10), carnation (ref. 11, Fig. 4), potato (ref. 12), tobacco (refs. 13 and 14), and strawberry (refs. 15 and 16). The data given above and other experimental results (ref. 1) support the hypotheses given in the previous section.

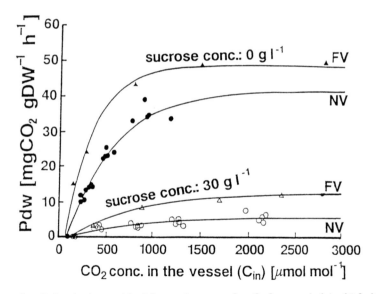

Fig. 3. Net photosynthetic rate per leaf dry weight (Pdw) of potato (Solanum tuberosum cv. benimaru) explant $\underline{in}$ $\underline{vitro}$ under forced (FV) and natural (NV) ventilation conditions (ref. 7.).

ADVANTAGES OF PHOTOAUTOTROPHIC MICROPROPAGATION

The advantages of photoautotrophic micropropagation are as follows (ref. 1):

1) The growth and development of plantlets are promoted under the conditions of high CO_2 and high PPF, 2) Physiological and morphological disorders can be reduced and relatively uniform

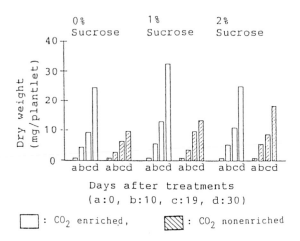

Fig. 4. Time courses of the dry weight of carnation (<u>Dianthus caryophyllus</u> L.) plantlets (ref. 11).

growth and development can be expected, 3) Biological contamination can be minimized with sugar-free medium; therefore, loss of plantlets due to contamination can be reduced, 4) A larger culture vessel can be used with a minimum loss of plantlets due to contamination, 5) Rooting and acclimatization <u>in</u> <u>vitro</u> and/or <u>ex</u> <u>vitro</u> can be achieved more easily, 6) Environmental control of the culture vessel can be facilitated so that the control of growth and development of plantlets can be conducted more easily, 7) Automation and/or robotization can be achieved by using a larger vessel containing normal and uniform plantlets, and 8) The use of plant growth regulators, vitamins and other organic substances can be minimized because some hormones will be produced endogenously by plantlets growing photoautotrophically.

The concepts of photoautotrophic micropropagation may give a novel idea for developing an automated micropropagation system to achieve a drastic reduction in the production costs. However, the advantages mentioned above must be examined carefully through a series of experiments before commercialization.

REDUCTION IN COSTS FOR HIGH CO_2 AND HIGH PPF

CO_2 enrichment for growth promotion is effective only under high PPF. There are several methods of increasing CO_2 in the vessel at reasonable costs (refs. 1, 2 and 3). The methods include: 1) use of gas permeable film as enclosures, 2) use of the film in a CO_2

enriched culture room, 3) use of a large vessel with a forced ventilation or a CO_2 supply system, and 4) use of chemicals to produce CO_2 in the vessel.

There are also several methods of reducing the costs for high PPF and for cooling the culture room. The methods include: 1) use of a water-cooled reflector, and 2) use of optical fibers and lateral lighting system (ref. 1, 2, and 3).

With these methods, the increase in costs for high CO_2 and high PPF can be minimized.

OTHER FACTORS AFFECTING THE GROWTH OF MICROPROPAGATED PLANTLETS

Numerous other factors can influence photoautotrophic growth of plantlets in vitro, e.g., inorganic nutrient composition, oxygen concentration in the air and in the medium, humidity of the air, water potential of the medium, air flow speed or diffusion coefficient of the air (ventilation rate under natural or forced ventilation conditions), air temperatures during photoperiod and dark period, length of photoperiod, light quality, preparation of explants transplanting density, etc.(ref. 1). These factors must be carefully assessed when considering the feasibility and practicality of photoautotrophic micropropagation. Influence of these factors have been examined experimentally to some extent but much more research needs to be conducted in future.

LARGE CULTURE VESSELS FOR PHOTOAUTOTROPHIC MICROPROPAGATION

The photoautotrophic tissue culture method makes it possible to use a larger culture vessel without the risk of increasing the loss of plantlets due to contamination. Fujiwara et al. (ref. 17) developed a photoautotrophic tissue culture system consisting of a culture vessel (28 cm wide, 53 cm long and 10.5 cm high) assembly, a gas flow assembly, and a culture solution assembly. This system was developed for shoots and/or plantlets at the rooting stage. A preliminary experiment showed that the dry weight and net photosynthetic rate of strawberry plantlets cultured with this system were nearly 2 times heavier and 10 times greater on day 21 than those cultured with a conventional tissue culture method.

The use of a large vessel facilitates the automation/robotization and automatic environmental control. Furthermore, if the plug tray, growth medium and transplanting machine being used for plug seedling production could be used also for micropropagated plantlets, the benefit would be substantial.

286

SOME PROTOTYPE ROBOTIC/AUTOMATED MICROPROPAGATION SYSTEMS RECENTLY
DEVELOPED IN JAPAN

In a recent review article of Aitken-Christie (ref. 18), it is
indicated that very few automation/robotization technologies are
being used in the commercial practice of micropropagation, with the
exception of some of the straightforward aspects such as media
preparation, handling of containers, and management of labor and
material requirements.

It is also stated in the article that significant progress in
research and development is being made towards the automation of
the entire micropropagation process. One of the more elaborate ways

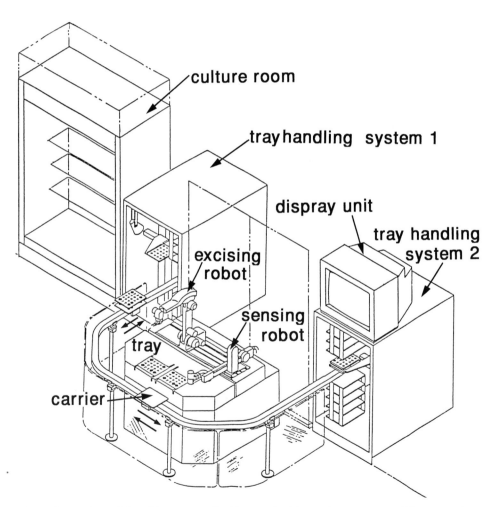

Fig. 5. Schematic diagram of the robotic micropropagation system
for plants with upright shoots and distinguishable nodes (ref. 19).

of automating operations in micropropagation is the application of robotic systems including a vision or sensor system and an automatic handling system.

In the following sections, three prototype robotic micropropagation systems recently developed in Japan are introduced. They were developed for exhibition the demonstration at the Industry/Technology Pavilion, Japanese Government Plaza, the International Garden and Greenery Exposition held in Osaka, Japan from April 1 to September 30, 1990. The author was the chairman of the technical committee on the demonstration of robotic micropropagation systems. The robotic systems were operated every day for 6 month for demonstration. In addition, an automated micropropagation system without vision capabilities recently developed in Japan is introduced.

Robotic Micropropagation system for nodal section method

This robotic system has been developed by Toshiba Corporation for automated micropropagation based upon axillary branching or nodal section method (ref. 19) which is the most commonly used method in the commercial micropropagation industry. The robot is developed for plant species which have one or two upright shoots and clearly distinguishable nodes and leaves, e.g., potato, carnation, Eucalyptus and redwood. The robotic system basically mimics a series of manipulations by a worker sitting in front of a laminar flow bench. A prototype of this system was first developed and demonstrated at an horticultural exhibition in Tokyo in 1988.

Fig. 5 is a schematic diagram of the robotic micropropagation system consisting of two robots. Fig. 6 shows the two robots installed in a clean room. One is a six-axis articulated robot with a scissor-like cutter for excising the plantlet and a forceps-like gripper for transplanting the explants. The maximum linear speed of the end effectors is 1 m/s. The robot is an industrial robot with the end effectors specially developed for micropropagation purposes. The other is a three-axis orthogonal robot attached with a laser emitter (semiconductor laser), a lens for producing a spot light, a laser scanner (movable reflective mirror), another lens for focusing the reflected light, and a position-sensitive device, for determining the node positions of plantlets on a culture medium (Fig. 7). Determination of the positions is conducted using the triangulation method with the image processing unit.

The two robots are controlled using a total control software system implemented on a 32-bit computer. One will be called "excising robot" and the other "sensing robot" hereafter.

Procedures employed by the system are as follows:

(1) A culture vessel containing 20 plantlets is uncovered and transported by an automatic conveyer system from the culture room within the robots work envelop. Each plantlet is cultured on a cylinder-shaped supporting material (plastic fibers, 1.5 cm in diameter, 4 cm in height) with nutrient solution. A vessel with only the supporting material is automatically transported beside the vessel containing plantlets from the other side.

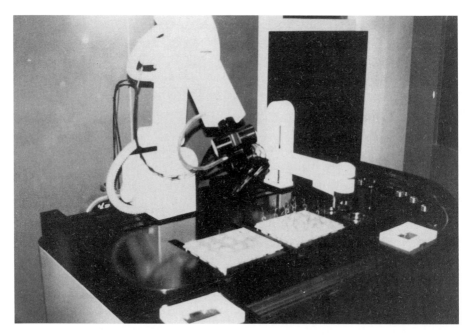

Fig. 6. A view of the two robots: excising robot (left) and sensing robot (right) (photograph: by courtesy of Toshiba Corporation).

(2) The sensing robot identifies the positions of the plantlet nearest to the position-sensitive device and then the node positions of the plantlet. Fig. 8 is an example of the node positions and cutting positions of carnation plantlet, recognized by the sensing robot and displayed on the screen.

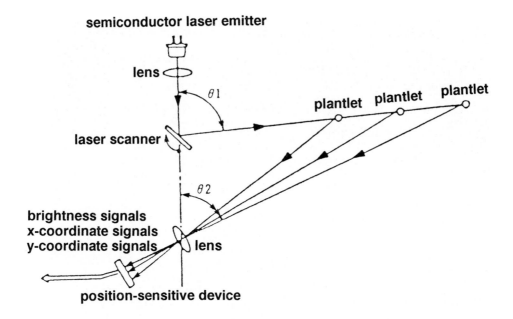

Fig. 7. Schematic diagram of the sensing system using the laser emitter and position-sensitive device (ref. 19).

(3) The excising robot grips the uppermost part of the stem softly (Fig. 9). The gripping force is controlled by a feedback control system with a strain gage, thin wire ropes coated with Teflon and a direct current servo motor, etc. (Fig. 10). The gripping force can be changed within a range between 0 and 200 grams with a resolution of 3 grams.

(4) The cutter which can orient itself to any direction approaches the stem and cuts the internode of the stem slightly below the gripped position to produce an explant (single node cutting with 2 leaves) 1-3 cm long. The gripper keeps holding the explant.

(5) The gripper transfers the explant and then inserts the stem of the explant into the supporting material at an appropriate location in the new culture vessel.

(6) The gripper and the cutter are cleaned up with a ultrasonic washer and are inserted into a container filled with alcohol for sterilization at predetermined time intervals. (The robot may sharpen the cutter by itself if it is so programmed.)

(7) The gripper returns to the plantlet from which the first

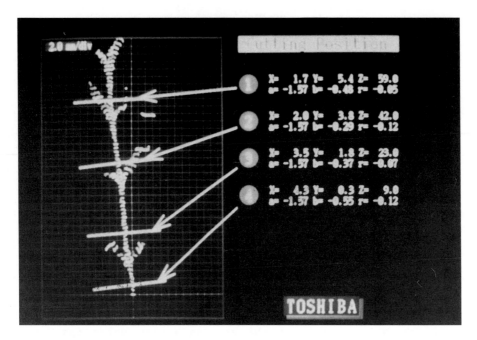

Fig. 8. An example of the node positions and cutting positions of carnation plantlet, recognized by the sensing robot and displayed on the screen (photograph: by courtesy of Toshiba Corporation).

explant has just been excised and grips the top of plantlet again to excise the second explant.

(8) The system repeats these procedures to obtain several explants from one shoot and then, starts excising the next plantlet to continue the procedures.

(9) The vessel with full of explants is covered and transported to the culture room automatically. An empty vessel is then transported in front of the robots. The vessel which was full of plantlets and has just been emptied is transported to the preparation room for washing and another source vessel containing 20 plantlets is entering the robot work envelope.

One cycle (from recognizing the position of the plantlet to returning to the same plantlet to excise the next explant requires approximately 15 seconds but the cycle time could be shortened to 10 seconds with a minor modification. If two sensing robots are attached to one excising robot, the cycle time can be halved.

The system is best suited for photoautotrophic micropropagation but it can also be used, of course, for conventional, heterotrophic or photomixotrophic micropropagation of any plants with upright shoots. The man-power needed for developing this system is estimated to be 3 man-years.

Robotic Micropropagation system for bulbous plants

Fig. 11 shows a schematic diagram of the robotic system for micropropagation of bulbous plants, developed by Mitsubishi Heavy Industries, Ltd. (ref. 20) under the guidance of Professor Miwa of Waseda University, Japan. The original idea (ref. 21) and a prototype of this system (ref. 22) were first presented by Professor Miwa. Fig. 12 gives a view of the whole system.

The system consists of the following subsystems:

(1) Delivery robot. This robot takes out the bulblet container from the container rack and places the container on the bulb scale separator; the container has 4 cuvettes (holes) each containing one

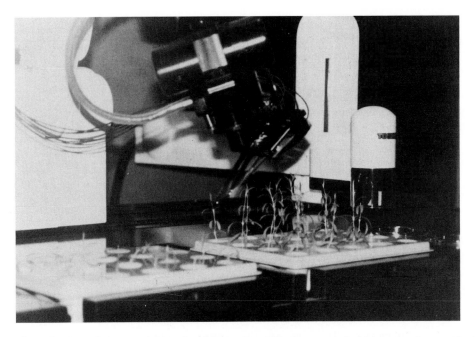

Fig. 9. A closer view of the end effectors and the sensing robot (photograph: by courtesy of Toshiba Corporation).

292

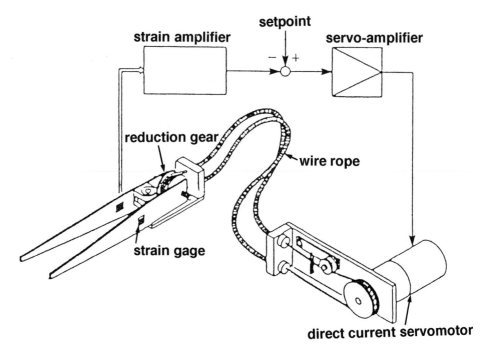

Fig. 10. Schematic diagram of the control mechanism of gripping force (ref. 19).

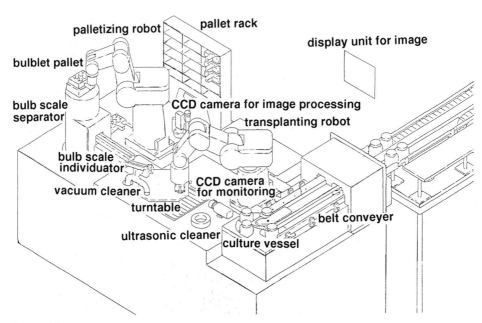

Fig. 11. Schematic diagram of the robotic micropropagation system for bulbous plants (refs. 20, 21 and 22).

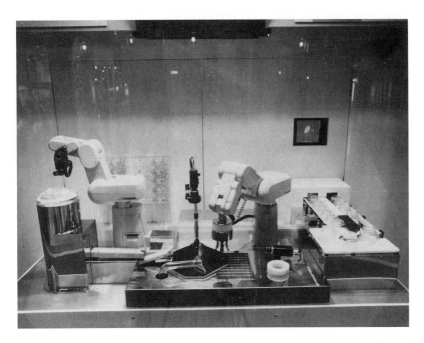

Fig. 12. A view of the robotic micropropagation system for bulbous plants (photograph: by courtesy of Mitsubishi Heavy Industries, Ltd.).

bulblet. The robot pushes down a button on the container to open the bottom of a cuvette and throw the bulblet down from the cuvette into the bulb scale separator. The robot transports the container back to the preparation room when all the cuvettes of the container are empty.

(2) Bulb scale separator. A rotating disk is installed at the bottom of the cylindrical separator. The rotating disk has small projections. The inside wall of the cylinder is covered with soft rubber. The disk is rotated at 2000 rpm for 2 seconds. During that time, the bulblet jumps up and down at random and hits the rubber wall due to the centrifugal force. The bulblet is thus separated into 10-15 bulb scales without any damage.

(3) Bulb scale individuator. This individuator, a vibrating, inclined V-shaped conveyer (also called parts feeder), makes the bulb scales into a line on the conveyer and sends them to the turntable separately.

(4) Image processing unit. A bulb scale is transported just below a CCD camera by the rotation of the table. The camera image of the bulb scale is processed with a microcomputer to determine the size, shape, orientation, centroid, etc. of the bulb scale. The image is displayed on the screen for monitoring by the human operator.

(5) Transplanting robot. This robot picks up the bulb scales from the table and transplants them onto the agar medium in the glass culture vessel (jam bottle) (Fig. 13). The stainless steal end effector has 2 vacuum tubes and 8 fingers (4 pairs of fingers). It picks up the scale with vacuum action one by one, holds 4 bulblets with 4 pairs of fingers, and transplants them onto the medium at the same time. This procedure is repeated once more; hence, 8 bulb scales are transplanted in one vessel. The depth of scales transplanted into the medium can be controlled by the robot. The transplanting speed is 18 seconds per bulb scale, 4800 bulb

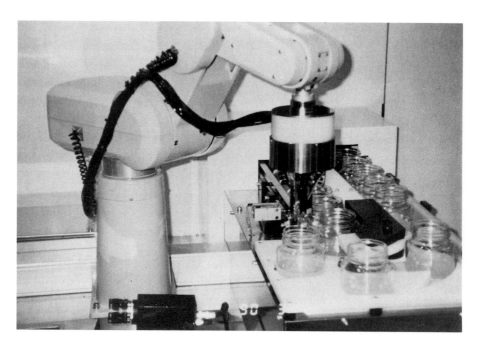

Fig. 13. A closer view of the transplanting robot of robotic micropropagation system for bulbous plants (photograph: by courtesy of Mitsubishi Heavy Industries, Ltd.).

scales per day. A bulb scale is usually grown to a bulblet within 90 days.

The end effectors are dipped in the ultrasonic washer at certain time intervals for cleaning . Refuse (broken pieces of bulb scales, etc.) remained on the turntable is cleaned up by a vacuum cleaner. This robotic system is sterilized in an air-tight box using ethylene oxide gas. The two robots used in this system, except the end effectors, are of commercial type for use in a clean room.

This system has now been tested for commercial micropropagation of Easter lily (Lilium longiflorum Thunb.) in Japan.

Robotic Transplanting system

This transplanting system is being developed by Japan Tobacco Inc. for transplanting micropropagated plantlets cultured in vitro on fibrous, cylindrical, plastic supporting material into cubic artificial substrates (rockwool cubes) with a hole at its central part of the top surface. Fig. 14 is a schematic diagram of the transplanting system consisting of a transplanting robot, an image

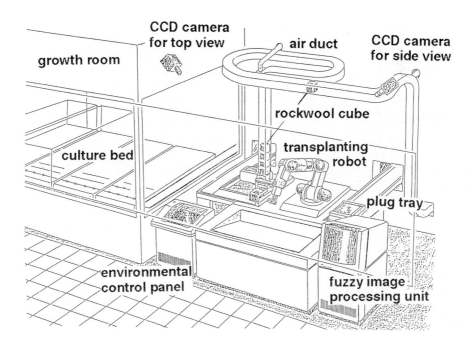

Fig. 14. Schematic diagram of the robotic transplanting system for micropropagated and acclimatized plantlets (ref. 23).

processing unit with 2 CCD cameras and an air duct system for transporting the rockwool cubes. The two cameras are used; one for obtaining a side view of the plantlet and the other for obtaining a top view of the plantlet. The robot arm is moved so as to obtain the views successively.

Fig. 15 is a view of the transplanting robot. The robot picks up the cylindrical substrate with a gripper from the plug tray (see, right hand side in the figure), recognizes the quality of the plantlet using fuzzy theory, then inserts the cylindrical substrate into the hole of the rockwool cube (left hand side in the figure) if the quality is satisfactory for transplanting. If not, the plantlet is discarded. Fig. 16 is a closer view of the robot. The plantlet will not be damaged by the gripper because the gripper does not touch the plantlet during transplanting.

Fig. 17 gives an example of the 2 images obtained by the cameras. The image processing unit determines the parameter values such as the number of leaves, leaf width, plant height, etc., based

Fig. 15. A view of the transplanting robot in operation (photograph: by courtesy of Japan Tobacco, Inc.).

Fig. 16. A closer view of the transplanting robot (photograph: by courtesy of Japan Tobacco, Inc.).

on the top and side images of the plantlet. Then, the unit calculates the grade of each parameter value by comparing them and their variations with those (membership functions) of high quality, using fuzzy matching method. Finally, the unit integrates the grade of each parameter and determines the integrated value, using fuzzy integral method. If the value is above a threshold value, it is determined to be a high quality plantlet.

A trial experiment using plantlets differing in size and shape showed that 96% of the judgment by the unit was acceptable. Thus, the fuzzy integral has proved to be a reasonable method for judging the overall quality of micropropagated plantlets as transplants. Features of high quality plantlets were selected from verbal expressions of human micropropagation experts by the persons who developed this system.

Automated masspropagation system

This system is being developed by Komatsu Ltd. for automated

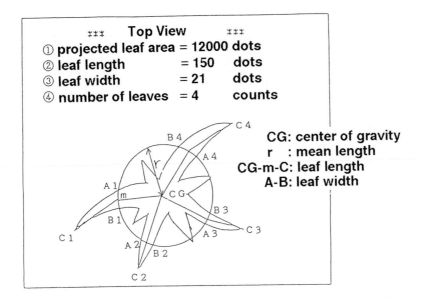

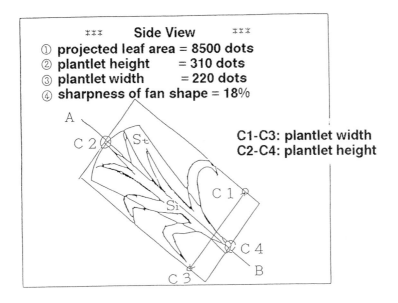

Fig. 17. An example of top and side views of micropropagated carnation plantlet, recognized by the image processing unit (ref. 23).

micropropagation using an axillary branching/nodal section and layering method (ref. 24). The design concept of this system is somewhat different from the other three systems.

Fig. 18 shows a schematic diagram of an automated masspropagation system for species with one or two upright shoots such as potato plantlets. The main part of the system consists of the excising and transplanting unit, the culture racks and the nutrient solution supply unit.

One culture rack which is installed with fluorescent lamps holds 5 culture containers (60 cm wide, 60 cm long and 15-20 cm high each). The container which is like a small, aseptic hydroponic propagation box has a punched metal sheet separating the aerial part from the root zone part to which the nutrient solution is supplied.

The excising & transplanting unit is attached with a sterilization unit at both sides. The sterilization of the container is conducted with a sprayer, producing mist of peracetic acid solution (0.4% conc.). It takes about one hour to complete sterilization using the spray system.

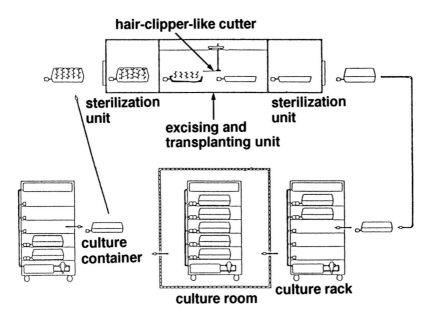

Fig. 18. Schematic diagram of the automated masspropagation system (ref. 24).

300

Fig. 19 shows the piping system connecting the container with the nutrient solution supply pipe, the nutrient solution drainage pipe and the clean air supply pipe. The joints connecting the container with the pipes are equipped with an electric heater for sterilization and cocks. The joints are closed with the cocks when removed from the pipes to prevent biological contamination.

In the excising and transplanting unit, there is a microcomputer-controlled, hair-clipper-like multi-blade cutter. The cutter moves horizontally at a certain height slightly below the plantlet height to obtain shoot tips or nodal cuttings as explants. Its motion is like trimming a lawn or pruning a hedge.

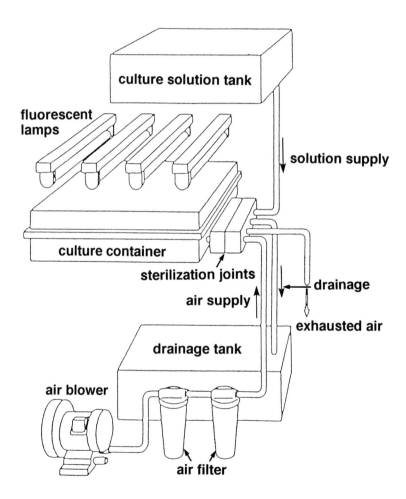

Fig. 19. Schematic diagram of the piping system in the automated masspropagation system (ref. 24).

The cutter holds a few hundred explants at one time on itself and transfers them to an empty container. Then, the explants are brushed off from the cutter, scattered and layered on a punched metal sheet in the container. The brush is operated in order to get an even distribution of explants over the metal sheet.

There is variation in the size of explants, mainly because there is no vision system. Besides, explants are not inserted into punched holes, therefore the orientation of layered explants on the sheet is random. Nevertheless, the growth of explants is relatively uniform.

Procedures employed by the system are as follows:

(1) A container with plantlets is transferred to the sterilization unit at the left hand side. An empty container is transferred to the sterilization unit on the other side. The two containers with covers are sterilized with peracetic acid mist and then rinsed with sterilized water.

(2) The containers are transferred into the excising and transplanting unit and the covers are removed.

(3) The cutter excises the upper part of plantlets to obtain explants in the form of nodal cuttings or shoot tips. The explants are then scattered (spread over) on the punched metal. Approximately 2,000 explants can be cultured in one container, therefore one rack holds approximately 10,000 explants.

(4) The cutter repeats the above operation at a lower height until it reaches the lowest part of the plantlets.

(5) The container is covered when it is full of explants. It is then transferred back through the sterilization unit to the container rack and connected with the pipes.

(6) The culture rack is transferred to either the culture room or multiplication room when it is loaded with 5 containers containing plantlets.

(7) Some racks are transferred from the culture room to the rooting and acclimatization room when plantlets are grown to a predetermined growth stage. The rest is used for further multiplication.

Fig. 20 is an illustrative diagram showing an overview of the masspropagation system. A feature of this system is its fast operation speed and a relatively complete sterilization method. This system will be suited for plants which require a very large cultivation area, such as potato, plantation crops, or trees for reforestaion.

302

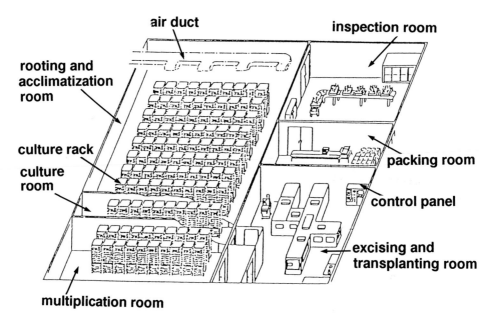

air duct

inspection room

rooting and
acclimatization
room

culture rack

culture
room

packing room

control panel

excising and
transplanting room

multiplication room

Fig. 20. An illustrative diagram of an overall view of the automated masspropagation system (ref. 24).

CONCLUDING REMARKS

A reduction of production cost by 90% or more is essential for worldwide commercialization of micropropagated plants in agriculture, forestry and horticulture. Automation/robotization in combination with environmental control or robotic micropropagation will be a key technology for realizing this drastic reduction in cost and thus for solving worldwide food shortage and reforestation problems in the 21st century.

In order to develop this key technology and to develop a robotic micropropagation system for commercial use, interdisciplinary and international cooperations are strongly required. Indeed, the micropropagation technology is an integrated high technology involving biotechnology, robotics, computer software technology, environmental control engineering, new material science, energy conservation engineering, ecophysiology, etc. We should make every effort to realize the system for our future.

Acknowledgment
The author wish to thank Ms. Jenny Aitken-Christie and D. K.C. Ting for their critical reading and stimulating discussion.

REFERENCES

1 T. Kozai, Micropropagation under photoautotrophic conditions, in: P. Debergh and R.H. Zimmerman (Eds.), Micropropagation: Technology and application, Kluwer academic publishers, Dordrecht, (in press), (1990).

2 T. Kozai, Autotrophic micropropagation, in: P.S. Bajaj (Ed.), Biotechnology in Agriculture and Forestry, Vol. 17 "High-tech and micropropagation I", Springer-Verlag, (in press), (1990).

3 T. Kozai, Controlled environments in conventional and automated micropropagation, in: R. Levin and K. Vasil (Eds.), Cell culture and somatic cell genetics of plants, Vol. 8 "Scale-up and automation in plant tissue culture, Academic Press, Inc., (in press), (1990).

4 K. Fujiwara, T. Kozai and I. Watanabe, Measurements of carbon dioxide gas concentration in closed vessels containing tissue cultured plantlets and estimates of net photosynthetic rates of the plantlets, J. Agric. Meteorol. Japan, 43(1), (1987), 21-30. (in Japanese with English summary)

5 T. Kozai, H. Oki and K. Fujiwara, Photosynthetic characteristics of Cymbidium plantlets in vitro, Plant Cell, Tissue and Organ Culture, 22, (1990), 205-211.

6 K. Watanabe, Y. Watanabe and N. Shimada, Effect of sucrose concentration in the medium on growth, apparent photosynthesis and Ribulose-1, 5-bisphosphate Carboxylase of Spathiphyllum plantlets in aeration culture, Plant Tissue Culture Letters, 7(2), (1990), 74-79. (in Japanese with English summary)

7 M. Nakayama, T. Kozai and I. Watanabe, Effect of sugar concentration in the medium and ventilation condition on the net photosynthetic rate of potato explants in vitro, Proc. of 11th Meeting of Plant tissue Culture Society, Okayama, Japan, (1989), 186-187. (in Japanese)

8 Y. Desjardins, A. Gosselin, R.C. jr. Beeson, Effects of photosynthetic photon flux and carbon dioxide concentration on activities of Ribulose-1, 5-bisphosphate Carboxylase $^{14}CO_2$ fixation in tissue-cultured plantlets of Fragaria x ananassa Duch., (submitted to Plant Physiology, (1990).

9 T. Kozai, H. Oki and K. Fujiwara, Effects of CO_2 enrichment and sucrose concentration under high photosynthetic photon fluxes on growth of tissue-cultured Cymbidium plantlets during the preparation stage, in: G. Ducate, M. Jacobs and A. Simpson (Eds), Plant micropropagation in horticultural industries, Arlon, Belgium, (1987), 135-141.

10 T. Kozai, Y. Iwanami and K. Fujiwara, Effects of CO_2 enrichment on the plantlet growth during the multiplication stage, Plant Tissue Culture Letters, 4, (1987), 22-26. (in Japanese with English summary)

11 T. Kozai and Y. Iwanami, Effects of CO_2 enrichment and sucrose concentration under high photon fluxes on plant growth of carnation (ianthus caryophyllusL.) in tissue culture during the preparation stage, J. Jap. Soc. Hort. Sci. 57, (1988), 279-288.

12 T. Kozai, Y. Koyama and I. Watanabe, Multiplication of potato plantlets in vitro with sugar free medium under high photosynthetic photon flux, Acta Horticulturae 230, (1988), 121-127.

13 M. Mousseau, CO_2 enrichment in vitro. Effect on autotrophic and heterotrophic culture of Nicotiana Tabacum (var. Samsun). Photosynthesis Research, 8, (1986), 187-191.

14 T. Kozai, A. Takazawa, I. Watanabe and J. Sugi, Growth of tobacco seedlings and plantlets in vitro as affected by in vitro

environment. Environ. Control in Biol. 28 (2), (1990), 31-39. (in Japanese with English summary)

15 Y. Desjardins and A. Gosselin, Effect of photosynthetic photon flux and carbon dioxide concentration on the activity of carboxylating enzymes and growth in tissue-cultured <u>Fragaria</u> <u>x</u> <u>ananassa</u> Duch. Plantlets following transplanting. Submitted to Annals of Botany, (1990).

16 Y. Desjardins and A. Gosselin, Effects of CO_2 concentration and photosynthetic photon flux on growth, leaf anatomy and chloroplast ultrastructure of tissue-cultured <u>Fragaria</u> <u>x</u> <u>ananassa</u> Duch. Submitted to Plant Science, (1990).

17 K. Fujiwara, T. Kozai and I. Watanabe, Development of a photoautotrophic tissue culture system for shoots and/or plantlets at rooting and acclimatization stages, Acta Horticulturae 230, (1988), 153-158.

18 J. Aitken-Christie, Automation, in: P. Debergh and R.H. Zimmerman (Eds.), Micropropagation: Technology and application, Kluwer academic publishers, Dordrecht, (in press), 1990.

19 H. Watake and A. Kinase, Robot for plant tissue culture, Robot, No.64, (1988), 74-79. (in Japanese)

20 Anonymous (Engineering Center, Mitsubishi Heavy Industries), Documentation on the robotic system for producing virus free bulblets of lily. (1990), 8pp. (in Japanese)

21 Y. Miwa, T. Yamamoto, Y. Kushihashi and H. Kodama, Study on automation of plant tissue culture process, J. of Precise Engineering, 54(6), (1988), 99-104. (in Japanese).

22 T. Murase and Y. Miwa, Automation of organ culture process of lily bulbs, Proc. of Japanese Society of Precise Engineering, (1989), 38-39. (in Japanese)

23 M. Nagaoka and M. Sei, Masspropagation system of tissue-cultured plantlets, Nogyo-oyobi-engei, 65(1), (1990), 87-92. (in Japanese)

I. Karube (Ed.) *Automation in Biotechnology*
Proceedings of the 4th Toyota Conference, 21–24 October 1990
© 1991 Elsevier Science Publishers B.V. All rights reserved

THE RUTHNER CONTAINER SYSTEM

Dr. E. RUTHNER

ASG AgroService GmbH, A 1010 Wien, Stubenbastei 12, AUSTRIA

SUMMARY
    An environmental controlled system for the continuos, all year
round production of fresh, living plants for the nutrition of men
in the same way as for the purpose of producing and maintaining
test plants for different research activities with combination
possibilities to modern robotic techniques and computeranalysers
in a modular containerized size is explained and discussed with
respect to economics.

INTRODUCTION

    In the last years there was an important change in production

methods of both, crops and ornamentals. The wish to harvest better

crops by intensive cultivation methods and equipment is as old as

horticulture is.

    In the same way a lot of investigations had been made to extend

the growing period against seasonal limitations. Even the use of

land changed and horticultural crops spread out into regions where

under normal conditions it was never thought to grow plants for

the needs of a fresh market.

    The developement of technology enhanced these efforts and due

to changing environmental conditions and air pollution, these

developements might never be finished, really.

    Depending on the increasing production capacities needed for

the different markets and due to changing environmental conditions

the development of new alternative growing techniques has to

ensure the high quality of products which are asked by todays and

future consumers much more,than before.

Today's problematics

The use of too many fertilizers and pestizides, the pollution by industries and city waste, the danger of radioactive fallout and catasrophes are the motivations for the development of systems for environmental controlled crop production.

Increased costs of labour, legal regulations on working place and lately as well ecologic limitations are the reason for a big rise in the production costs, which could not be brought into relation to the consumers market any more.

The shortage of energy with related increase of energy costs demands for a continuous improvement of both, traditional and new technologies in search for better efficiency.

All mentioned above, together with the problems of bad infrastructure and corresponding transport systems have to be seen as a vicious cycle. To get rid of it should be one of our main duties in future.

Solutions

The development of these, new, and future technologies for fresh plant production should solve all these problems, but the experience of the past has demonstrated that a unique answer, with universal respect to all the needs, can only be a solution with compromises.

Traditionally thought, the main goal to achieve is to decrease the production costs by
a) lowering the costs for crop protection as building construction and cladding,
b) lowering heating and ventilation costs,
c) using crop techniques wich permit minimazation of labour,
d) selecting better adapted seeds - introduction of hybreds,
e) more efficient use of nutrients,

f) use of crop adapted pesticides.

We think that the goal has to be seen in establishing an all year round crop production, which is well known as the only chance for some specific areas in the world, areas where they can`t get a second or a third crop within one year especially with outdoor agriculture and for growing a very uniform genetic material like tissue cultures and clone propagation.

The Organisation of a real all year round crop production in arctic and arid areas is somewhere still in an experimental stage. A lot of efforts have been made around the world, so in Japan, in the USA and UdSSR, in Israel, in the UK and the Netherlands and out of this some productive techniques have been developed and are in use and it is possibel to row names and ideas, all together with the background of many scientific institutes.

With that experience and our own philosophy created by Prof.Dr.Othmar Ruthner in the 60th'ies, we tried to concentrate all results to make proposals on that matter which harms human beings most in future:

the confrontation
of
fresh plant production with the endangered environment

First of all it is no rule to have a crop production soil-bounded, what has to be seen in two ways:
On the one side there is no other use of natural soil than as a substrate with buffer capacity, and therefore it can be substituted best by hydroponics really,
and on th other side, it is a fact a plant has not to be grown in one stationary place. It is possible to move plants while they are growing - in specific situations it is a big advantage, especially for the most efficient use of installations for providing a controlled environment such as lighting equipment, ventilation, nutrient supply and transportation system.

Now it is a fact too, that traditional greenhouse systems are very wide, flat buildings what results in having a large contamination surface - and this bears a double risk:

- 1) pollution from underground (groundwater, soilbound by city waste, wastewater from neighbouring industries and agriculture ...)

- 2) pollution given by surrounding air (trafficexhaustion, industry-smoke, pesticides-dust, fall-out etc.)

To avoid this, our goal is to grow as much as possible in a closed and controllable environment with high volume and small surface. It is the same to compare geometric figures of a flat one looking like a traditional greenhouse and a cubic-like building. If both have the same volume, the total surface of the cube is much smaller than that of the flat figure.

In other words it does mean that any building most similar to a cube has less contamination problems with endangered environment. (Ideally seen the optimum figure is a globe)

In a traditional constructed greenhouse people have to grow plants on benches and therefore they have to move them for handling and cultivation work - and that results in a lot of work. Finally many of unfavorable microclimatic effects are the result of fixed growing places in such traditional systems too.

Our strategic is to grow plants without these disadvantages. Following the principle of the Ruthner philosophy, growing the plants on continuous moving hangers in a closed environment and its own ecosystem. we started to design and to construct production units to face all these problems.

Our principle of solution had to be realized with following criterias if possible to be enclosed. For items too difficult to be included, the best compromise had to be worked out:

* unchanging high quality of products

* all year round production

* production does not harm the environment

* screening of pollution and nuclear radiation

* continuous or batchwise production

* independence of seasons and climate

* less ground works

* low construction and building costs

* spacesaving

* modular design

* easy to be transported

* relocatable

* integrationable to traditional greenhouses

* order/contract production

* good working conditions

* efficient energy throughput

* precalculable production costs

The system being developed also has to be set up on the experience with modern equipment for horticulture, has to use the advantages of three-dimensional growing systems and to avoid their disadvantages as shown in the past. It has to be costefficient and should be designed to compete traditional greenhouses.

The new system has to follow the modular-design principle, and a combination with equipments bought in addition and well known to a gardener should be possible.

For the use of natural sunlight, translucent walls should be foreseen but in the case of special production requirements a hundred percent closed unit with a most efficient light radiation system should be delivered as an option.

To maintain the controlled environmental conditions according to any growing pattern program, the unit walls have to be isolated. For climatisation every airflow should be filtered, the input and the output of water, materials and crop should be controllable.

The system has to be layed out in that way that the optimum growing dimensions are given for every crop to be produced. If there is a need to connect the modular units with existing greenhouses or to incorporate them, it should be possible.

The production of locally desired crops should be possible in the most areas of interest of the fresh products markets.

Realisation:

There are may transport container on the market. Out of these the 40`feet container can be used especially as the basic volume for the installation of a moveable three-dimensional crop growing system. It represents a optimum use for the point of view of modular design.

Containers can be set up side by side, one on the top of the other and one after another. The plants are cultivated in mobile hangers. These hangers are led by a horizontal conveyor system within the container.

Depending on the type of plants there can be installed a certain number of cultivation trays in each hanger. Both, hangers and the conveyor system have to be installed that way, that the integrated irradiation system can illuminate the whole container with all the mounted trays and can be used as supplementation to daylight, if needed, as well as an entire replacement of daylight.

The hydroponic system for feeding of plants with nutrient solution and/or water is installed for each cultivation tray seperately.The feeding can be made either at one or more places.

The airconditioning system depends on the needs of location, climatic conditions and crop planned and is assembled and installed individually.

Connecting load for the completely isolated system (incl. irradiation system and airconditioning system) is max.30 kVA, out of this, the use of the artificial irradiation system is 15 KVA

THE RESULT

The result of our new investigations is an advanced principle using the RUTHNER Container System with its single growing unit, with the brand name, the ÖKOTAINER Typ Cont/60g and with the technical description within the definition for Middle Europe.

The basis of a ECOTAINER is a 40 feet steelcontainer with the outside dimensions:

	length	12,0 m
	width	2,4 m
	height	2,6 m

Walls incl. isolation: 60 mm

Windows: 2 x 5 Thermoglass screens   each   2000 x 800 x 24

Fig. 1. Photograph showing the outside view of a ECOTAINER

312

The mechanical installation: (the conveyorbelt system) consists of a rolling track, the hangers with the traylevels, the trays and the driving unit for moving the hangers on the rolling track.

Fig. 2 . Photograph showing the conveyorbelt system

Rolling track: 2 parallel tracks, each linked at the end with the
other one   lenght   22,5 m
                                                 radius   0,6 m
Hanger: 60 hangers with 1 - 6 levels each to support the
        cultivation trays
        max. load per hanger   80 kg
Tray: Size   600 x 330 x 50 mm
      Distance of levels (6 trays/hanger)   380 mm
      Total number of trays (6 trays/hanger)   360
Driving unit: pneumatically with plunger and air compressor
                        compressorinput   1,2 kW
                        rotating speed   1-2 m/min variabel

Nutrient supply: System consisting of tanks, pumps, outlets and
a gutter for recirculating the nutrient solution.
Each tray has a seperate in and outlet and a
separate leveling device to be regulated.

                              Tank volume: 2 x 800 l
                              Pump delivery: 360 l/min

Fig.3 . Photograph showing the nutrient supply system

Airconditioning system: 4 compact cooler conn.load each 3,1 kW
                        2 heating units conn.load each 2,0 k
                        day/night - cycle programmable
                        temperature range   16⁰ - 30°C

Irradiation system: 36 Highpressure Sodiumlamps   400 W each,
                    day/night period programmable
Electric device system: 3x 220/380V, 50Hz with neutral conductor
                        max.connecting load 30 KVA

The ECOTAINER will be delivered inclusive electric control system and turn key installed. Water, electric current and a gully has to be foreseen by the customer. Besides levelling there is no need of foundation constructions.

MARKETS

In consideration of making compromises it is obvious, that there are ideal markets defined for the RUTHNER Container System. Nevertheless all countries with the following characteristics are proper markets:

* consumer-consciousness of quality

  (interest in qualitatively outstanding,

  healthy plant products)

* pollution and/or radiation endangered areas

* existence of conurbation

* existence of industrial centres

* ambition for autarky

* low energy costs

* high labour costs

* high property costs

* lack of space

* no traditional horticulture possible

* unfavourable infra-structure

* high transportation costs

* dependence on imports

* arid climatic zones (arctic, subtropical arid)

(If only about 4 criterions mentioned before apply, a marketing of RUTHNER Container System might be of interest.)

Considering the markets according to the clientel, there are mainly young-plants-producers, reproducers, florists, producers of vegetables and herbs, and nurseries among horticulture, who will be interested in this system. For agriculture it is interesting for breeders of highly performanced hybrid seeds, institutions and companies, which are engaged in resistance tests and contamination tests.

Within the range of industrial manufacture the use of the Container System is interesting for the continous production of medicinal plants, pot-herbs, baby-food, as well as fresh vegetables with low nitrate content.

PROFITABILITY

One unit of the RUTHNER Container System with the brand name ECOTAINER including the completely isolated container with all equipment as the complete conveyorbelt system, the irradiation system, the airconditioning system and the nutrient supply costs

appr. AS 850.000,-- (excl.VAT, ex works)

1. Operating costs for one single ECOTAINER of the RUTHNER Container System for Austria, Far East and countries with low costs for energy (Gulfstates, USA, UdSSR, Canada, ...)

example:

investment: AS 850.000,--

interest rate: 8,5 %

amortization: 10 years

man-hours: 520 hours per year and ECOTAINER

version I: ECOTAINER with use of natural daylight
plus 100 days additional irradiation
version II: ECOTAINER with 100 % irradiation only

1.1. Austria:

 est. costs for energy: AS 1,30 per kWh

 est. costs for man-hour: AS 80,-- per h

operation costs per year incl. annuity, energy, man-hour, spare

parts approximately: version I : AS 250.000,--

            version II: AS 340.000,--

1.2. Far East:

 est. costs for energy: AS 0,80 per kWh

 est. costs for man-hour: AS 80,-- per h

            version I : AS 240.000,--

            version II: AS 310.000,--

1.3. Countries with low energy costs:

 est. costs for energy: AS 0,15 per kWh

 est. costs for man-hour: AS 80,-- per h

            version I : AS 200.000,--

            version II: AS 230.000,-- "

2. Production capacity of one ECOTAINER vs different crops:

2.1. Production of fresh herbs:

 * chives, parsley, dill, chervil, marjoram, basil,

  thyme, mint, citronella, rosemary, etc.

   - specific to the variety of herbs:

      appr. 3000 - 5000 trays/year/ECOTAINER

      or appr. 300.000 - 400.000 portions

        (3 - 5 to/year)

  * cress: appr. 400.000 boxes/year/ECOTAINER

       (box: 120 mm X 70 mm)

2.1.1. rentability - example "fresh herbs" in Austria:

Production standards:

* equipment: 60 hangers, with 6 cultivation trays each,
             what results in 360 trays
             (size of tray: 600 mm X 300 mm)
* growth of herbs: 35 days (excl. germination)
* operation: 365 days with artificial light
* program for irradiation system: 18 x SON400W,
                        appr. 12 hours light, appr. 5.000 lux
* temperature day/night: appr. 24/16 °C
* nutrient solution
* substratum: expandet clay granulate , size 2 mm
* sowing in "multiplant-trays"

Annual production per container:
   360 X 365/35 = 3.754 trays fresh herbs

Annual production costs:
----
costs for seed   appr. AS 10.000,--
material   "   "   5.000,--
spare parts   "   " 15.000,--
energy   "   " 60.000,--
man-power   "   " 60.000,--
----
production costs   appr. AS 150.000,--

Production cost per tray: i.e. AS 40.-- /tray

Fig. 5. Photograph showing the cultivation trays with seedlings

Est. proceeds per tray: wholesale trade   appr. AS 150.-- /tray

                          retail trade   180.-- /tray

                          gastronomy   180.-- /tray

Est. proceeds per portion *):   AS   5.-/portion

to the ultimate consumer, with 51 portions

per tray it sums up to   AS 255.-- /tray

*)based on selling prices for ultimate consumers for single

portions: in supermarkets: AS   8.--

              in delicatessen stores: AS 12.--

In other words,

- if the whole annual production can be sold directly to the

ultimate consumers (e.g. own delicatessen stores,..), and the

proceeds are calculated at AS 5,00 per portion with an annual

production of i.e.3754 trays with 51 portions each, there is a realized profit before taxes of AS 807.110,-- per year, or if the complete production is sold at the wholesaler trade, profit is sinking to an amount of AS 412.960,-- before taxes, from one to the other version the invested funds bear a 48 % - 95 % profit and is a proof for an excellent way of investment.

This model can be taken also for the production of all other herbs, it is also possible to cultivate all the herbs mentioned before at the same time in one ECOTAINER, in this case risk of selling is becoming less too.

2.2. Production of Ornamental plants:

The capacity in ornamental plant production is dependent on the size of pots, which is used respectively, on the duration of the planned growing period in the system and the kind of plants to be produced.

- Referred to pots size 5 this means a capacity of 19.800 pieces with growingperiod of 8 weeks 128.700 new plants can be produced a year.
- If pots size 8 are used, there is a capacity of 10.800 pieces and with a stay of 16 weeks there will be a production of 35.640 new plants.
- For pots size 10 there is a capacity of 5.400 pieces. With a term of growth of 16 weeks 17.820 plants will be produced.

If you put in the usual proceeds for the annual production, a realistic marketing chance for the RUTHNER Container System will result for countries with a horticultural products demand similar to the Austrian/European market.

For example:

Production of Syngonium - young plants:

Out of an annual production of 128.700 young plants 15cm high, delivered in potsize 5, you can get a selling price per plant of AS 8,-- and thereof a turnover of   AS 1,029.600,-- With production costs per plant of   AS   4,-- the profit before taxes per year is   AS   514.800,--

As well as in case of the production of delicate vegetables, when starting from seed, the vegetation phases: germination, cultivation of young plants and maturing to harvest are very important.

These phases also apply to the production of ornamental plants except there is a vegetative multiplication, what means, that instead of seed multiplication, before production starts, there is first a clonecutting propagation.

In case of propagation from tissue-culture first you have a phase of adaption, whereby the tissue-culture will be bought or produced in separate and special laboratories.

Apart from the possibility to do all necessary phases of production in one ECOTAINER, you just have to use only some special hangers.

In the following study we always combine one ECOTAINER for cultivation with one or more for production to obtain a better efficiency.

2.3. Production of fresh vegetables:

With regard to the different market situations in the world, we like to mention only the production capacities of one ECOTAINER. We understand that for well organised markets, as we have them in Europe, the use of such systems for this kind of vegetables to produce is less economic.

That there are some specific markets in the North and in desert climates is realistic and the figures about production capacities may help to evaluate different feasibilities:

tomatoes	5.000 kg/year/ECO.
cucumbers	8.000 kg/year/ECO.
hot pepper	3.500 kg/year/ECO.
lettuce	25.000 pieces/year/ECO.
kohlrabi	50.000 pieces/year/ECO.
radish	375.000 pieces/year/ECO.

With RUTHNER ContainerSystem it is possible to combine one or more ECOTAINER to reach a certain size of vegetable production output.

2.4. Production of Test plants:

Especially in the field of research in plant physiology, in breeding and genetics as well as in resistance and contamination tests for a lot of many very sophisticated institutes for agriculture and industrial research labs, there is a need for having all the year round certain quantities of testplants for tests in vivo and vitro.

With the RUTHNER Container System, using the ECOTAINER it is possible to grow testplants under desired environmental factors from January to December just on side of the labs where they are needed.

It is easy to decrease cost assumptions for budgeting plant research projects if you know the possibilities of shortening the demand of time for the whole research project.

Usual research people has to calculate minimum five growing seasons for reasons of statisic realability for a certain result, now they have the chance to get it within one year.

The reproduction of testplants under controlled environment

is possible by continuous daily planting and harvest or batchwise with regards to   follow   a   certain   environmental   programm automatically.

Due to the closed system of the ECOTAINER it can be used for studies   on   the   uptake   and   placement   of   radioactive   marked nutrients with isotopes. The system can be handled just from one location with all the safety devices like handling openings and etc., all typical for such work.

We think, that it must be possible to connect the ECOTAINER with some kind of robotic workcells even for automatic handling and computer management.

2.5. Adaption of tissue culture plants and clone propagation:

The   problems   of   growing   tissue   cultures   in order of clone propagation had been expierenced before and we found a real good solution by using the Ruthner Philosophy.

To   achieve   uniform   young   plants   out   of   tissue   cultures techniques mean to us not only to keep the clones alive. It seems to us very important to have uniform plants of a certain phenotype.

In our own adaption work with tissue cultures of special ornamentals, as there were Syngoniums, Spathiphyllums, Ficus and Musa, we got excellent propagation rates due to a special environment programm possible by the Ruthner Technology.

A well defined humidity decreasing programm in combination with programmed irradiation ensured best quality of resulting young plants, best prepared for being transplanted outside. Therefore the ECOTAINER should be used on line after the tissue cultivation.

The percentage of succsessful adaption was best with Syngoniums (97 %), with Spathiphyllum 92% and with Ficus 91%. Musa had been very sensitive to the age of the cultures.

We noticed that in our system the adaption was less problematic since the tissue cultures had been more developed within their sterile plastic box.

Starting the adaption program with 95% relative humidity in the surrounding air for the first week, followed by a decreased value of 85% for the next week and after a final week with 75%, the adaption from tissue culture sensitivity to a normal young plant is finished sucsessfully. Now the so produced tissue culture grown plants can be handeld in the usual way and are ready for delivery. We are shure that this is just a smale scale mentioned, where the development of the Ruthner Container System bears a realistic chance to be used for.

SUMMARY

With respect to the company`s know-how having developed the Container System there is an exciting chance for the economical use of the Ruthner Container System.
1) providing domestic gardeners with this new technique to compete nursery- and young plants imports producing independent of season and being able to be on time with orderdeliveries and contracts.
2) to become a serious partner to the exclusive markets of hotellery, restaurants and delicatess-shops by delivering fresh and healthy herbs in the range of fresh vegetable production with the new possibility of continuous uniform and high quality plant production, and
3) to provide research activities by realizing a continuos and uniform testplant production with the aim to achieve a chance for more efficient budget planning due to get statistical results much earlier and to give automation and robotic techniques additional input to be used in plant physiological research,
4) and to achieve a possibilty of adapting tissue plants from sterile media to greenhouse environment by dehumidification steps program inside the ECOTAINER
We see a very strong possibility in selling licences to interested groups of foreign countries in the same way as in the installation of joint ventures.
The development of the Ruthner Containersystem with its small production units (ECOTAINER) and their modular layout with the possibility of being combined with traditional horticultural facilities (greenhouses in the same way as Ruthner Tower Systems and scientific laboratories) is the consequent result of many years work.
The prospection of the feasibility of one ECOTAINER vs. different crops demonstrates the unique chance such systems will have in future too.

I. Karube (Ed.) *Automation in Biotechnology*
Proceedings of the 4th Toyota Conference, 21–24 October 1990
© 1991 Elsevier Science Publishers B.V. All rights reserved

REQUIREMENTS AND TECHNOLOGIES FOR AUTOMATED PLANT GROWTH SYSTEMS ON SPACE BASES

T.W. TIBBITTS[1,2] R.J. BULA[1] R.C. MORROW[1] R.B. COREY[3] AND D.J. BARTA[1]

[1]Wisconsin Center for Space Automation and Robotics, University of Wisconsin-Madison, Madison, WI 53715

[2]Department of Horticulture, University of Wisconsin-Madison, Madison, WI 53706

[3]Department of Soil Science, University of Wisconsin-Madison, Madison, WI 53706

SUMMARY

Plant growth systems involving the use of higher plants, and possibly algae, appear to be a necessity for long duration habitation in space. Use of plants in a bioregenerative life support system would minimize the cost of removing carbon dioxide and providing food, oxygen and pure water to maintain humans in space. The requirements for such life support systems will involve technologies that are quite different from those of systems being used for short duration space missions. The plant growing system, including the support equipment needed to sustain growth, harvest the crop, process the useful edible product, and recycle the waste products, must be of minimum size and weight to reduce the cost of transport to the space base. The system must incorporate a high level of automation and robotics so that the astronaut-hours required for plant maintenance are kept to a minimum, but have provision for astronaut interaction if system malfunction occurs. The system must be constructed to sustain growth and productivity of several different plant species simultaneously, to provide diversity in the diet, and redundancy in case of loss of one or more of the species. The growing area should be compartmentalized so that the individual units can be isolated for separate maintenance, cleaning and sanitation. All chamber and plant culture equipment must be constructed from materials that can be effectively sanitized at appropriate intervals. Another important requirement for space bases will be the need to keep power consumption at low levels. Effort is being directed toward the development of new technologies for plant growth in space. Progress has been made in the development of an improved plant lighting unit, a nutrient delivery system that can supply water and nutrients to plants in microgravity, and a nutrient composition control system utilizing ion exchange materials for maintaining nutrient concentrations and nutrient balance for plants. There are many additional technology needs that will require resolution for the effective operation of a plant growing system in space. Technologies need to be developed for effective gaseous exchange between the plant growth units and the human habitation areas, control of pathogenic microbes in the nutrient media, identifying and controlling contaminants that accumulate in the atmosphere and nutrient solution, and for monitoring the productivity of the plants.

INTRODUCTION

The establishment of permanently inhabited space bases will very likely involve the use of plants to reduce the cost of providing life support for the inhabitants. Plants have the potential of not only providing food, but they can provide oxygen, remove carbon dioxide, and recycle water. Their value for regulation of oxygen, carbon dioxide and water is greater than for food because the daily requirements to support one person are 0.9 kg of oxygen, 1.1 kg of dry filter to remove carbon dixoide, and 4.1 kg of clean water for drinking and hydration of food, whereas only 0.6 kg of dry food are needed. Also, it has been reported that 13 m² of growing plants can provide the daily, per person life support requirements for oxygen, carbon dioxide and water whereas at least 20 m² of space is required to meet the food needs (refs. 1-3).

At the time of the exploration of the moon, both the U.S.S.R. and the United States initiated considerable effort to study the potential of using plants for life support. These efforts have been directed toward the use of both higher plants and algal systems, both of which can provide a useful oxygen supply, carbon dioxide removal, and water recycling. However, as a food source, higher plants provide significantly more flexibility and usefulness than algae because algal foods are not as palatable and can be toxic if large quantities are ingested. Most of the comments in this report will be centered around the use of higher plants.

A plant growth system in space will be most valuable when it is a component of a total life support system that recycles the inedible plant parts and the human wastes through processing components that regenerate carbon dioxide, water and other elements in forms that can be utilized again by plants.

This report will review a number of general requirements that will be necessary for the development of the plant growing component for bioregenerative life support and then discuss specific technologies that need to be developed if we are to have an operating and effective bioregenerative life support system.

REQUIREMENTS

Minimum size and mass

A primary requirement of an operational life support system at a space base is that the plant growth component must be as small as possible and of minimum mass. The size requirements are dictated primarily by the productivity of the plants, expressed as grams per cubic meter per day (g $m^{-3}d^{-1}$) of digestible food. Certain low growing plants, such as potatoes and wheat, have been found to have high productivity and would be particularly suitable for a bioregenerative life support system (refs. 4-6). Volume requirements for the plant growing unit could be reduced by careful regulation of plant spacing during early plant enlargement. Mass requirements could be reduced through utilization of thin water film growing procedures, and development of minimum weight production, harvesting, food processing, and waste recycling hardware.

Maximum recycling

The major requirement of a bioregenerative life support system is that a high proportion of the life support elements be recycled because the portion not recycled must be provided from storage or resupplied from earth. In addition, the non-recyclable 'waste' portion will have to be stored or returned to earth at a significant cost.

Water, which makes up the greatest percentage of the life support waste materials, must be recycled. Technologies are required to effectively obtain highly pure water for drinking and hydration of food. Organic compounds, such as cellulose, hemicelluloses and lignins, make up essentially 90% of the solid portion of the inedible plant parts and human waste. These require oxidation to produce CO_2 and H_2O which can be recycled in the system. Mineral elements, even those essential for plant and/or human nutrition, are required in small amounts and will be recycled in certain space bases and may not be recycled in other bases. This includes potassium, nitrogen, phosphorus, calcium, magnesium and sulfur as well as the significant amounts of sodium chloride utilized in human food preparation. Other minerals present in very low concentrations in the waste material, and which are required in these low amounts by plants and humans, may not need to be recovered since resupply costs for such small amounts of material would be low.

Low power consumption

Power consumption will be of critical concern in space bases and long duration missions that rely on solar collectors for electrical energy production. It will be of less concern on space bases or missions that would be powered by nuclear generators. The demands for power are large in a bioregenerative life support system. Power for the plant growth lighting system is a major concern because a large amount of lighting is needed to grow plants effectively. A minimum of 300 W m^{-2} of electrical power is needed for 16 hours each day. In addition, essentially all of this electricity, at least 90%, becomes heat and must be removed by conductive and convective cooling systems along with additional energy to operate the motors and fans required to circulate the air over heat exchangers so as to maintain a favorable temperature in the growing area. Also, the power consumption needs for the waste processor are significant because large amounts of energy are required to evaporate water in waste processors and collect it by condensation on cold coil systems.

Automation and robotics

Automated procedures for operating the bioregenerative life support system are needed to minimize the amount of astronaut time devoted to life support. The system, however, should not totally exclude human interaction because this may be desireable for solving problems that arise and also for enhancing the daily routine of the astronauts. There is a need for developing automation and robotic systems that are as simple as possible so that they can be repaired and maintained with minimum effort. Robotic devices need to be developed that can manipulate tender plants and plant materials easily damaged by mechanical systems.

Several different plant species

The plant growing component must be designed to permit the growth of several different crop species to provide variety in the diet and to provide redundancy in the event that a particular crop fails. It will be desirable to select species that have similar growth requirements in terms of temperature, lighting, and nutritional demands so that they can be grown together in the same growing area. Therefore, it would be desirable that the growing system be constructed with several separate compartments so that each can be individually operated and controlled. This will also provide the opportunity to clean, sanitize, and perform maintenance on one unit without interfering with the productivity of the other units. Planting of each species will need to be staggered in each of the sections spaced only a few weeks apart, so that a nearly continuous harvest is maintained. This will provide more uniform oxygen production and minimize food storage requirements.

Non-toxic materials

The system must be constructed of materials that will not release compounds that are toxic to the plants or to microorganisms used in biological waste treatment. Of principle concern, are volatile organics released into the air and heavy metals released into the liquid streams from the structural materials used in the various components. Volatile organics from materials, such as plastics, paints, sealants, can seriously affect plant growth in closed recycling environments. Isobutyl phthalates have been identified as one of the more troublesome gases and cause injury to plants at approximately 1 PPB (ref. 7). Heavy metals, such as nickel and cadmium, released from piping and containers that contain and supply the nutrient solution for the plants, can also be a problem. Generally, these are released slowly and will likely only pose problems after extended periods of operation as they accumulate in the recycling nutrient solution.

NEW TECHNOLOGIES

Light Emitting Diodes

An interesting new technology that we are studying is the use of light emitting diodes for photosynthetic lighting of plants (ref. 8). These are particularly useful because they can be fabricated with a peak emission in wavelengths of 630 to 690 nm which is of maximum photosynthetic usefulness for plants. The spectrum of these red radiation emitting LED is shown in Fig. 1. They have essentially no longwave emissions compared to other lamps being used for plant lighting, thus reducing requirements for convective cooling in the plant growing area (ref. 9). They also are solid state devices and thus have a high level of reliability, plus not being subject to the many hazards of gas filled tubes.

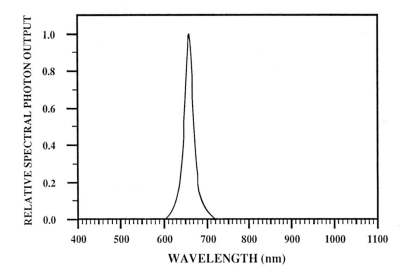

Fig. 1. Spectral energy distribution of a gallium-aluminum-arsenide light emitting diode having a peak emission at approximately 660 nm.

We have grown several species of plants, including lettuce and wheat, under LED's supplemented with 10% of photosynthetic photons from blue fluorescent lamps and found growth similar to that obtained under cool-white fluorescent and incandescent lamps (Table 1).

TABLE 1

Growth of lettuce plants, cv. 'Grand Rapids', maintained under a LED plant irradiation system and under cool white fluorescent (CWF) plus incandescent (Inc) lamps (From: ref. 10).

	LED system[z] (mean ± SE)	CWF + Inc lamps[y]
Fresh weight, (g)	19.00 ± 2.76	16.7
Dry weight, (g)	1.11 ± 0.16	0.97
Stem length, (mm)	9.30 ± 0.63	10.2
Number of nodes with leaves >1 cm long	11.00 ± 1.73	9.6
Length of 5th leaf, (mm)	111.70 ± 6.60	123.3
Width of 5th leaf, (mm)	117.70 ± 11.50	117.5

[z]Data are mean values of two plants from each of three separate growth periods.
[y]Data from Hammer et al. 1978, SE values not available.

It has been found that some broad leaved plants require a small percentage (5-10%) of blue photons to obtain plants with normal characteristics (Table 2). At present, blue LED's of sufficient output have not been available and the red LED arrays have been supplemented with the required amount of blue wavelengths from fluorescent lamps.

TABLE 2

Lettuce seedling hypocotyl extension at different levels of incident blue photon (400-500 nm) levels (From: ref. 12).

Blue Photon Flux (μmol m^{-2}s^{-1})	Hypocotyl length (mm)
0.0	27.3
4.0	18.8
7.5	17.0
10.0	13.0
15.0	13.1
22.5	7.7
30.0	5.6
37.5	3.1
45.0	3.3
52.5	2.5
60.0	2.1
300.0	1.0

Nutrient Delivery System

Special technologies need to be developed to provide water, and nutrients, effectively to plants in microgravity. Special procedures are needed because in the absence of a significant gravity vector, methods for prevention of dispersion or escape of the liquid into the atmosphere must be developed. Also, in weightlessness, liquid added to roots or a rooting matrix likely would fill all pores so that oxygen supply to the roots might become limiting resulting in restricted plant growth.

Research is underway at several NASA Centers to develop a membrane nutrient delivery system first proposed by Wright et al. (ref. 13). This system circulates the solution under a low level of negative pressure beneath or within a membrane with micron size pores. The root system is maintained to the outside of the membrane. Efforts at the Kennedy Space Center are directed toward utilizing tubular membranes enclosed within a larger rigid plastic tube. The liquid is contained within the inner membrane tube and roots permitted to grow between the outer and inner tube (refs. 14-15). At the University of Wisconsin, we have developed a system utilizing porous stainless steel tubes having a pore size of $\sim$ 30 microns and imbedding these porous tubes in a 1-2 cm depth of media consisting of particles of arcillite (calcined clay). The nutrient solution is circulated through these tubes at -5.0 to -10.0 cm of water pressure. The media and tubes can be heat sterilized as required after each growth cycle. We find the use of a rooting medium necessary for growth of potatoes and feel this will enhance productivity of most plants. Growth of potatoes, (Solanum tuberosum), lettuce (Lactuca sativum), and Brassica species using this system has been equivalent to that observed in peat vermiculite, whereas it appears growth is restricted some by the tubular membranes without media (ref. 16).

Nutrient Composition Control

Maintenance of nutrients at a desirable balance and concentration in plant growing systems in space is a major problem since it would be essentially impossible to take into space the large and complex analytical equipment utilized for this purpose on earth. We have been developing technology to obtain nutrient balance and concentrations with a series of ion exchange materials would have high reliability with a minimum of analytical and electronic regulation (refs. 17-18). The ion exchange materials (resins, zeolite, etc.) are loaded with a mixture of plant nutrient ions so that as plants remove these ions from the nutrient solution, and the ion balance is altered, there will be a regulation within the ion exchange materials to maintain the original ion balance. These exchange materials may be stacked in layers in a single cylinder as shown in Figure 2, or coupled in series in separate containers. At the inlet and outlet of the series of materials an additional ion exchange material will be included for controlling the pH of the nutrient solution. The ion ratios loaded on the ion exchange materials will be based upon the desirable ion balance for the crops being grown and the amount of ion exchange material incorporated in the system will be determined by the anticipated nutrient uptake prior to recharging.

As ion uptake by the plants occurs, nutrient solution conductivity will decrease. A conductivity meter is installed in the system, that will control the addition of concentrated nutrients from storage containers to restore the conductivity. The nutrient balance in this solution will be proportional to the concentration of nutrients present in plants of the species of interest as established in preliminary research.

This unit will have particular usefulness in small space bases or in space experiments where it would be very difficult to provide and maintain complex analytical equipment.

Oxygen/Carbon Dioxide Exchange

A major aspect of a bioregenerative life support system will be the absorption by plants of the carbon dioxide given off by people in the space enclosure and the release of oxygen from plants for supply to people. The habitats must be maintained within certain CO_2 and O_2 concentration ranges to avoid toxicities and depletion of each of these gases as required by both plants and humans. Thus, technologies need to be developed to ensure that gaseous concentrations of the habitat atmosphere can be kept within the proper ranges. This will likely require separate technologies to remove excess concentrations from plant growing compartments and transfer them to human habitation compartments and vice versa, or, store excess quantities in pressurized cylinders to provide a reserve that would maintain the desired concentration of these gases in the atmosphere of the human habitation and plant growing compartments. The area of actively photosynthesizing plant tissue at any particular time, and the number and activity of astronauts present, will influence the balance of production versus consumption of each of these two gases.

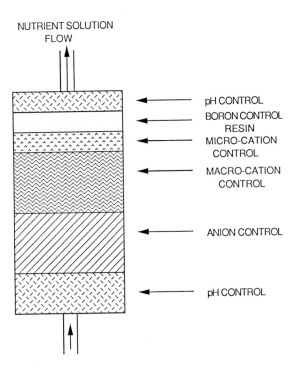

Fig. 2. Ion exchange materials for maintaining the mineral ion composition and pH of the plant nutrient solution.

Microbial Growth in the Nutrient Solution

New and effective technologies are needed to ensure that pathogenic microbial populations in the nutrient solution do not significantly limit plant growth. Scientists generally agree that microbes will be present in the recirculating nutrient solution because it would be extremely difficult, if not impossible, to keep the solution sterile. The use of chemicals to control pathogenic microbes is strongly discouraged because of the costs for resupply and because of potential accumulation in the recirculating systems and toxicity to plant and humans. Exposure of the nutrient solution to ultra-violet radiation also might be effective in killing the microbes, however, side effects of such a treatment have generally limited its use on earth. Micropore filters will also remove microorganisms but technologies would have to be developed to keep them from plugging and restricting nutrient flow.

It is felt that primary effort should be directed toward maintaining a balance of microbial populations that maintains sufficient beneficial organisms to limit the excess multiplication of pathogenic microbes.

Control and Monitoring of Plant Vigor

Automated technologies are needed to effectively monitor the vigor of plants without any significant perturbation of the plants. Monitoring is needed for determining plant size, leaf and stem morphology, chlorosis of leaves, photosynthetic rates, transpiration rates, and maturity of the harvestable portion. Combinations of both visual and physical measurements are needed.

A carefully developed artificial intelligence (AI) program is needed to ensure that environmental conditions in the plant growing area are maintained within a tolerable range and as close as possible to optimum operating levels for operation of the total bioregenerative life support system including the waste processors and astronaut habitation areas. This is of paramount importance when unexpected perturbations occurs, as for example failure of an exhaust fan, or heavy metal toxicity of one crop species.

Removal of Atmospheric Contaminants

Technologies need to be developed to remove contaminants that accumulate in the atmosphere within the space base. Of principal concern are hydrocarbon compounds released from structures and facilities of the bioregenerative system and from plants growing in the bioregenerative life support system. Many of these, such as ethylene (refs. 19-20) and isobutyl phthalate (ref. 7), are much more toxic to plants than to humans. It has been shown that plants can absorb and remove certain contaminants but other additional means for removing these contaminants are needed. Different forms of activated charcoal can be used for absorption of many hydrocarbons but technologies will need to be developed to effectively regenerate the activated charcoal at appropriate intervals. Catalytic combustion can be effective in removing most hydrocarbons but this procedure is limited by the large electrical requirements to provide the high temperatures required.

334

Providing an economical, reliable, and safe life support system for long term presence of humans on bases in space, Moon or Mars presents unique challanges. Technical developments will involve the expertise of a wide spectrum of scientists. It will require the interaction of mechanical, electrical, civil, and computer engineers along with plant, soil, food, and nutritional scientists. We have a long way to go to accomplish these goals, but the payoffs for space and very likely for earth activities will be great.

REFERENCES
1. B. Bugbee and F.B. Salisbury, Wheat production in the controlled environments of space, Utah Science Winter (1985) 145-151.
2. I.I. Gitelson, I.A. Terskov, B.G. Kovrov, G.M. Lisovskii, Yu.N. Okladnikov, F.Ya. Sid'ko, I.N. Trubachev, M.P. Shilenko, S.S. Alekseev, I.N. Pan'kova, and L.S. Tirranen, Long-term experiments on man's stay in biological life support system, Adv. Space Res. 9(8) (1989) 65-71.
3. T.W. Tibbitts and R.M. Wheeler, Utilization of potatoes in bioregenerative life support systems, Adv. Space Res. 7(4) (1987) 115-122.
4. R.M. Wheeler and T.W. Tibbitts, Utilization of potatoes for life support systems in space: III. Productivity at successive harvest dates under 12-h and 24-h photoperiods, Amer. Potato J. 64 (1987) 311-320.
5. B.G. Bugbee and F.B. Salisbury, Studies of maximum yield of wheat for the controlled environments of space, in: R.D. MacElroy, N.V. Martello, and D.T. Smernoff (eds.), Controlled Ecological Life Support Systems: CELSS '85, Ames Research Center, Moffett Field, CA, TM-851388, 1986, pp. 447-485.
6. I.I. Gitelson, B.G. Kovrov, G.M. Lisovsky, Yu.N. Okladnikov, M.S. Rerberg, F.Ya. Sid'ko, and I.A. Terskov, Problems of Space Biology, Vol. 23, Experimental Ecological Systems Including Man, Moscow: Nauka, 1975.
7. R.C. Hardwick, R.A. Cole, and T.P. Fyfield, Injury to and death of cabbage (Brassica oleracea) seedlings caused by vapors of dibutyl phthalate emitted from certain plastics, Ann. Appl. Biol. 105 (1984) 97-105.
8. R.W. Ignatius, T.W. Martin, R.J. Bula, R.C. Morrow, and T.W. Tibbitts, Method and apparatus for irradiation of plants using optoelectronic devices, U.S. Patent Application 07/283,245, 1988.
9. D.J. Barta, T.W. Tibbitts, R.J. Bula, and R.C. Morrow, Application of light emitting diodes for plant irradiation in space bases, Submitted for publication in Adv. Space Research, (1990).
10. R.J. Bula, R.C. Morrow, T.W. Tibbitts, R.W. Ignatius, T.S. Martin, and D.J. Barta, Light emitting diodes as a radiation source for plants, Accepted for publication in HortScience, (1990).
11. P.A. Hammer, T.W. Tibbitts, R.W. Langhans, and J.C. McFarlane, Base-line growth studies of 'Grand Rapids' lettuce in controlled environments, J. Amer. Soc. Hort. Sci., 103 (1978) 649-655.
12. M.E. Hoenecke, R.J. Bula, and T.W. Tibbitts, Lettuce seedling response to blue photon levels. Submitted for publication in Amer. Soc. Hort. Sci., (1991).
13. B.D. Wright, W.C. Bausch, and W.M. Knott, A hydroponic system for microgravity plant experiments, Trans. Amer. Soc. Agr. Eng. 31 (1988) 440-446.
14. T.W. Dreschel, The results of porus tube plant growth unit experiment T6B, NASA Tech Memo 100988, John F. Kennedy Space Center, FL, 1988.
15. T.W. Dreschel and J.C. Sager, Control of water and nutrients using a porous tube: A method of growing plants in space, HortScience 24 (1989) 944-947.
16. D.L. Bubenheim, T.L Reschel, and C.A. Mitchell. Comparison of plant growth in a tubular membrane hydroponic system with that in conventional hydroponic culture (Abs) HortScience 22 (1987) 1051.

17. R.T. Checkai, L.L. Hendricksen, R.B. Corey, and P.A. Helmke, A method for controlling the activities of free metal, hydrogen, and phosphate ions in hydroponic solutions using ion-exchange and chelating resins, Plant and Soil, 99 (1987) 335-345.
18. R.B. Corey and S.M. Combs, Control of nutrient concentrations in plant growth media, in: Gabelman, W.H. and B.C. Loughman (eds.), Genetic aspects of plant mineral nutrition, Martinus Nijhoff Pub., Boston, MA, 1987, pp. 591-601.
19. F.B. Abeles, Ethylene in plant biology. Academic Press, New York, 1973.
20. E.C. Sisler and S.F. Yang, Ethylene, the gaseous plant hormone, BioScience 34(4) (1984) 234-238.

I. Karube (Ed.) *Automation in Biotechnology*
Proceedings of the 4th Toyota Conference, 21–24 October 1990

337

Automation in Space Life Sciences

Masamichi Yamashita
Institute of Space and Astronautical Science

SUMMARY

Advanced concepts of automation are reviewed for its application to space. Telescience for space life sciences are discussed in terms of evolution of automation for scientific researches. A telescience testbed for space biomedical studies were conducted to define critical issues at its application.

INTRODUCTION

Space is a place of adventure for humankind. Outer space has severe environment for lives. There remain many constraints for manned activities in space. It drives efforts to replace man with abiotic systems which survive under vacuum, cosmic radiation and weightlessness. Fully automated unmanned probes have been sent to extraterrestrial space on behalf of scientists. Automated tools work as extended eyes and hands of scientists. Preprogrammed sequences are executed to achieve mission objectives. This stage of automation is characterized by "feasibility of science" without presence of scientist on the site. Many efforts have been devoted to improve capability of automation, such as adoption of intelligent functions for the system.

A concept of telescience is originated with a rather different context to automation. It has been proposed to conduct interactive experiments in space. Interactive capability of the system takes an advantage of mobilizing intellectual parts of scientist at its execution. Telescience is a scheme of methodology for science which utilize various tele-tools for rather wide area of scientific endeavors. It ranges over full span of activities, such as experiment planning, management, communication, data acquisition, analysis and so forth. Automation of sciences in a part of definitive routines is the target of telescience. It enables to execute experiments under physical isolation of a scientist from his operation site. Since investigator is believe to be the most expensive part among the space system, telescience reduces cost of experiment by without sending them to orbit. At having no presence of scientists on site, quality or outcome of science might be less compared to ones obtained in ordinary ground laboratories. Even this demerit is counted, the ratio of quality/cost could be increased by telescience. This ratio is a quantity related to "productivity of science" with which rationale of telescience is measured. Telescience explore the feasibility of automation applied to the most vivid acts of humankind, i.e. science. Experimental set-up itself should be re-configure at unexpected findings during the mission on orbit. Flexible and adaptive system is requested to assure less de-qualification of scientific achievements.

Telescience testbed is an engineering stage to develop telescience. There have been reported several testbeds for space life sciences. At Nagoya University, two series of testbed were conducted with biomedical experiments as their constituent objectives. Critical issues for application of telescience were defined at this testbed.

Unmanned vs Manned

There had been certain confronting aspects of automation against 'man'. Industrial revolution in 19th century accompanied luddite movement as its opponent. It has been accepted that a part of man's labors is replaced with industrial machines or robotics in case operation is tedious and repetitive, or recognition and judgment are relatively simple and definitive to make. Furthermore, machines and automation may expand human capability beyond the biotic limitation in many scenes at industrial factory, research laboratory and even at home. At the same time, humankind is thought to be more creative than robotics. Welfare of mankind is expected to be improved by shifting time and other resources to more creative works at expense of automation for less creative activities. Human activities might be replaced more and more by abiotic system. Human heritage would be transplanted to the abiotic part in some content. Evolution of automation should lead liberation of mankind, even if we cannot imagine ultimate figure of happiness for mankind at its utopia.

For space activities, there have been a hot debate on "manned or unmanned" among space engineers. It may rise some sort of space ludditism at the development of space automation. It relates to the fundamental query of why we expend our efforts for space programs.

Manned mission expense enormous resources to sustain life on orbit. Developmental cost of manned space crafts is extraordinary higher than that of unmanned. Many constraints are imposed on space system and mission payloads. Heavy safety requirements originate in carrying man on space crafts. Labor cost in space activities gives a strong driving force for robotics and automation. Even there are so many draw back for manned space pointed out, we cannot imagine unmanned space programs could drive space activities in past and future.

In another word, 'presence of man in space' itself had been a major target in many space programs, regardless of whether it is spoken out clearly or implicitly. Decision comes from the level of national prestige or political motivation, backed by a dream of exploring space. Once astronauts are defined as pilots rather than passengers watching automated flight, they should be doing flying. We should give them a chance to do better job. Similar feature is seen in antarctic science program. Many scientific researches planned might not require presence of man all through a year at an antarctic base instead of leaving unmanned probe during severe winter season. Geopolitics is a driving force in some contents for either antarctic or space program.

It is well accepted that space is now related to core subjects of modern sciences. Space physic has an intimate tie with elementary particle and high energy physics. History of galaxies has its chapter on origin and evolution of life on mother earth. Although importance of studies on these fundamental issues could not be denied, quite amount of investment on big sciences including space would not reflect to fair grading of its value against other sciences. People who defend 'man in space' have made their excuse to run expensive programs as follows. Since early days of manned space flight, manned operation in outer space has been verified to be feasible to conduct. Man can do something in space. Advantages to have man on space craft have been shown at various scenes. There were several occasions in which crew resolved unexpected troubles onboard to save mission and life of himself. If we consider robotics having capabilities similar to man in those cases, developmental cost of such an automated system is estimated much higher than that for building and operating manned mission. Fully unmanned system is surely cheap at sacrifice of variety of functions and capabilities which are provided by man. Questions can be stated as; Is there any driving force to develop robotics or automation cheap and functionable enough to replace man with it? Even various requirements or tasks for space activities might be evolved in future, will the ratio of cost for manned and unmanned system remain same as the conventional one?

Certainly cost depends on specification of the system and its operation based on the requirements defined for a mission. Productivity is another quantity to measure economical aspects of the system. Although quantitative evaluation of merit of having man on orbit is hard to make, it is worth to draw a conceptual chart of productivity for manned, unmanned and combination of the two. Productivity here is the ratio of achievements and cost. If we argue that sciences are a part of culture and hence they are priceless, we cannot apply the terminology of economics to scientific endeavors.

For unmanned system with automation, development of a system having capability same as man is postulated to result in extraordinary high cost and also to be not feasible with technologies off the shelf. Man without automated system pays a different penalty of expensive cost and restrictions mentioned before. It is supposed that achievements might certainly fruitful by manned operation, especially for space life science. In terms of productivity or cost efficiency, combination of manned and unmanned system may have the maximum point as shown in Fig. 1-1 for a typical case of space life sciences. If achievement should be kept at a definite level of manned operation, as shown in Fig.1-2, cost is increased drastically for fully automated unmanned case. At non-zero presence of man on orbit, design of reliability and verification of safety for space systems gives an almost same load on developmental and operational cost. There are missions which require specifically pure-unmanned or intensive-manned. Voyager type outer space exploration, and space human physiology experiment are good examples of those type of mission. Figure 1-3 explains relation of three parameters for such a case.

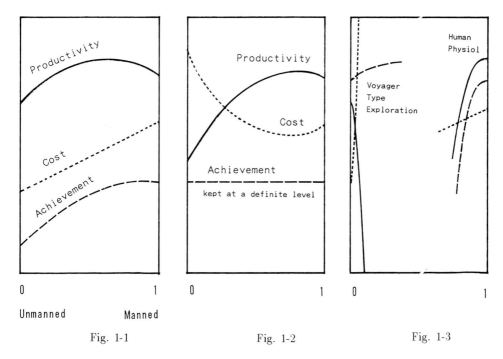

Fig. 1-1   Fig. 1-2   Fig. 1-3

Then, questionnaire comes as follows. How should we allocate functions among manned and unmanned part of space system? There are at least two kinds of answer. One is this way; man to be a component of system functioning as its part. Cost to replace that part with automated subsystem is compared to cost of having man. If latter stays cheaper, man can keep a seat on space crafts. The other way is that man does man's part on orbit, in stead of sitting on ground to send command in remote. Unmanned system assist this activity in this case.

Space Robotics

Conceptual studies on space robotics have been conducted by several groups in Japan. Key technologies for future robotics and its development scenario are surveyed with several aspects concerning functions and tasks imposed on it. Robotics is somewhat an interface between man/operator and target of work. Functions or subsystems of robotics are spread over a spectrum in between man and target of work. Work function of end effector and acquisition of data or information are a part facing to the target. Control, power and actuator systems are in the middle between. Data and information processing and trouble-preventing system are close to operator of robotics. Abilities to be developed in these subsystems are;

Effector: Autonomous operation and multi-freedom of mechanical structure.

Data acquisition: Database style acquisition.

Control: Autonomous control by applying artificial intelligence.

Power: New energy source rather than fuel cell and conventional ones.

Actuator: Actuator driven by the new energy source, or analogy of muscle.

Data processing: Real time 3-dimensional image processing.

Trouble prevention: Self diagnosis and repair with redundancy and AI.

Task supervisor: Accept natural language.

Almost all targets of technological development mentioned above are common for robotics in general. Emphasis on points specific to space application might be hidden under the general formulation of robotics technology. More or less, critical factors at implementation of robotics for space are different from those for ground. Followings are features which concern space application.

Preventive Maintenance Concept is the one which in particular relates to a limited capability to make access to the system on orbit. Operation of robotics requires high durability for a long duration without fatal trouble. Expected troubles are forecasted to prevent them by maintenance and effective measures. Subsystems for diagnosis and preventive maintenance are planned to be installed as a part of the system. Status of the system is monitored at defined sequence and situations, without breaking its normal operation and within a period allocated to maintenance. Inputs from sensors are evaluated to assess a margin of safety limit. It plans maintenance with consideration of cost. Cost for preventive maintenance should be lower than loss of merit at loosing functions plus cost for repair without maintenance. Subsystem to keep system life, at defined duration, is designed to fill safety requirements and to achieve cost effectiveness. Purpose of the preventive maintenance is to reduce propagation of trouble and its impact on core mission objectives. It does not mean reduction of trouble events themselves. Design of system reliability should be made properly to avoid over-specification. Redundancy and other countermeasure against troubles are kept at a level which can accept in terms of cost efficiency. Planning of maintenance concurrent to normal operation based on reliability analysis is the heart of advanced preventive maintenance concept.

Adaptability of system for various tasks is another advanced feature to be developed. Adaptive system is capable to re-configure itself at different operational purpose. It is somewhat an analogy of metamorphosis or flexibility shown in living creatures. Adaptive structures are defined as they can purposefully vary their physical and mechanical properties such as shape and movement. This concept can be applied to other domain of robotics including information systems. Function of intelligence is a next step to improve capability of robotics by introducing sub-autonomous operation of each part with a hierarchy of control. Going beyond intelligent system, intellectual features of the system might bridge automated operation to man's supervision on it.

Telescience

Science is a vivid act of humanity. No one speak, explicitly at least, that automated sciences without participation of scientists give better results compared to sciences led by scientists. Telescience is an attempt to conduct science by the introduction of tele-systems and some kinds of automation. Experimentation objected by telescience ranges its whole process. From defining objective of experiment, certain steps are required to be executed. Experiment strategy is followed by selection of facility, which might require modification of facility or strategy itself some time. When facility and mission are selected, various constraints come up for experiment planning. Functional verification of the system certifies final definition of the experiment. During the experimental execution, a variety of tele-tools is mobilized to make scientist's participations in his/her experiment possible, to manage experimental operation and scheduling, and so forth. Telescience is a concept of utilizing tele-tools in those processes.

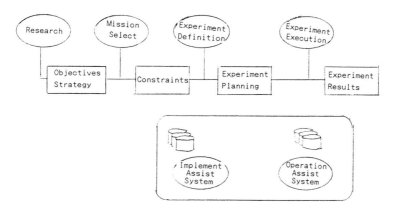

Fig. 2

Space stations and other manned space crafts will offer opportunities for conducting various scientific experiments onboard. Space experiments are subjected to many different constraints compared to laboratory work on ground. Being contrary to the nature of science, space experiments tend to estrange investigators in various unprecedented scenarios. At the initial planning phase, the scientific requirements for space research must be stated according to standard forms which are inclined to be stereotyped to suit engineering conveniences, rather than relying on inspiration or A-ha events. Several scientific objectives and procedures are integrated to constitute a multipurpose mission. Principal investigators are rare to both fabricate flight hard and software and also execute experiment in space hands-on. Developmental and operational cost of manned space systems can in part be reduced by replacing man with an unmanned system. Cost are lowered by keeping fewer astronauts in orbit. On the other hand, this excludes or minimizes the use of specified skills and abilities inherent in the crew in case the onboard crew operator cannot conduct the experiments on behalf of on-ground investigators. Those factors and the non-presence of the principal scientist onboard might result in less qualified science being performed in space.

It is important to judge impacts of such an arrangement on scientific outcome, Δ (quality), and reduction of cost, Δ (cost). The ratio of these two factors, Δ (quality)$/\Delta$ (cost) relates to the productivity of science. Improvement of this ratio, (quality)$/$(cost), is a key to open new avenues of space utilization for various sciences which otherwise could not use space. Figures 3-1 and 3-2 explain a shift of the point A, fully unmanned, to the point B, where productivity is higher than A and get into the affordable region by introducing telescience.

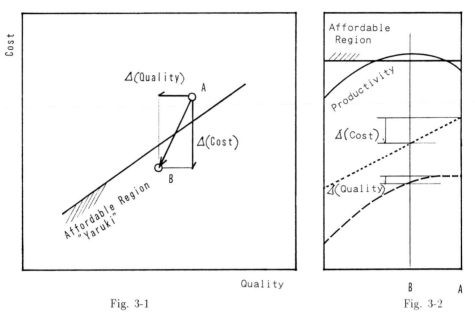

Fig. 3-1   Fig. 3-2

Telescience is expected to solve some of the problems which originate from the physical separation of scientists from their space experiments. Tele-Tools have been powerful constituents in many modern sciences. They expand man's abilities to approach on the subjects. Tele-Tools for telescience are motivated in the opposite way. They enable to enlarge distance between investigators and their sciences, without loosing their controls in touch.

Telescience testbed is an engineering stage to show advantages of telescience and feasibility of tele-tools. The quantity, Δ (quality)/ Δ (cost), is analyzed at the testbed for typical applications of telescience. It can also be a useful model for implementation processes of integrated missions serving different scientific objectives superimposed. Implementation of experiments is subject to constraints stemming from both science and engineering aspects. Those constraints come from the state of art technology. Issues which during the test-bedding are found critical in fulfilling scientific requirements or in reducing costs might reflect upon future technological developments.

Telescience has a wide spectrum in terms of functional allocation of tele-tools and crew commitment. It ranges from a fully automated system for unmanned experiments, as one extreme. There may be tele-operation with command/data link between orbit and ground. In this scenario onboard crew operators are supposed to participate in a restricted way, such as maintenance and trouble shooting. Another extreme of telescience is the intense manned operations with the intervention of investigator on ground through tele-command.

Space physiology and biomedicine is a discipline which requires telescience in the last configuration as mentioned above. Medical and biological experiments depend heavily on the crew's activities and its presence. They usually consist of rather complicated and delicate maneuvers. Automated unmanned systems for these purposes will cost obviously higher than a configuration in which the crew conducts major operations on orbit. Furthermore, the crew is supposed to be the human subjects for the biomedical study. In this situation, there is also a request for the crew to ensure the safety of the subject on site. Telescience will greatly enhance the level of safety because of the surveillance of the experiment from the ground.

We conducted two testbed experiments. At the first testbed, water immersion experiment was performed to study the physiological effects of simulated weightlessness on human. In the second testbed, rat was selected to an subject of biomedical experiment which includes surgical operations applied on it. Various scientific objectives were overlaid on these telescience testbeds. The experimental items integrated for the first testbed were; electrocardiography, blood pressure measurement, skin flowmetry, blood sampling and analyses, and echocardiography. For the second rat experiment, the experimental items were; embedding a telemetric sensor in body cavity to monitor cardiac activity and body temperature, blood sampling to analyze hormones, and sampling of muscles. One operator executed the experimental sequences in both case. Investigators sent instructions from a remote control room.

Experimental objectives and sequence were integrated to form a single mission with a process similar to those for planning of space experiments. In this paper, planning and execution of the first testbed experiment is summarized. At the beginning, the overall scientific aim was stated as to study cardiovascular and hormonal effects of blood and fluid shift during simulated weightlessness. The conditions imposed during planning and execution of the testbedding were chosen to be as close as possible to a real space mission. Certain constraints and criteria, considered at the initial stage, were feasibility to conduct it with one experimental subject and one operator whose skill would be at a limited level.

Five items selected as the constituent experiments were; electrocardiography (ECG), blood pressure wave recording (BPW), laser Doppler skin flowmetry (LDF), blood sampling (BLS) and echocardiography (ECH). Principal investigators (PI) were assigned for each item. They gave a short explanation of the scientific objectives and methods applied.

In order to define the experimental requirements for the testbed system and its operation, PIs made their primary inputs in a descriptive form. These inputs explained the safety and scientific aspects, the experimental methods, the principal protocol and sequence, the requirements for data acquisition, and the specific instruments. Based on these informations collected from the PIs, the coordinator composed an out-line of the testbed. ECG was considered as the basic reference to analyze the physiological state and was monitored all through the experiment for the safety reasons. Instruments were checked at the initial stage of experiment. Electrodes, catheter, and other probes were installed on the subject and their functions were confirmed to be normal. Measurements were performed with the subject standing during three conditions, 1) dry, 2) water immersion to the navel and 3) water immersion to the nipple. Measurements were repeated after reduction of water level. The interval between measurements or samplings, and the time necessary to stabilize the physiological state after a change of water level were considered to be managed in the later implementation process. Such an interpretation of the requirements was sent back to PIs from coordinator.

The experimental sequence of the overall integrated experiment was planned as a combination of a series of Functional Objectives (FO). Each FO had a definite purposes and consisted of a set of operation units with a certain sequence from start to end. PIs filled in a standardized FO form. This form contained information about; name of FO, preceding conditions to start this FO, list of micro-procedures, and expected duration time of those. Other items in the FO-form were; requirements for data acquisition, command and communication links, priority of this FO, and other comments. An example of FO forms is shown in Fig. 4 for venous catheter insertion FO. In addition to scientific FOs, check-out, shut-down, emergency, and stand-by FO were described from an operational point of view. Priorities for the execution of FOs were agreed between coordinator and PIs. The highest priority was given to the FO related to safety issues. Optional experiment items had low priority. Crew time is a major restricted resource shared by PIs in space. One issue focused upon in the present study was time line management for integrated missions.

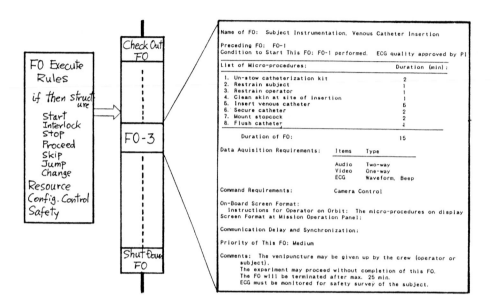

Fig. 4

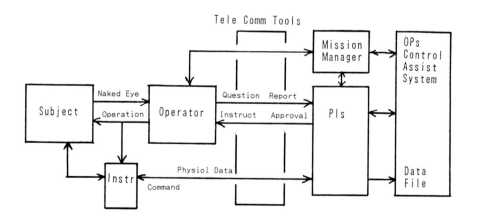

Fig. 5

Experimental System of Telescience Testbed was installed at Research Institute for Environmental Medicine at Nagoya University. The water immersion facility was utilized in the first experiment. Both subject and operator were standing from the beginning in the water immersion tank which simulated an on-orbit experiment facility. Physiological data were sent to PI in the control room, simulated an operation control center on ground. Two video cameras sent visual images of operator and subject to the control room. The view angle and zooming of one of the cameras were remotely controlled. At the second testbed, operator conducted the rat experiment using a safety bench installed in an isolated room.

Data and command links back and forth between the operating site and the control room were through cables. Communication links provided were; the one way video link mentioned above, a bit-link between personal computer with key board and monitor screen, and a two way audio link. At the second testbed, up-link video line was installed to send still images to operator. Effectiveness of instruction with a mark on down-linked in-situ images were evaluated by two-way video line. The functional block diagram is shown in Fig. 5.

To facilitate the operational management, FO activation and its performance were controlled and displayed by the computer. At the beginning of each FO, instructions and micro-procedures came up on the monitor screens provided for operator and PI. The PIs were requested to confirm or eventually modify these build-in set of instructions and elapse time. The operator followed the micro-procedures and informed the PI about their completion step by step. Certain steps proceeded under the guidance or approval of the PI. The PI could send messages through the keyboard in case the audio link would failed. A computer in the control room stored filed data and recorded all important events. The data handling system for ECG was designed to fulfill safety requirements. ECG and BPW signals were processed and their trend curves were displayed on a monitor screen for quick look and alarm in abnormal situations.

The constraints imposed upon this testbed were simulating the limitations foreseen in the type and capacity of communication links for space experiments. Other problems investigated were; delay in audio communication and data acquisition, and de-synchronization among transmission of video, audio and physiological data. The transmission delay in the video link was performed by using two video tape recorders, VTR. The video signal was recorded by the first VTR on tape which was loaded into the second VTR for play-back. This gave a delay of about 30 sec, determined by the length of the tape between the two magnetic heads of the record and play-back VTRs. At the second testbed, delay of video transmission was made by frame memory with 0 to 4 second delay capacity. The audio signals were delayed by a commercially available device. The delay was adjustable within a range of 0 to above 4 seconds. In the case of the physiological data, a shift register data recorder was utilized so that the PI observed the measurements on his oscilloscope with the set delay. A software developed for this testbed took care of the delay in command and data communication between the two PCs used for this purpose.

With regard to the echocardiography, frames were frozen in digitized mode and sent through a modem at a preset bit rate. The more spatial resolution of the image requested, the longer transmission time required. The moving image of echocardiograph was presented only to the operator. The PI instructed the operator by observing the frozen images of ECH, and the delayed video image of the operator's hand positioning the ECH probe held to the chest of the subject. At the second testbed, down-linked video images of surgical operation were hard-copied by the PI. It was sent back to the operator with marks which instructed position to operate and so on. Photographs and schematic drawings were also transmitted to the operator by this up-link video line.

Organizational Structure and Operational Management were also evaluated in testbed. One of features of this testbed was to rehears the roles played by scientists and engineers during a simulated space experiment. The total number of participants was about 20 at each run. The supervising scientist and the chief engineer were responsible for controlling and monitoring the whole experiment from scientific and engineering view points, respectively. The safety manager was authorized to activate the emergency FO at any notification concerning safety of the subject and the operator. The principal investigator had the role of communicating with the operator. Only the responsible PI was on the line of communication. The PI instructed and guided the operator to perform FO and assured data quality. In the case of decrease in quality or complete loss of signals, PI could warn the operation control manager a change in time line. The operator performed the tasks requested by the person on the line, either PIs or operation control manager. The start and completion of major procedures had to be confirmed by the PI. The operator surveyed that the subject was safe and well. The operator reported all significant observations, related to safety, scientific, technical or operational aspects. An assistant was assigned to the operator for practical reasons as his/her extended arm. The operation control manager gave permission to activate the FOs and to intervene if necessary. During the intermittent periods between FOs, the operation control manager took the communication line to the operator. The operation control assistant assumed performance of the testbed system. The staffs for measurement and data logging, and the assistant for communication worked under the operation control assistant to maintain functions of the system. A time keeper recorded in detail events happening all through the experiment. The reports from the time keeper were important in the later evaluation of the testbed. Performance of software to assist experiment management was evaluated in the course of experiment runs.

Evaluation of Telescience Testbed was made at two testbed experiments. One way to evaluate results is to observe the differences between actual time elapsed and that planned for each FO. For an example, catheter insertion was performed relying upon the operator's own skill. The PI was passive during this FO, except for watching video images of the operation for confirmation. Other FOs proceeded under much communication between the operator and the PI. The latter case had much overtime compared to pre-planned elapse time. It was shown that active and adequate communication habits and tools are key elements in telescience.

Difficulties of conducting experimental operation for operators did not correlate to the degree of complications in sequences. It was found that if operation was well defined, operator could conduct those steps easily. On the contrary, if operation needed basic knowledges of the discipline, such as anatomy of muscles, it was hard to educate operator on the site. Skill level and career of operators is an important parameter. Operators were selected with a variety of background concerning to research experience, space engineering, and even nationality. Operators were trained and familiarized to the experiment at some content before the testbed. Lesser experience of operator in the field of physiology and in space experiments made incomplete execution of experimental objectives. Mental mood of operator and psychological relation between operator and PI are another factor. At the second testbed, electrocardiograms of operator and PIs were measured during testbed run. Trend of pulse rate was analyzed to refer mental stress and degree of fatigue of either operator and PIs during testbed runs. The effects of training was shown apparently after several runs of testbed. All the team was familiarized with the experimental equipments, communication links, and the rules of the game. The team also had been used to perform conversation with delays, etc. and to use the right communication links in each procedure and situation.

Evaluation of the telescience testbed was made in several ways. Complaint and trouble reports were gathered after each run. When the series of runs were completed, a questionnaire form was given to all participants together with a summary of problems and complaints occurring during the experiment. Questions, regarding the aspects of the scientific outcome, were concerned with if the quality of science was affected by the experimental system and operational management which were desired according to constraints for space experiments. With regard to the implementation process, some questions were concerned with the adequacy of following items; selection of experiments, selection of time lining, documentation by the FO forms, engineering understanding of the science, and work interface between scientists and engineers in general. Suggestions were solicited for additional requirements for training and familiarization procedures for the objectives in the experimental situation, and for means to give proper instructions to operator and operation control manager. There were questions regarding the performance of the operation system software and hardware, and the quality and quantity of information transmitted.

Concluding Remarks

Testbedding is an educational process for scientists and engineers. Scientist PIs are trained to communicate their science to engineers in a well defined and logic way. The engineers accumulate experience in interpreting and understanding the scientific objectives and requirements. How can the scientists participate and control their experiments while conducted by the crew operator? This is a general issue in space physiology. The roles of crew operator versus scientist should be defined in this context. Resources, such as the timeline, allocated to the individual experiments in an integrated mission should be allocated to increase the scientific returns. Ways and means for a flexible rescheduling procedure of a timeline and so forth are important issues in space physiology.

The present simulation of a series of space medical and physiological measurements showed the value of using telescience for space experiments. The testbed also permitted the study of implementation processes in a mission. Both scientists and engineers can extract and define their mutual objectives for study and accumulate their experiences to share. It will drive the efforts to develop proper tele-tools and telescience strategy. The fruits to be harvested from testbedding studies will be an increased scientific productivity in space.

Telescience testbed experiments reported in this paper were performed by the leadership of Prof. Satoru Watanabe of Nagoya University and by the participation of his research group. Its engineering part was managed by Messrs. Takatoshi Shoji and Hideo Sudoh of Kawasaki Heavy Industries. International participation was made by Drs. Flemming Bonde-Petersen and Niels Foldager of Danish Aerospace Medicine Center. Dr. Takehisa Matsumoto of National Space Development Agency of Japan made his contribution to this project. The author would like to express his sincere thanks to all of the persons who joined the telescience team.

I. Karube (Ed.) *Automation in Biotechnology*
Proceedings of the 4th Toyota Conference, 21–24 October 1990
© 1991 Elsevier Science Publishers B.V. All rights reserved

AUTOMATED MEASUREMENTS FROM LABORATORY TO INDUSTRIAL SCALE
A TOOL FOR BETTER UNDERSTANDING OF FERMENTATION PROCESSES

JOHN VILLADSEN

Department of Biotechnology and Research Center for Process Biotechnology, The Technical University of Denmark, DK-2800 Lyngby, Denmark

SUMMARY

New analytical tools have become available for on-line measurement of key fermentation variables. Flow injection Analysis (FIA) is used as an illustration, and it is shown that stable and very accurate measurements are obtained over a wide range of concentrations. The experimental data are used to verify models which far better than conventional fermentation kinetic models describe the microbial physiology.

INTRODUCTION

The vast possibilities of Modern Biotechnology are being realized right now, and the impact of Information Science, New Materials Science and Biotechnology is certain to change our daily life. Every week scores of genetically engineered microorganisms are registered in patent offices or described in the open literature. Lower working temperatures and much higher stability of detergent enzymes, extremophilic microorganisms which convert dangerous industrial waste to harmless substances or remove lignin in the paper pulp industry without the need for an ecologically undesirable chlorine treatment. These are but a few of the news items which assert to the vitality of Biotechnology. Some of the subjects which are treated in fundamental and applied biology and biochemistry are frightingly complex. Studies of the sequencing of the human genome, the folding geometry of proteins, and the localization and description of receptors on the surface of bacteria demand not only ingenious theoretical research, but also a host of extremely complicated and expensive instruments.

Turning to the industrial implementation of biology and biochemistry one is struck by the disparity in the level of sophistication between fundamental bio-research as conducted in the renowned cell biology laboratories and the engineering science approach on which the commercial success of the bioprocess is ultimately going to depend.

The gap may unfortunately widen since the ever increasing demands of front line research in protein chemistry or gene regulation will leave less and less time for the young scientist to become familiar with techniques which were long since described and utilized in classical chemical engineering research - in rheology, mass transfer, reaction engineering, mathematical modelling and process control - but never really transplanted to the bio sciences or properly understood by the microbiologists who are the research leaders in what is commonly thought of as the Modern Biotechnology.

The bioreactor is an expensive piece of equipment which often works with quite expensive materials. When running a penicillin fermentation one requires more than 200 hours of process time with complicated feeding strategies and at least some expensive substrates - notably the precursor. The contents of a 50 m^3 fermentor tank is far too valuable to allow any mistakes to happen. Still, penicillin fermentation is basically run on empirical know-how only. One may know from experience the best level of glucose, ammonia or precursor and the best switching time from batch to fed batch operation, and the penicillin strain may be selected after countless screening experiments or even be the product of r-DNA studies. One does, however, not quite understand why the fungus produces penicillin; the precursor uptake mechanism and the influence of fermentation conditions on the rate limiting metabolic step are certainly not understood. And the influence of mixing and of proper dispersion of both oxygen and liquid substrates in the tank has hardly been studied. Considering that penicillin production is extremely competitive with an overcapacity worldwide in fermentation equipment, it is, of course, imperative that all major producers try to reduce the direct production costs - and this requires an intensive research effort.

The same rather lamentable story can be told for other major industrial fermentations: bakers yeast, SCP, polysaccharides, citric acid. Empiricism plays a dominant role, perhaps disguised by process computers which work on a limited set of easily monitored process variables such as pH, temperature and dissolved oxygen and with a controller based on a "black box" model of the process.

What really is lacking is a true model for the process and for the process equipment. A model which is expressed in terms of key metabolic variables, each giving independent information on the state of the microorganism, and a model which incorporates the influence of key reactor variables: gas dispersion, mixing pattern, liquid-gas contact time, and the like.

Why are such models not available - considering the tremendous power of modern process computers to handle the necessary calculations?

One reason may be the traditional reluctance of the experimental microbiologist to describe his results in a mathematical guise. Another - and more acceptable reason is that reliable on line, semi on line or at least fast off line analytical methods were until recently not available. Even the problems of withdrawing samples for analysis from industrial bioreactors have seemed to be unsurmountable. Very recently several on line sampling devices (ref. 1-3) have been marketed. Also during the last few years model considerations applicable for laboratory to plant size fermentations have been published, and the models are based on results obtained with a remarkable line of new experimental techniques.

The purpose of my lecture is to illustrate by means of examples how informative the model guided experimental study can be. Also some new techniques, in particular those based on the Flow Injection concept, will be discussed.

On line measurement of broth composition

The performance of the microorganism is completely determined by its environment, the fermentor broth. We have no means of directly interfering with the processes inside the cell, but we can manipulate the concentration of any substrate or product in the broth provided that it can be measured accurately and in "real time". If finally the cell state is satisfactorily modelled in terms of substrate and product variables we have the means to operate the process in an optimal fashion. True in-situ measurement methods are available only for a few fermentation variables. Classical measurements include pH, temperature and dissolved O_2 and CO_2, and these measurements are rarely useful for modelling purposes whereas they can provide alarm signals. Some cations are measured by ion-selective electrodes, the cell density by light absorption, and recently very good fluorometric methods have become available for measurement of the state of "activity" of the cell (the interpretation of fluorometric signals is still not quite certain – perhaps the NADH/NADPH level is measurable by the method).

Semi on line measurements are more valuable for modelling purposes since variables directly related to the process kinetics and the yield are measured. The most versatile measurement concept is possibly the High Performance Liquid Chromatography (HPLC) which employs adsorption, reversed phase, liquid/liquid partiton and gel or affinity chromatography, sometimes after a derivatisation of certain constituents of the liquid sample. Amino acids, volatile fatty acids, alcohols, sugars and most of the substrates, intermediates, products and byproducts involved in the penicillin fermentation have been determined by HPLC (ref. 4 and 5). An intricate sample preparation, low analysis frequency and high apparatus cost + maintenance are the major drawbacks. High performance capillary electrophoresis (HPCE) may be both more accurate and have higher resolution than HPLC. Very recently a whole symposium (ref. 6) was devoted to this analytical method. On line or semi on line applications are, however, only just emerging. Membrane inlet mass spectrometry (MIMS) has been very successfully used to determine volatile liquid phase components (methanol, methane in SCP fermentations) besides O_2 and CO_2. Some beautiful transient experiments which reveal major parts of the kinetics of the key enzyme methane monooxygenase (converting methane to methanol) are reported in (ref. 7). Sometimes mass spectrometry must be combined with gaschromatography to get sufficient information and the relatively poor permeability and long diffusion times of larger molecules do create difficulties for MIMS, at least if the instrument is placed some distance away from the fermentor. It is claimed that ion chromatography (based on ion-exchange, ion-exclusion or ion pair formation) is the most versatile semi on line method. A comprehensive instrument is, however, prohibitively expensive and sample preparation is again quite complicated.

We shall now give a more detailed description of Flow Injection Analysis (FIA), a rather inexpensive and truly versatile semi on line method which is now used in many variants and for thousands of applications (ref. 8).

Flow Injection Analysis

FIA is perhaps originally conceived as an improvement of the well known auto-analyser concept where liquid plugs of e.g. fermentation broth separated by gas are pumped through diverse mixing units, reaction coils etc. to a detector. Once the advantages of FIA - the method is based on injection of a sample pulse in a flowing liquid stream of either buffer or reagent - had been fully realised there was no doubt about the superiority of the much better defined transportation and dispersion offered by this method.

Figure 1 shows a typical set-up for FIA. 3 fermentation variables are measured: cell concentration (by optical density, OD), glucose, and lactate, the main product of a fermentation with Streptococcus cremoris, a lactic acid bacterium. All valves and pumps on figure 1 are activated by a process computer. Sample is withdrawn from the fermentor by an automatic syringe. It passes a small (0.5 ml) chamber where it is cooled to 2°C by a Pelltier element to stop cell activity. Thereafter the sample is transported either to the OD or to the lactate and glucose injection valves. The sample volumes are typically 25 μL and 100-200 μL respectively in the OD and in the lactate/glucose valves.

Optical density is determined by light absorption at 585 nm using $KMnO_4$ as standard. When the injection valve is activated (figure 2a) a plug of cell containing broth is pushed into the small, well stirred mixing chamber by a carrier stream of distilled water. The output signal from the mixing chamber is a spike with an almost ideal exponential tail (figure 2b). The peak height is proportional to the cell concentration and the sample frequency can be as high as 120 h^{-1} depending on the choice of mixing chamber volume. The standard deviation is less than 0.95%, and the analyzer is used for 100 hours to monitor lactic acid fermentations (biomass concentrations up to 4 g/L without any fouling problems from cells or from deposits of MnO_2 in the tubings or in the detector. For higher biomass concentrations (e.g. bakers yeast fermentation) a second injection valve can be fitted into the analyzer. As described in connection with figure 3 this device allows the linear range of the detector to be expanded to about 50-100 g/L.

The lactate and the glucose analyzers are both based on enzymatic conversion of the organic compounds after removal of the cells in a membrane module.

$$L\text{-lactate} + O_2 + H_2O \longrightarrow \text{pyruvate} + H_2O_2$$

$$(1)$$

$$\beta\text{-D-glucose} + O_2 + H_2O \rightarrow \text{gluconic acid} + H_2O_2$$

Lactate oxidase and glucose oxidase respectively are immobilized on the inner surface of 1 m, 1 mm i.d. nylon tubes using standard immobilization techniques. A minute amount of enzyme is sufficient (2 mg enzyme gives 92% conversion of the 200 μL lactate sample passed through the coiled nylon tube by a phosphate buffer). The reaction coil retains its activity even after several months on stream.

Detection is by chemiluminiscense after reacting the H_2O_2 with luminol using Fe^{+3} as a catalyst.

As shown in figure 1 several chemical analyzers can be set up to work sequentially. The only difference between the analyzers lies in the choice of enzyme in the reaction coil. Consequently a series of calibrations interspersed with measurements on samples can be programmed on the process computer.

Figure 4 is a standard curve which relates peak height to lactate concentration. The measurements were made in the equipment shown on figure 3, but without the on line dilution and second injection valve (indicated in the box on the figure). It is remarkable that peak height is linear in the concentration (correlation coefficient 0.9997) from 0.1 to 100 mg lactate/L. Measurements can be taken every two minutes and the simple (inhouse built) apparatus appears to be nearly ideal both for monitoring fermentations and for routine analysis of clinical samples. Note that at the lowest concentration (90 μg lactate/L), a sample volume of 200 μL in the injection valve and 92% conversion in the coil, the instrument is able to measure accurately as little as $1 \cdot 8 \cdot 10^{-10}$ moles of H_2O_2 without even correcting for dispersion in the reaction coil.

The mixing chamber and the second injection valve shown on figure 3 allows measurements to be made on samples with concentrations of glucose and lactate in the range 0.5 to 20000 mg/L without exceeding the linear range of the detector. On figure 5 the output from the mixing chamber is shown as a function of time elapsed since the first injection valve was activated.

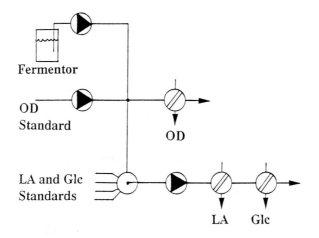

Figure 1a : Manifold for FIA.

Figure 2a : Biomass analyser.

Figure 2b : Sequential signals from biomass analyser.

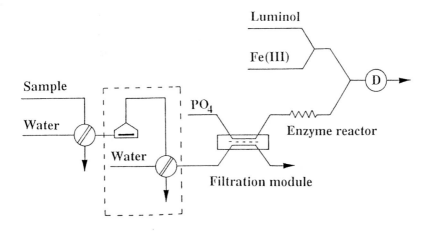

Figure 3 : FIA set-up for a number of organic compounds.

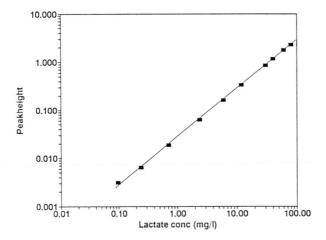

Figure 4 : Standard curve relating lactate concentration to FIA peakheight.

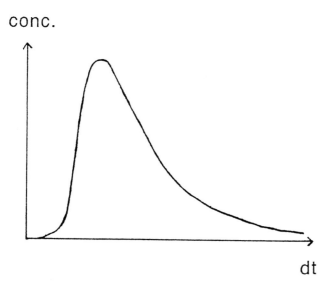

Figure 5 : Output signal from mixing chamber in FIA.

If a sample of the output is collected in the second injection valve at a precisely measured time after injection of the original sample into the mixing chamber one may operate anywhere on the exponential tail of the output. The timing between the injection into the chamber and the capture of the diluted sample in the second valve is of course crucially important. The operation is performed by the process computer which is programmed to change the delay time based on previous measurements of concentrations. Hence, the analyzer works automatically, and always in the linear range of the detector. It can be used both for batch fermentation (glucose level changing from e.g. 10 g/L to 10 mg/L) or it can be used to monitor transient experiments in a chemostat where equally large changes in the outlet concentrations may be found when the dilution rate (D = volumetric flow/fermentor volume) is drastically changed. - see e.g. figure 6.

All the analyzers described here were developed by graduate students in our department (references 9,10 and 11).

Figure 6 collects results from monitoring of a lactic acid fermentation over 17 hours. Outlet concentrations from a laboratory size (1.0 L active volume) chemostat with cell free feed of 10 g/L glucose and sufficient N source to make sugar the limiting substrate. At the low dilution rate (D = 0.12 h^{-1}) up to 2.7 h 97% of the glucose is converted to lactate (3% is used in synthesis of biomass). After the change of D to 0.39 h^{-1} the biomass and lactate concentrations decrease, and simultaneously the glucose concentration in the outlet increases. Throughout the experiment the total carbon balance seems to be satisfied (very little byproduct formation) and it is apparent that the almost noiseless output signals can be used with great confidence to test various growth models.

It appears that FIA is a versatile analytical concept - less connected with a particular instrument than with a certain general technique.

Transportation of a liquid sample - inserted in a flowing buffer or reagent stream - from an injection valve, passing various separation devices, mixing units and reaction coils to a detector is the key concept of FIA.

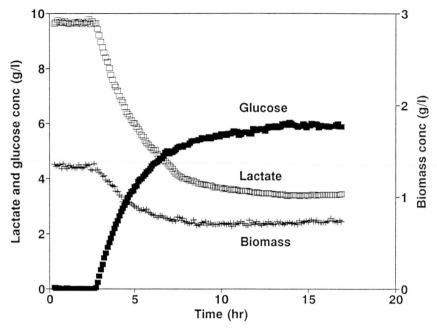

Figure 6 : FIA measurements of lactic acid fermentation variables during a transient chemostat experiment. The dilution rate is changed from 0.12 1/hr to 0.39 1/hr.

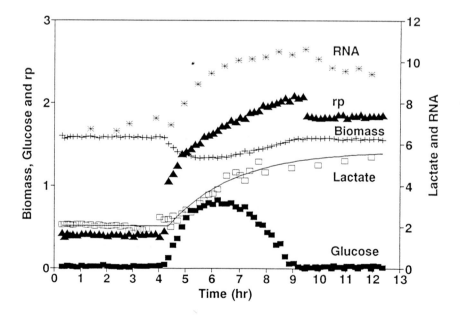

Figure 7 : Result of a transient experiment. Here the dilution rate is changed from 0.075 1/hr to 0.409 1/hr using another strain of Lactococcus than in figure 6.

The dispersion of the sample is carefully reproduced from one analysis to the next, and at regular intervals standards are sent through the equipment and treated exactly as the samples. Thus base line drift or other slowly developing systematic errors are almost quantitatively absent.

FIA will not work unless timing of the sample transport is perfect. A process computer is essential - it is programmed to activate valves, start chemical treatment of samples etc. at exactly the right time. The hardware should also be of the best quality. Membrane modules must not leak, valves must open correctly and tubing must not age. If certain obvious automatic cleaning procedures (e.g. of flow cuvettes) are programmed to occur at regular intervals, and if membranes, pumps, tubing and other minor items are serviced properly the equipment may function without any problems for weeks or even months.

Applications of FIA for chemical and biochemical analysis run into the hundreds. Sugars, cations (PO_4^{-3}, SO_4^{-2}, NO_3^-), NH_3, urea, ethanol, fatty acids, amino acids, penicillin, 6-APA, certain enzymes (proteases and dehydrogenases) have all been analyzed by FIA. Units for analysis of analogous compounds (e.g. lactate, glucose, galactose) can be prefabricated and used in a common framework which includes a suitable computer software to operate the complicated analytical procedure.

Modelling of fermentation kinetics

The analytical tools described in the last section are used to enhance our understanding of the metabolism of the microorganisms through a proper mathematical modelling of the fermentation. The ultimate goal is, of course, to improve the process yield or the productivity of the microorganism.

Batch fermentations are of very limited use for modelling purposes, although they may serve a useful purpose in screening experiments. Chemostat experiments where a steady state can be maintained for sufficiently long time to get a detailed, pointwise picture of the influence of substrate components on growth rate and product formation are essential.

The most powerful experimental technique may be a combination of steady state and transient experiments: subjecting the steady state obtained in a chemostat to pulses of substrate or to step changes in dilution rate. The major variables do not necessarily have to change very much during the transient, but there can be a wealth of information in the transients from one steady state to another (or back to the previous steady state in a pulse experiment).

If transient experiments shall be meaningful one must have access to many, accurate measurements which are free from distortions due to time delays. This is where the new automated analytical techniques demonstrate their power.

Figure 7 shows a transient experiment similar to that of figure 6, but with more state variables.

Again one considers the formation of lactic acid by a lactic acid bacteria (a different strain from that of figure 6). The feed concentration of glucose is 7 g/L and for the first 4.2 h the chemostat works at the extremely low dilution rate of 0.075 h. Obviously much of the glucose is converted to byproducts (ethanol, CO_2, acetic acid) since the lactate concentration is only about 2.2 g/L. The rate of acid formation r_p is 0.4 g/h per g biomass/L. (=0.66 g lactic acid/Lh). The glucose level is very low, 0.5 - 0.8 mg/L, practically at the limit of detection.

When the dilution rate is changed to 0.409 h^{-1} at 4.2 h there is an instantaneous increase in acid production rate whereas the lactate concentration slowly increases. The jump in r_p can only be explained by the presence in the starved cells of some active machinery (enzymes etc.) which immediately springs into action when the glucose level exceeds a few mg/L. It is, however, important that r_p continues to increase for the next 5 hours - i.e. as long as the glucose level stays above the mg/L range.

The microorganisms do, however, not have enough active machinery to consume all glucose and there is a transient accumulation of glucose in the chemostat between 4.2 h and 9 h.

The increase of D to 0.409 h^{-1} leads to an incipient wash out of the cells which at their initially starved condition cannot properly utilize the glucose for growth purposes. However, after 2 hours the cell concentration begins to increase and the biomass level reaches almost the same level at the new state as it had at the much lower dilution rate. The explanation of this unexpected shape of the biomass concentration x vs. time t profile is likely to be that more "effective" cell machinery is formed in the relatively rich environment.

The figure also includes measurements of one <u>internal</u> cell variable, the RNA content (g/100 g cells). The RNA level is low (7%) at the low D-value and it increases (almost parallel to the r_p curve) until the accumulated gluose is consumed and the glucose concentration at the new steady state (4.8 mg/L) is reached. Now the RNA content slowly decreases but its steady state level is still much higher (9.5%) than in the original steady state.

It is known that the cell activity is proportional to the RNA level and the measurement of the single internal variable nicely confirms the experimental results concerning the influence of external variables - glucose concentration, biomass concentration and r_p.

Models

Based on many experiments such as those described in figure 7 it is possible to collect enough information to set up a model which accounts for the influence of processes which have time constants greater than about 5 minutes. Figure 8 is a schematic representation of lactic acid formation. The cell mass is divided into two compartments, A and G. It is the active compartment A which is responsible for production of lactic acid, and the structural compartment G (mostly inactive material such as cell membranes, polysaccharides etc.) is also formed from A. To form compartment A one needs an energy source (glucose) and a nitrogen source, and the rate of formation of G from A must depend on the internal concentrations of energy source and nitrogen source. In the external medium the concentrations of glucose and nitrogen source are s and s_N respectively. The substrates are transported into the cell by fast transmembrane enzymatic processes to form the intracellular substrates S and S_N. The concentrations of A (X_A), of G (X_G), and of S and S_N are measured in units of g/g cell. RNA constitutes the major part of the active component A, and X_A is proportional to the RNA content. The internal concentrations of S and S_N are close to zero at all operating conditions and the internally formed lactic acid complex P is transported to the external medium (concentration p). Figure 7 clearly reveals the key features of the kinetic model.

Thus, at small values of s the rate of formation of lactic acid must be proportional to s whereas the influence of s is small for higher s values. Otherwise the jump in r_p when the glucose level increases above about 1 mg/l cannot be explained. Also r_p must increase with the size of the active compartment X_A. A simple and yet completely reasonable model is

$$r_p = k_1 \frac{s}{s + K_1} X_A \quad (h^{-1}) \tag{2}$$

where K_1 must be in the mg/L range.

Similar relations can be set up for the rate of formation of A and G, and measurements of the transient increase of RNA and of the biomass concentration can be used to fit kinetic parameters.

The resulting model (ref. 12) does, of course, only give a very simplified picture of the metabolism of lactic acid bacteria. Many different enzymes are pooled into the active compartment A, and compartment G also combines a multitude of components. Still, the model with only two cell compartments fits nicely with the existing level of experimental information - only one internal cell compartment (A) can be followed experimentally, whereas the major external variables are well taken care of by the FIA measurements. As soon as reliable analytical techniques become available to measure more cell components the model can be expanded.

The strikingly simple picture of the cell offered by figure 8 can be used as a starting point for other experimental/theoretical fermentation studies.

Thus we have recently (references 13, 14) proposed structured models for other fermentations. Figure 9 shows the compartment model for thermally induced plasmid runaway of r-DNA microorganisms. A slight temperature increase from 37 to 42°C induces a rapid increase of the number of plasmids per cell. Hence, the production of a plasmid encoded enzyme is boosted. Unfortunately the cells will die if the runaway continues for several cell generations and after some time at the high temperature the cells must be transferred back to the lower temperature in order to recover.

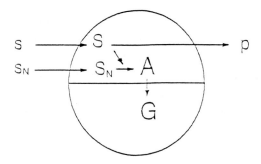

Figure 8 : Structured cell model for lactic acid fermentation.

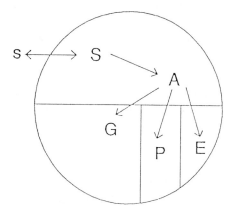

Figure 9 : Structured model for genetically engineered E. coli.
P = plasmids, E = plasmid encoded enzyme.

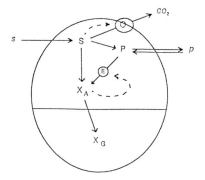

Figure 10 : Structured model for Saccharomyces cerevisiae.

Figure 10 shows a compartment model for bakers yeast fermentation. The model is based on saturation of the glycolytic machinery when the sugar concentration exceeds an upper limit. Intracellular glucose is converted to either CO_2, to ethanol or to building blocks in the active compartment. Energy for the cell building process is obtained by oxidative phosphorylation, but when this machinery has become fully occupied the cells start to gain energy by conversion of glucose to ethanol. At low sugar concentrations ethanol from the external medium can also be used to form building blocks in A, but an enzyme E formed from A is necessary, and this enzyme is only operable when the glucose concentration is low.

The bakers yeast model has been used to predict the results of a number of experimental studies by different research groups. Kinetic parameters can be found from steady state experiments using only the oxygen uptake rate, the carbon dioxide evolution rate, the biomass concentration, and the concentrations of sugar and ethanol in the external medium.

To quantify the model for the genetically engineered microorganism one must, of course, also have measurements of the plasmid content and the enzyme production rate.

Further perspectives

Although some very encouraging results can be obtained using the automated analytical techniques described in this paper there is, however, a long way to go before bioreactions are understood as well as classical organic and inorganic reactions.

When it comes to identification and quantification of proteins produced by fermentations - perhaps excreted to the external medium, but often retained inside the cell membranes - one has to rely on tedius off line analytical models. Certain key variables such as the ATP level of the cell are not even available by analytical methods.

Another important point is that fermentation kinetics, however well they may be described, will only constitute one part of the mathematical description of a bioreactor. The inhomogeneities of the microbial environment caused by inadequate substrate dispersion, by complicated mixing patterns in the reactor, and by coalescence and redispersion of gas bubbles have to be measured by techniques which are not discussed here.

References

1. K. Schügerl, On line analysis and control of production of antibiotics, Anal Chim Acta 213(1988) 1-9.
2. K-H. Kroner and N. Papamichael, Continuous sampling technique for on line analysis, Pro Bio Tech 10 (1988) 3-10.
3. M. Garn, M. Gisin, C. Thommen and P. Cevey, A flow injection analysis system for fermentation monitoring and control, Biotech Bioengr 34 (1989) 423-428.
4. J.C. Mothe, X. Monsear, M. Termonia, M. Hofman, G. Alaerts, Computer monitoring of sugars, acids and volatile compounds in fermentations, Anal Chim Acta 163 (1984) 275-280.
5. J. Möller, Penicillin Produktion mit einem Hochleistungsstamm von Penicillium crysogenum, Diss, Univ. Hannover (1987).
6. Second International Symposium on HPEC (High performance capillary electrophoresis). San Francisco, California, January 1990.
7. L. Jørgensen and H. Degn, Mass spectrometric measurements of methane and oxygen utilization by methanotrophic bacteria, FEMS Microbiology Letters 20 (1983) 331-335.
8. J. Ruzicka and E.H. Hansen, Flow Injection Analysis, 2nd. ed., John Wiley Series on Analytical Chemistry vol. 62, N.Y. 1988.
9. J. Nielsen, K. Nikolajsen and J. Villadsen, FIA for monitoring important lactic acid fermentation variables, Biotech Bioengr 33 (1989) 1127-1134.
10. J. Nielsen, K. Nikolajsen, S. Benthin and J. Villadsen, Flow injection analysis applied for on line monitoring of sugars, lactic acid, protein and biomass during lactic acid fermentations, Anal Chim Acta, accepted (1990).
11. S. Benthin, J. Nielsen and J. Villadsen, Growth and product formation model for recombinant lactic acid bacteria, 3rd Symposium on Lactic Acid Bacteria, Wageningen (Holland) 17-21 Sept. 1990. FEMS microbiology reviews 87 (1990) 107 (G2).
12. J. Nielsen, K. Nikolajsen and J. Villadsen, Structured modelling of a microbial system, part 1: A theoretical study of the lactic acid fermentation, Biotech Bioengr, acepted (1990).
13. J. Nielsen, A.G. Pedersen, K. Strudsholm and J. Villadsen, Modelling fermentations with recombinant microorganisms: Formulation of a structured model, Biotech Bioengr, accepted (1990).
14. J. Villadsen and J. Nielsen, Modelling of Fermentation Kinetics, review paper ECB5, Copenhagen 9-13 July, 1990. To be published in Transactions ECB5, December 1990

I. Karube (Ed.) *Automation in Biotechnology*
Proceedings of the 4th Toyota Conference, 21–24 October 1990
© 1991 Elsevier Science Publishers B.V. All rights reserved

TREND OF AUTOMATION IN BIO-INDUSTRY

M. HACHIYA

Systems Engineering Division, Hitachi Ltd., 6, Kanda-Surugadai 4 chome, Chiyoda-ku, Tokyo 101 (Japan)

SUMMARY

All sorts of machines which are used for manufacturing various bio-products are required by users to provide high efficiency and high quality, so manufacturers have been continuing to develop high efficiency automating machines. The research and development of bio-products are generally made in the R&D stages. Among these bio-products thus developed, marketable ones are produced by bio-machines in the production stages. High efficiency and automation are required not only in bio-machines to produce bio-products but also in instruments and apparatus to develop bio-products. In this paper, we will introduce the recent trend in the automation equipements and bio-machines in the R&D stages and the production stages. Discussion will also be made on future direction in automation of bio-machines.

1 INTRODUCTION

Molecular biotechnology is unraveling organisms and even life itself, and remarkable advances have been made in this field in recent years. We are pursuing the automation in the R&D (research and development) stages for biotechnology which will be applied to such new fields as pharmaceuticals, basic cosmetics, foods stuffs, animal feeds, and energy.

In the production stages of bio-industry, we have been producing antibiotics, amino acid, physiologically active substances, enzymes etc, and proceeding to the automation of fermentation factory, and also producing pharmaceuticals by new bio-machines.

Now I will introduce automation in the R&D stages and in the production stages.

2 AUTOMATION OF MAIN BIO-MACHINES USED IN BIO-PROCESSES (ref. 1)

Nowadays, automation of bio-machines used in bio-processes in the R&D stages and the production stages are proceeding at a fast pace. They are shown on Fig. 1.

368

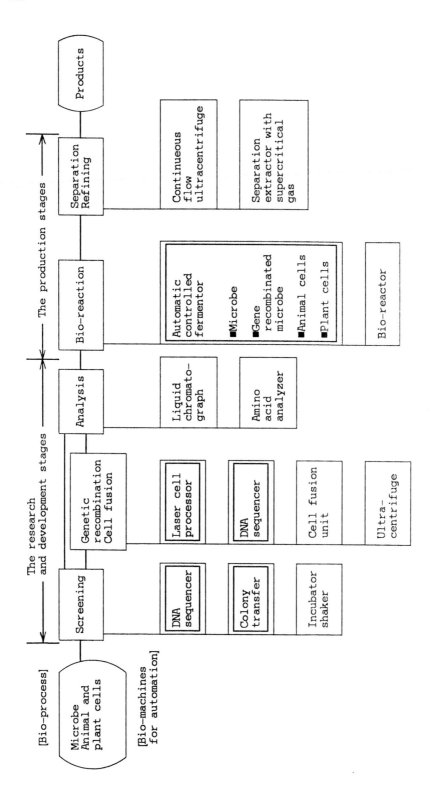

Fig. 1. Bio-process and bio-machines for automation

3 AUTOMATION IN THE R&D STAGE

(1) Micro-injecting cell-processor by laser
   (the laser cell processor) (ref. 1)

Fig. 2 shows the general method of injecting foreign substances (for example, genes, protein and etc.) into the cells. Because of injecting into very small objects, operation efficiency of injection is low, and skilful operation is necessary.

The laser cell-processor makes high performance gene transfection. The laser cell-processor works in such a way that a fine laser beam is irradiated through the object lens of a microscope upon a particular cell in the culture medium in which the desired genes are suspended. (see Fig. 3, 4) The laser makes a very small self-heating perforation in part of the cell membrane, enabling the gene to be introduced into the cells. As a result, the laser cell-processor brings about a dramatic improvement in transfection rate and speed, as well as in the efficiency of cell modification. Moreover, by use of the laser microbeam, microsurgery for subcelluar organelles, embryos, and so on, has been put into practice.

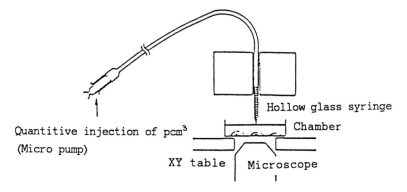

A foreign substance (DNA, protein, etc) is forced
to inject into the cells.

Hollow glass syringe

Quantitive injection of pcm³
(Micro pump)

Chamber

XY table   Microscope

Fig. 2. The general method of injecting into the cells

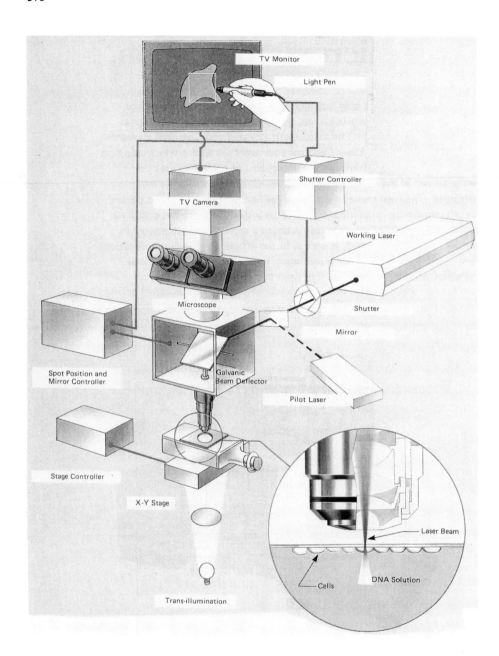

Fig. 3. The system configuration of the laser cell-processor

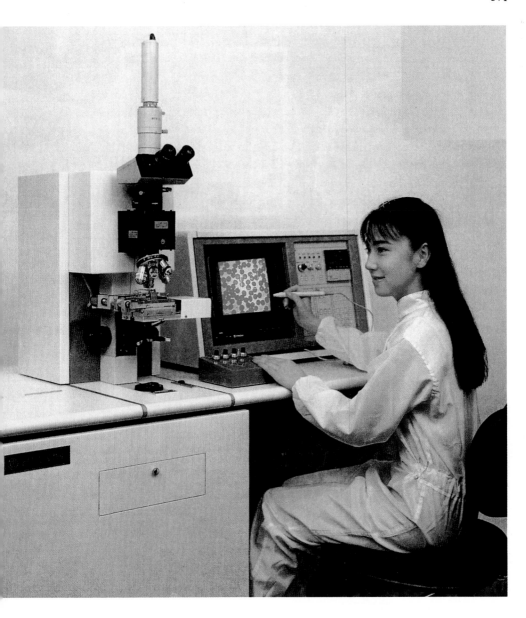

Fig. 4. The laser cell-processor

372

(2) <u>Automated colony transfer system</u> (ref. 2)

For getting useful microbes, screening various microbes is necessary.
Fig. 5 shows the operation of screening.

The automated colony transfer system is designed to transplant a
colony to a plastic petri dish, a microplate or a tube. The automated
colony transfer system recognizes a colony on a petri dish and
transplants the colony automatically to the specified point with high
accuracy through the use of a microcomputer.

Fig. 6 shows the configuration of the automated colony transfer
system. Control unit consists of image processor, main controller,
transfer controller and transplating controller. Image processor
transfers from catching image to the monitor TV. Main controller controls
transfer controller and transplating controller. Transfer controller
controls dish supply and storage. Source dishes and object dishes are
transferred to conveyer. Transplating controller consists of XYZ table,
transplating unit, needle check unit and ITV camera. Needles are
transferred from the colony on the source dish to the indicated position
on the object dishes. When finished, each dishes are transferred to the
storage side.

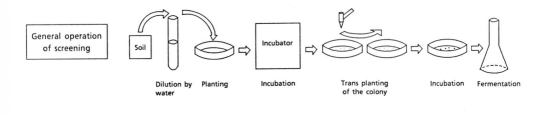

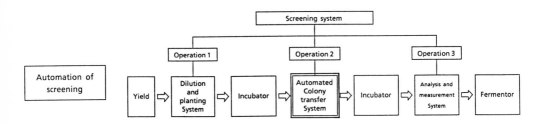

Fig. 5. The operation of screening

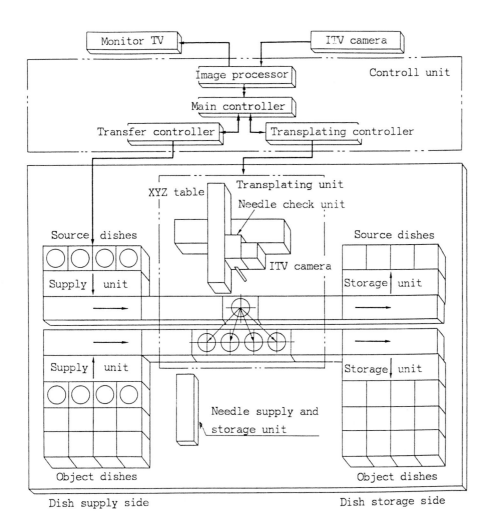

Fig. 6. The automated colony transfer system

(3) Fluorescence DNA sequencer (ref. 3)

If the DNA sequence of an organism could be deciphered, various genetic information contained in that organism could be analyzed and the mechanism of growth and aging of organism, cancer generation, genetic disorders could be clarified and biological species improvement and other researches would make great strides.

① Fig. 7 shows the generally followed method of reading and determination of DNA sequence. These operations are handiwork. Fluorescence DNA sequencer permits automatic electrophoresis and analyzing of DNA sequence.
Fluorescence labeled DNA is subjected to an electrophoresis and is simultaneously detected and analyzed by a computer, thereby saving labor and time required for sequence analysis.

② Fig. 8 shows the detecting principle.
Each fluorescence labeled DNA samples whose terminal is A, C, G or T are charged in each lane of the gel.
Samples are subjected to electrophoresis separation under the gel in the ascending order of length.
When DNA fragment reaches the detecting position of a high sensitivity camera, a fluorescence is excited by laser beam which irradiates the gel inside at all times.
The beam emitted from a fluorescent body is detected by a high sensitivity camera, its brightness is converted by a computer into an analog signal and is immediately indicated on a display.
All data input in the computer is stored in a hard disk. Therefore, the data remains safe regardless of whether DNA electrophoresis in buffer solution past the detecting position.

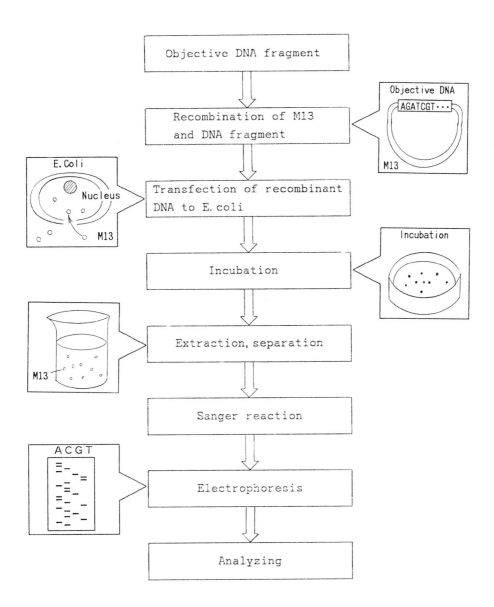

Fig. 7. The method of reading and determination of
DNA sequence

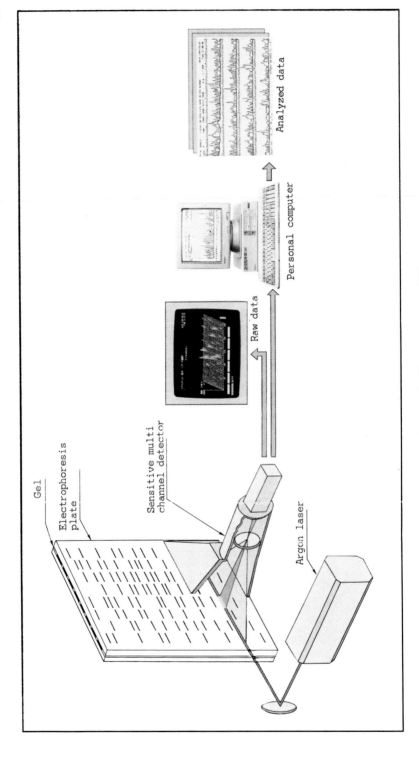

Fig. 8. The detecting principle of fluorescence DNA sequencer

4 AUTOMATION IN THE PRODUCTION STAGE (ref. 5, 6)

(1) Fermentation plants

Fig. 9 shows the general pharmaceuticals manufacturing process, including the fermentation process.

The fermentation process in the bioindustry consists of seed fermentation and the main fermentation, each of which has its own facilities for regulation of the substrate, sterilization, etc. The final products are made by taking what is produced in the fermentation process and future isolating, refining, systhesizing, drying, and packing it.

Microorganism fermentation plants have come to be used in many fields in recent years, and their manufacturing process varies depending on the microorganism and the products to be produced. A typical example is presented in Fig. 10, which is a flow chart outlining how antibiotics are manufactured.

Our company supplied many fermentation plants, until today about 100 plants, and supplied fermentors are about 500.

In Japan, high growing GNP ages, about 20 years ago, we designed and supplied large fermentors for antibiotics, that is 300m^3 volume to Meijiseika Co. in Japan and 1500m^3 volume fermentors for protein of animal foods. (see Fig. 11, 12)

(2) Examples of automation in the production stage

i) Computer-controlled fermentation

In general, the fermentation process is accomplished by first inoculating a small amount of the microorganism onto a nutriment-containing culture medium which is then prepared in the fermentor. Next, it is propagated with air blown in to obtain the desired microorganism or its matabolized substances.

In most cases, production is performed under given conditions at room temperature and at standard atmospheric pressure.

There are the key points involved in the control of the fermentation process.

① Process is by batch or semi-batch control.
② Correlative interface occurs due to the complicated reaction of the microorganism in the nonlinear system.
③ Response time is slow and in minutes.
④ The nunber of quickly measurable objects is limited.

378

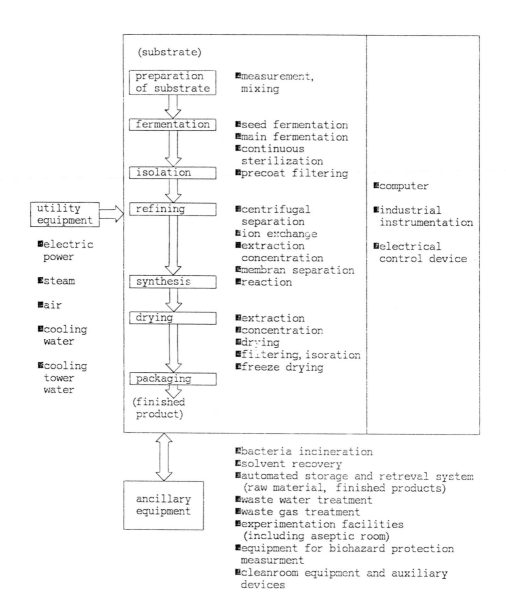

Fig. 9. The general pharmaceuticals manufacturing process

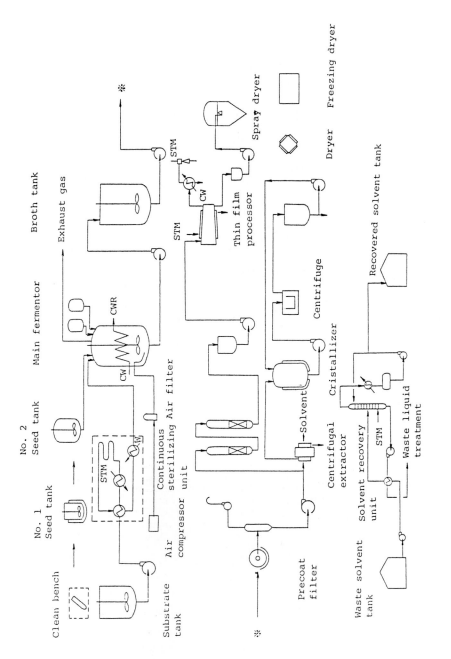

Fig. 10. The fermentation plant flow diagram

Fig. 11. The 300m^3 fermentor

Fig. 12. The 1500m^3 fermentor

In order to obtain the maximum yield, a function is needed which optimizes the microorganism conditions for the fermentor control systems. (Refer to Fig. 13 for the control system for a fedbatch culture.) The values of pH dissolved oxygen (DO) in the fermentor are fed into the CPU of the microcomputer through various sensors, and then the calculation results are output to the operation section for the fermentor whereby control of the various factors is accomplished.

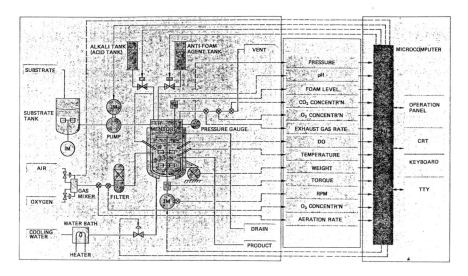

Fig. 13. The fermentor control system

ii) Automation of fermentation factory

■ Concept

Recently, a strong movement is under way which asserts that this is an opportunity to review the basic conventional thinking in regard to automation.

For example, in future production activities, needless to say, it will be necessary to reduce costs through rationalization and labor- and energy-saving measures. A future important matter, however, is how to provide with efficiency products of stable quality while making the transition to the small-lot production of many different types of products that will be required by the diversification of market demands.

Accordingly, factory automation systems have now newly appeared to
serve as trump cards in support of tomorrow's production activities.
These systems point the way to flexible manufacturing systems by
organically combining CAD (Computer-Aided Design), CAM (Computer-Aided
Manufacturing), and CAT (Computer-Aided Testing) in a system which is
based on consistent concepts and extends from product design to
production design, manufacturing, and inspection.

■ Information-processing systems for fermentation plants

Three important features may be cited when considering the subject of
information processing in fermentation plants;

① When applied to raw material manufacturing processes, information
   processing can occupy an important position as a part of a company's
   overall production control system.
② Information-processing systems can be used to store and control
   manufacturing records over an extended period of time.
③ Information-processing can be employed to analyze process data for the
   purpose of yield improvement.

Because each of these three features involves the necessity of
processsing voluminous data, it is desirable to employ a control computer
instead of processing at the instrument-monitoring level. When considered
as a part of company's overall production control system (Feature ①),
the ares of application for computers used for plant control in computer
systems are shown in Fig. 14.

iii) Plant bio-systems

An example of a plant bio-systems is shown in Fig. 15, from the R&D
stage to the production stage. There are automated machines in each
stages.

In the R&D stage, an automated transfer system for screening of carus,
a cell fusion unit and a gene injector are used.

In the production stage, automated seeding nursery and plant factory
are being developed.

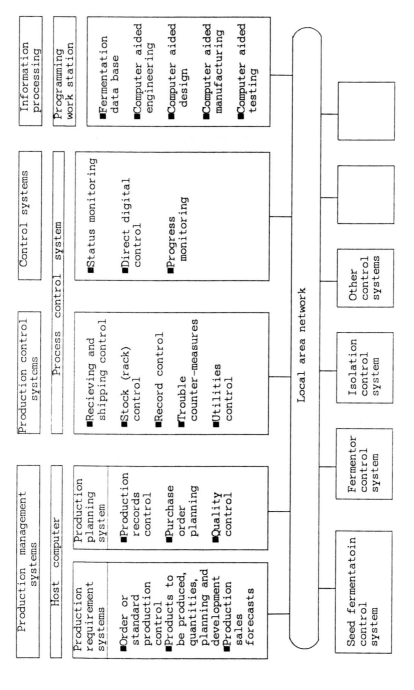

Fig. 14. The configuration of information system for fermentation

384

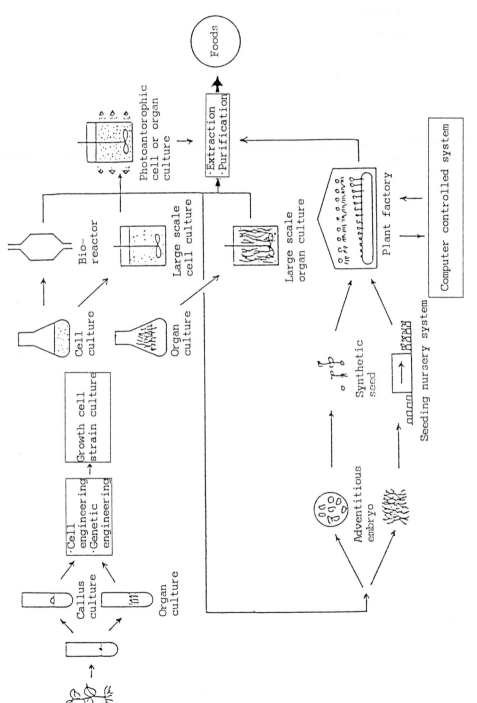

Fig. 15. The plant biotechonology process

5 FUTURE OF AUTOMATION IN BIO-INDUSTRY

The automation of bio-machines in Japan in recent years is shown as follows:

(1) Orientation of automation in bio-industry

	1979	1989	1992
① Automation ratio of bio-machines	20% ——→	80% ——→	90%
② Automation ratio of bio-factory by CIM		·········· 4% ——→	55%
		Accomplished CIM	Probability

(CIM: Computer Integrated Manufacturing)

(2) Introduction of state-of-the-art technology in bio-industry will promote the planning of automation and high efficency of productivity. (see below)

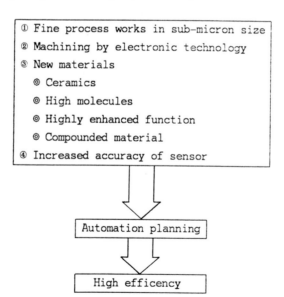

① Fine process works in sub-micron size
② Machining by electronic technology
③ New materials
　◎ Ceramics
　◎ High molecules
　◎ Highly enhanced function
　◎ Compounded material
④ Increased accuracy of sensor

Automation planning

High efficency

REFERENCES

1 Hitachi Ltd., Hitachi Biotechnology Systems (1983)
2 N. Kimura, N. Hosomi, S. Murakami and T. Kasuya, The Hitachi Hyoron, 69, 307-312 (1987)
3 M. Nukumi, F. Iwaya, M. Okuma and Y. Odawara, The Hitachi Hyoron, 69, 313-319 (1987)
4 H. Kambara, T. Nishikawa, Y. Katayama and T. Yamaguchi, Bio Technology, 6, 816-821 (1988)
5 M. Hachiya, Kagaku Kikai Gijyutsu, Vol. 13, 109-121 (Society of Chemical Engineering, Japan, 1961)
6 M. Hachiya and Y. Odawara, Bio Process Engineering, 1-10 (CMC, Japan, 1985)